Springer Aerospace Technology

For further volumes:
http://www.springer.com/series/8613

Tarit Bose

Airbreathing Propulsion

An Introduction

 Springer

Tarit Bose
Aerospace Engineering
Indian Institute of Technology Madras
Chennai 600036, India

ISSN 1869-1730   ISSN 1869-1749 (electronic)
ISBN 978-1-4614-3531-0   ISBN 978-1-4614-3532-7 (eBook)
DOI 10.1007/978-1-4614-3532-7
Springer New York Heidelberg Dordrecht London

Library of Congress Control Number: 2012938708

Printed on acid-free paper

Springer is part of Springer Science+Business Media (www.springer.com)

Preface

This book is an outgrowth of my lectures to undergraduate and postgraduate students in aerospace engineering at the Indian Institute of Technology, Madras, India. Before the first course in aircraft propulsion is offered, it is expected that the student has already undergone a course in both thermodynamics and gasdynamics and has a sufficient understanding of the laws of fluidmechanics. The actual course material is divided into two parts. In the first part, the aircraft propulsion systems, including ramjet, straight turbojet, small bypass jet, and large bypass fanjets, and piston engines are treated in systems approach, in which certain amount of fuel mass and energy are introduced without really considering the flow inside the blade rows. The equations are written in terms of nondimensional parameters for analysis in ideal cycles, or in terms of dimensional performance parameters in a real gas system. This part has been analyzed with the help of one cycle analysis code written in Fortran and given along with the text in this book. The engine parameters of a large number of piston engines, turboprops, and turbojets are also included as an appendix in this book, and will be made available online as spreadsheets on the book's webpage at www.springer.com. It is expected that the reader will use them in deepening his or her understanding of the subject and update also the data from time to time from the Internet. The files may be copied for personal use, but commercial exploitation is not allowed. In the second part, the subject of the design of aircraft gasturbines is treated more exhaustively and the results are compared with the engines, which either were designed earlier or currently exist. While preparing the second part, however, I had difficulties in referring to books on aircraft engines that would be suitable for teaching the subject, which is the prime motivation for writing this book. However, some of the books that I consulted during the preparation of the course lectures and the manuscript of this book are given in the Bibliography, and any reproduction of figures has been acknowledged. For example, the figures prepared by reproducing figures in Traupel's famous work, *Thermische Turbomaschinen*, are given after explicit permission was granted by the publisher, Springer.

Much of the information was taken from unknown authors on the Internet. Thanks to all of them. Readers of this book are strongly encouraged to browse the Internet to collect information.

To a large extent, I wrote this book (text and graphics) myself on a personal computer. In the early stage of writing the manuscript, while I was still at the Indian Institute of Technology in Madras, some of the text was typed on a PC by Jayanthi and some graphics were made by a number of my students; my hearty thanks to them. At that stage I used the Corel Ventura program. Unfortunately, at a later date, Corel Ventura's subsequent version was not compatible with the PC's operating system and so I had to convert my work line by line into the WINWORD editor. This process involved difficulties in putting the engine database in ASCII format into Word format, and hence it was decided to put the matter separately. Unfortunately, all that is new in WINWORD is not good, and what is good there is not new. Thanks go to Dr. M. Ramakanth for help with various publication issues and to the anonymous reviewers for making very useful suggestions. Thanks go also to my wife, Preetishree, for maintaining a congenial environment for writing the book, and also to my three children, Mohua, Mayukh, and Manjul, for bearing with me while writing this book, especially Manjul for procuring some necessary software.

I also would like to thank Springer New York for taking lot of interest in publishing this book following their exacting high standards.

I am now retired from the Indian Institute of Technology in Madras. I worked for some time in California but am now back in Kolkata. My e-mail address is bose.tarit@gmail.com. I look forward to suggestions, reactions, or discussions with readers through e-mail.

Kolkata, India   Tarit Bose

Contents

Chapter 1
Introduction

1.1 Historical

The earliest efficient *reversible thermodynamic cycle* was proposed by Nicolas Léonard Sadi Carnot (1796–1832) with two isotherms and two isentropes in a cycle, but it remained mostly a curiosity until today. One of the earliest engine types used for aircraft applications was piston engines running on *Otto* and *diesel cycles*. The inventor of the *Otto engine*, Nicholas A. Otto (1832–1891), built a successful engine in 1876. These engines helped drive the Industrial Revolution in Europe. However, the workshops had low roofs, and some early engines built vertically required holes in the roof. In the absence of spark-plugs, ignition was done by positioning a flame near the top of the cylinder and a sliding valve would open to ignite the air–fuel mixture. There was no *crank shaft*, and the force of the cylinder was transferred through an arrangement of linear gear and some "catcher" pins to transfer the linear motion to a rotary motion. Sometimes the "catcher" would fail and the linear rod would go through the hole in the roof. Later these engines were built with a horizontal axis. Such early engines are on display at the German Museum in Munich, Germany. The inventor of the diesel cycle, Rudolf Diesel (1853–1913), was born in Paris to German parents who later moved to London because of Franco-Prussian War in 1870. He built the first engine in 1893 in MAN's German factory in Augsburg. Diesel planned initially to build an engine based on an earlier proposal of the reversible Carnot cycle, which was to have the best thermodynamic efficiency within a given temperature ratio. He realized very quickly, however, that to realize a *Carnot cycle*, one would have to run the thermodynamic process of two isotherms as slowly as possible, but in an actual engine the two adiabatic processes must be run as fast as possible, with the result that the two opposite requirements cannot be satisfied. Further, he realized that the Carnot cycle, in spite of the best thermodynamic efficiency, has a very small specific work output. On the other hand, an Otto cycle, because the air and fuel are premixed, could have a very low compression ratio with the resultant low thermodynamic efficiency.

T. Bose, *Airbreathing Propulsion: An Introduction*,
Springer Aerospace Technology, DOI 10.1007/978-1-4614-3532-7_1,
© Springer Science+Business Media, LLC 2012

Diesel therefore proposed a cycle consisting of two adiabatic processes: one constant-pressure and one constant-volume. Diesel disappeared in 1913 while crossing the English Channel during a storm.

The first engine for an aircraft used by the Wright brothers in 1903 was a water-cooled, inline, 4-cylinder engine of weight 890 N (90 kg) delivering a power of 8.95 kW (12 hp). In 1908 the French Gnome air-cooled rotary radial engine was developed, in which the propeller rotated along with the cylinder, and it developed about 70 hp. The power-to-weight ratio for such an engine was better than the water-cooled engine, but there was a gyroscopic effect. After 1920, the stationary radial piston engine was built, and one such engine, the Wright Whirlwind, built by Wright Co., with 9 cylinders and a weight of 1,000 kg developing 220 hp, was the engine used by Charles Lindbergh in 1927 to cross the Atlantic in his monoplane, the *Spirit of St. Louis*. During World War II, a typical piston engine, such as the Rolls-Royce Merlin, was an ethyleneglycol–cooled V-type 12-cylinder engine with 6,360 N (640 kg) and a power of 8,434 kW (1,130 hp) at 3,000 rpm.

Among the large-sized air-cooled radial piston engines, mention must be made of the Pratt & Whitney engine WASP of 2,891 N (290 kg) weight developing 400 hp at 1,900 rpm followed by the twin WASP-R-1830. Another WASP-type (cyclone-powered) engine was used to power one of the most successful passenger transport aircraft, DC-3 (Dakota), about 10,000 of which were built. Another air-cooled radial engine, the WRIGHT R-3330-54, with 18 cylinders developing 3,300 hp at 2,900 rpm, was used to power the four-engined passenger aircraft of the 1950s, the Superconstellation G.

Among the more successful piston engines having diesel cycles, we should mention Junkers' (Germany) two-stroke diesel motor JUMO 205E developing 600 hp and used for regular transatlantic flights between Lisbon and New York around 1938 run by the German airliner Lufthansa. The British company Napier and Sons developed a two-stroke, multiple-cylinder diesel motor for aircrafts with power over 3,000 hp.

Among the aircraft gas-turbine engines, we should mention the turbo-supercharger developed in 1918 by Sanford A. Moss. In 1930, Frank Whittle received a patent for a turbojet engine, and 5 years later, Hans von Ohain also received a German patent for a turbojet engine. The first turbojet engine was flown in June 1939 on the German aircraft Heinkel He178, which developed 4,900 N thrust. Whittle's engine was flown in England in May 1941, and the first jet-propelled plane flew in 1942. By 1944, both Junkers and BMW had operational engines for the German *Luftwaffe* (Air Force) and the world's first jet-fighter aircraft, *Messerschmitt Me262*, was flown in 1944 (and later the world's first jet-bomber aircraft, the *Arado Ar234*). However, these fuel-guzzling jet aircrafts (debuting during a shortage of petroleum-based fuel) appeared too late to make any significant contribution to the outcome of the World War II. The first commercial jet-propelled aircraft, the Comet, was introduced in 1949, but the flight of this new aircraft was very short-lived because of the unexpected appearance of the fatigue phenomenon. But only 5 years later, Boeing aircraft powered by four Pratt & Whitney USA JT-3 engines was successfully introduced.

In the late 1940s, the French company Rateau developed a jet engine in which a part of the incoming air was bypassed around the high-pressure compressor stages, combustion chamber, and turbine stages to be mixed with the hot air before the mixed air was sent through the nozzle. The *bypass ratio*, which is the mass ratio of the cold bypassed air to the hot air, had to be small. When the flight speed of an aircraft is in the subsonic or transonic region, one can have a high-bypass-ratio engine, where the air coming from the low-pressure compressor is split, with one half going through the route of hot air, and the other half expanding in a nozzle. The method has the advantage of having a high thrust and good propulsive efficiency, without losing the simplicity of a turbojet engine. The first such large-sized turbofan engine was introduced in 1967 by Pratt & Whitney (P&W JT9D) for the Boeing 747 aircraft. Another high-bypass-ratio fanjet engine (P&W JT3D) was developed to replace the JT3 engine of the Boeing 707 aircraft.

The same consideration of a large thrust at a high propulsive efficiency is valid also for turboprops. The most successful of these to fly on a commercial aircraft is the DART by the Rolls-Royce Company. In recent years, a via media between two- to four-blade turboprops and turbojets with a large number of blades have been designed, called *propfans,* or UDFs (*unducted fans*), which have a very large number of blades in one or two rows, and in the latter case, counterrotating against each other.

At this point it is of historical interest to note some of the specific fuels used in some special aircrafts. In the 1950s, Lockheed (USA) built the U2 spy plane, for which a reliable engine was required to fly at about 22-km altitude with very little oxygen and an outside temperature of −55°C with a range of 8,000 km without refueling. At that time, Pratt & Whitney built the highest-pressure-ratio J57 engine, which was adopted for the U2 plane. However, in the environmental conditions, the standard JP-4 fuel would have frozen or boiled off due to low atmospheric pressure. Thus, a special low-vapor kerosene, designated LF-1A, was developed for high altitudes. It smelled like lighter fluid, but a match could not light it. It is believed that the stuff was actually very similar in chemistry to a popular insecticide and spray known as Flit, since the latter company belonged to the petroleum company that delivered the fuel. For about 4 years, the aircraft flew unchallenged over the Soviet Union on spy missions until it was shot down in 1959, but it continued to fly elsewhere in the world for a few more years on surveillance operations.

A few years later, another high-performance titanium body airplane, called *Blackbird* because of the black paint to reduce the skin temperature by radiation, was developed to fly at the sustained speed of Mach 3 and an altitude of about 30 km. The fuel, put in five noninsulated fuselage and wing tanks, which would heat up to about 170°C during supersonic flight, had to be a high-flashpoint fuel that would not vaporize or blow up under tremendous heat and pressure. The fuel had to remain stable, on the one hand, at −65°C, which was experienced when pumping the fuel in air, and, on the other hand, at 170°C when it was fed to the engine; it was also used as a coolant. As an added safety precaution, nitrogen was added to the fuel tanks to pressurize them and prevent an explosive vapor ignition. The engines chosen were two Pratt & Whitney J-58 afterburning bypass turbojets that had a movable spike-shaped cone inlet and allowed an astounding 84% propulsive efficiency at Mach 3. Another problem that occurred here is called an *unstart*, when air entering one of the two engines was blocked due to separation of air by the angle of pitch and yaw, and the efficiency dropped to a very low value in milliseconds, which was much faster than the time required by the movable spike-control system to control the problem. The problem was finally tackled by designing an electronic control, which saw to it that when one engine was hit with an "unstart," the second engine dropped its power to it, and, finally, both engines were relit automatically.

1.2   Comparison Among Piston Engines, Turboprops, and Turbojets

We start our discussion by making a sketch of a propeller-driven (piston-engine or turboprop) and a propeller-less engine (turbojet, with or without bypass, fanjet, or ramjet) (see Fig. 1.1). In the first group, the thrust is developed by large propellers across which the flow is slightly accelerated and the pressure changes very little. In the second group, a comparatively smaller mass flow rate is accelerated by a large value.

In Fig.1.1, $\dot{m}_a$ and $\dot{m}_f$ are the respective air and fuel flow rates, and u_∞ and u_e are the respective approaching flow velocity and exit velocity. For the present, we neglect the fuel mass flow rate, that is, $\dot{m}_f = 0$. Now, the thrust in both the cases is given by the relation

$$F = \dot{m}_a(u_e - u_\infty) \qquad (1.1)$$

By multiplying the thrust with the flight velocity u_∞, we get the developed power (based on thrust)

$$P_F = Fu_\infty = \dot{m}_a u_\infty(u_e - u_\infty) \qquad (1.2)$$

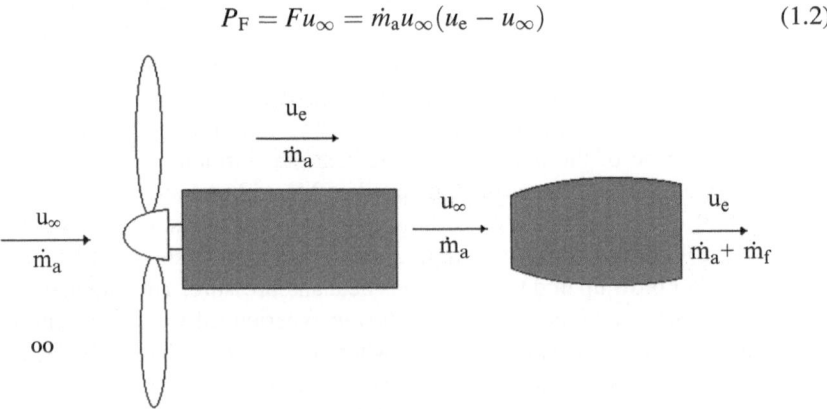

Fig. 1.1 Schematic sketch of a turboprop engine

On the other hand, the work extracted from the engine can be evaluated from the specific work from a thermodynamic cycle. This is simple and will be shown in chapter 2 as being equal to the increase in the kinetic energy of the air mass flow in the jet. For propeller-driven engines, the relations are not quite straightforward, but taking an analogous relation, the extracted work from the engine is obtained from the increase of the kinetic energy of the propeller-driven air mass. Hence, for all these engine types,

$$P_E = \dot{m}_a (u_e^2 - u_\infty^2)/2 \qquad (1.3)$$

Now the *propulsive efficiency* is the ratio of P_F/P_E:

$$\eta_p = \frac{P_F}{P_E} = \frac{2}{\left[1 + \left(\frac{u_e}{u_\infty}\right)\right]} \qquad (1.4)$$

From (1.1), we can also write that

$$\frac{F}{\dot{m}_a u_\infty} = \left[\left(\frac{u_e}{u_\infty}\right) - 1\right]$$

Hence, the alternate definition of the *propulsive efficiency* is

$$\eta_p = \frac{1}{\left[1 + \frac{F}{2\dot{m}_a u_\infty}\right]} \qquad (1.4a)$$

Thus, for the approaching flow velocity $u_\infty \to 0$, we get the maximum thrust, $F/(\dot{m}_a u_\infty) \to \infty$ and $\eta_p \to 0$. On the other hand, for the approaching flow $u_\infty \to u_e$, $F/(\dot{m}_a u_\infty) \to 0$ and $\eta_p \to 1$. Since in the cruise there has to be a finite thrust to affect the drag, $u_\infty < u_e$ and the propulsive efficiency η_p must necessarily be smaller than 1. This is shown schematically in Fig. 1.2.

It is now evident that for a given thrust, the propeller-driven engines have u_∞/u_e less than, but very near, 1, which gives a small value of $F/(\dot{m}_a u_\infty)$, but a very good propulsive efficiency, whereas for propeller-less engines, u_∞/u_e is very much less than 1, with large $F/(\dot{m}_a u_\infty)$ finite and a very small propulsive efficiency.

Since for propeller-driven engines (piston, turboprop), the shaft power developed by the engine and given to the propeller is an important criterion, for jet engines the developed thrust is an important criterion. Hence, a comparison of these various engine types is done by comparing P_E for a propeller-driven engine with $P_E = P_F/\eta_p$ for a jet engine. For example, for $\eta_p = 0.85$, 1 kN thrust for a flight speed of 460 km/h is equivalent to the power of 150.3 kW. At 920 km/h, the same thrust with the same η_p is equivalent to 300.6 kW. However, because of the compressibility effects at such flight speeds, η_p cannot be more than 0.65, and 1 kN here is equivalent to 393.2 kW. Another consideration when comparing various

Fig. 1.2 Airplane drag and
engine thrust and propulsive
efficiency

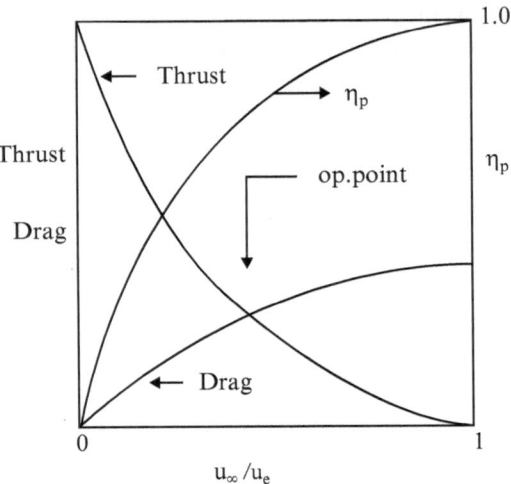

engine types is the weight consideration of the engine per unit of developed power
on thrust. From an analysis of various engine data, the following ranges of data are
evaluated for various engine types:

Aircraft piston engine: weight/power 0.63–1.52 kg/kN
Turboprop engine: weight/power = 0.17–0.57 kg/kW
Turbojet engine: weight/thrust = 0.015–0.035 kg/N

Hence, one can see that the turboprop engines are much lighter than the piston
engines, and the piston engines, used for low-power ratings, are justifiable only on
the basis of lower manufacturing costs for small piston engines.

Now, for a turbojet engine,

$$\frac{W}{P_E} = \frac{W\eta_p}{Fu_\infty} = \frac{W}{F}\left(\frac{\eta_p}{u_\infty}\right)$$

Taking an average weight-to-thrust ratio of 0.025 kg/N, we get the weight-to-
power ratio:

$$\frac{W}{P_E} = \frac{W\eta_p}{Fu_\infty} = 25\left(\frac{\eta_p}{u_\infty}\right), \frac{kg}{kW}$$

when u_∞ must be taken in m/s. Taking an average weight-to-power ratio for an aircraft
piston engine to be 1.0 and for a turboprop to be 0.3, and assuming the propulsive
efficiency of a turboprop equal to 0.35 and for a turbojet equal to 0.05, the minimum
flight speed for an aircraft piston engine against the turboprop is 16.25 m/s, or
58.5 km/h, and for a turboprop against the turbojet is 54.25 m/s, or 195.3 km/h.
In actual practice, however, such a comparison is not as simple as described above.

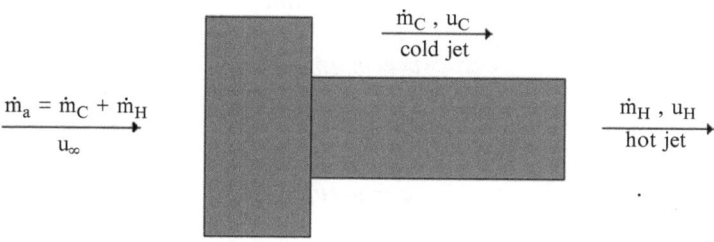

Fig. 1.3 Sketch of a high-bypass turbojet (fanjet)

It is well known that for a propeller engine for an aircraft, the thrust decreases with the flight speed, but the thrust changes only marginally for a turbojet engine, and thus the propulsive efficiency of a turbojet engine is much smaller than that of the propeller-driven engine.

We could now make another comparison between a straight-jet or a low-bypass turbojet engine on the one hand with a high-bypass turbojet (Fig. 1.3) engine (fanjet) on the other. In the latter, a very large volume of cold air is ejected out of the fanjet nozzle, and only a very small part of the air is sent through the combustion chamber and turbine. Both the fan and the low- and high-pressure compressor are driven by the turbine. While in the former case the propulsive efficiency is given by (1.4), in the latter case we have to consider a two-stream exhaust. For this purpose, we again neglect the introduction of fuel.

For the above engine (bypass turbojets have a hot gas stream through the combustion chamber designated with the index "H" and a cold gas stream bypassing the combustion chamber designated with the index "c"), the thrust and power are given by the relations

$$F = \dot{m}_H(u_H - u_\infty) + \dot{m}_c(u_c - u_\infty) \tag{1.5a}$$

and

$$P_F = Fu_\infty = u_\infty[\dot{m}_H(u_H - u_\infty) + \dot{m}_c(u_c - u_\infty)] \tag{1.5b}$$

The kinetic energy increase of the fluid is

$$P_E = \frac{1}{2}\left[\dot{m}_H\left(u_H^2 - u_\infty^2\right) + \dot{m}_c\left(u_c^2 - u_\infty^2\right)\right] \tag{1.5c}$$

From the definition of *propulsive efficiency*, $\eta_p = P_F/P_E$, we therefore write

$$\eta_p = 2\left[\frac{u_\infty\{\dot{m}_H(u_H - u_\infty) + \dot{m}_c(u_c - u_\infty)\}}{\dot{m}_H\left(u_H^2 - u_\infty^2\right) + \dot{m}_c\left(u_c^2 - u_\infty^2\right)}\right] \tag{1.6}$$

Introducing the definitions of the *bypass ratio*,

$$b = \dot{m}_c / \dot{m}_H \tag{1.7a}$$

the *jet speed ratio*,

$$\varsigma = u_c / u_H \tag{1.7b}$$

and the *hot jet speed-to-approaching flow speed ratio*,

$$\varphi = u_H / u_\infty \tag{1.7c}$$

we get from (1.6) the relation

$$\eta_p = 2 \left[\frac{(\varphi - 1) + b(\varphi\varsigma - 1)}{\varphi^2 - 1 + b(\varphi^2\varsigma^2 - 1)} \right] \tag{1.6a}$$

The above equation is indeterminate if $\varphi = 1$ and $\varsigma = 1$ independent of b. On the other hand, the definition of ς is meaningless if $b = 0$.

For a given b, $\varphi = u_H / u_\infty$, and the choice of ς for the maximum η_p can be obtained by differentiating η_p with respect to ς and setting it equal to zero. This leads to the result for the optimum ς as follows:

$$\varsigma_{opt} = \left[\frac{b(1 + \varphi^2) + (1 - \varphi^2)}{\varphi\{2b + (1 - \varphi)\}} \right] \tag{1.8}$$

The results of calculating (1.8) appear in Fig. 1.4, which shows that, generally, $(u_c / u_H)_{opt} \to 1$, but since for a high-bypass fanjet engine, $(u_\infty / u_H) \approx 1$, the optimum (u_c / u_H) is also just above 1. It has therefore been suggested that for fanjets, the condition $(u_c / u_H) = 1$ can usually be considered adequate.

1.3   Thermal Efficiency, Overall Efficiency, and Specific Fuel Consumption

In (1.4) we defined the *propulsive efficiency* as the ratio of the actual energy of air due to the aircraft's motion [energy = force (thrust) × flight speed of aircraft] to the difference in the kinetic energy of the jet air. This ratio is given by the expression

$$\eta_p = P_F / P_E$$

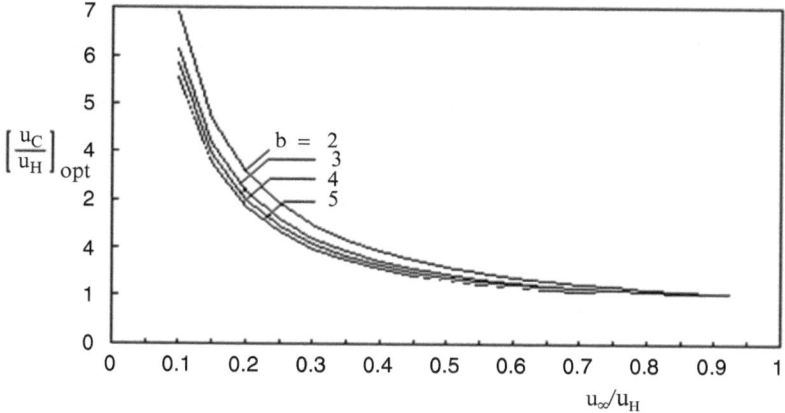

Fig. 1.4 Optimum cold-to-hot velocity ratio vs. ratio of velocity of approaching flow to hot jet with bypass ratio as parameter

We will show later that P_E is equivalent to the overall work that one can extract in a thermodynamic cycle. For a single-jet engine (ramjet, straight jet, or low-bypass jet), connecting (1.3), we can write

$$P_E = \dot{m}_a \left(u_c^2 - u_\infty^2 \right)/2 = \dot{m}_a w \tag{1.9}$$

where w is the *overall specific work* (per unit mass of air) in kJ/kg. Now *the thermodynamic efficiency* of a complete thermodynamic cycle is defined as

$$\eta_{th} = \frac{P_E}{\dot{m}_a q} = \frac{w}{q} = \frac{P_E}{\dot{Q}} \tag{1.10}$$

where q is the *specific heat addition* (per unit mass of air) in kJ/kg. We'll define the thermodynamic efficiency in terms of specific work. The specific fuel consumption is unambiguous; it is independent of whether the work is supplied to a propeller or through an increase in the kinetic energy of the jet stream. The alternate definition of work is

$$w = \frac{P_E}{\dot{m}_a} = \frac{1}{2} \left(u_e^2 - u_\infty^2 \right) \tag{1.11a}$$

and is meaningful for jet engines only.

The *overall efficiency* is defined as the product of the propulsive efficiency and the *thermodynamic efficiency*:

$$\eta = \eta_{th.} \cdot \eta_p = \frac{P_F}{\dot{Q}} \tag{1.11b}$$

where $\dot{Q}$ is the heat input in kW.

   While the thermodynamic efficiency η_{th} is an important criterion for any thermodynamic cycle process, some other performance parameters, such as specific work or specific fuel consumption, are also very important. For jet engines, the *specific fuel consumption* (SFC) is defined as the fuel flow rate $\dot{m}_f$ per unit of thrust kg/N·s; that is,

$$\text{SFC} = \frac{\dot{m}_f}{F}, \frac{\text{kg}}{\text{Ns}} \tag{1.12a}$$

which for a *ramjet* and *straight turbojet* with or without reheat is

$$\text{SFC} = \frac{\dot{m}_f}{\dot{m}_a(u_c - u_\infty)} = \frac{f}{(u_c - u_\infty)} \tag{1.12b}$$

where $f = \dot{m}_f/\dot{m}_a$ is the *fuel–air ratio*. Note that

$$q = \dot{m}_f \Delta H_p = \dot{m}_a f \Delta H_p \tag{1.13a}$$

where ΔH_p is the *heat of reaction* of fuel (per unit mass of fuel). For the specific fuel consumption (per unit of thrust), we get the relation

$$\text{SFC} = \frac{f}{(u_e - u_\infty)} = \frac{\dot{Q}}{\dot{m}_a \Delta H_p(u_e - u_\infty)} = \frac{\dot{Q}}{F\Delta H_p} = \frac{P_E}{F\Delta H_p \eta_{th}} = \frac{\dot{Q}u_\infty}{P_F\Delta H_p} = \frac{u_\infty}{\eta\Delta H_p}$$

where $\dot{Q} = \dot{m}_f \Delta H_p$ is the heat added per unit of time, and the heat addition per mass of air and per unit of time is

$$\dot{q} = \dot{m}_a q = \dot{m}_a f \Delta H_p \tag{1.13b}$$

   Thus, for $\eta \to 1 : (\text{SFC})_{max} = u_\infty/\Delta H_p \to (\text{SFC})_{min}/\text{SFC} = 1/\eta$.

   Thus, for jet engines, including the ramjets, the specific fuel consumption is immensely proportional to both the thermodynamic efficiency, η_{th}, and the *overall efficiency*, η. Equation 1.12b also gives the limit of the minimum SFC possible (for $\eta = 1$) for a particular flight speed. For example, for a typical straight-jet engine like the JT3C-2, the given SFC is 62.30×10^{-6} kg/Ns, whereas the theoretical minimum SFC for an aircraft flying at 880 km/h and using JP-8 fuel ($\Delta H_p = 43,264$ kJ/kg) will be SFC $= 880/(3.6 \times 4.3264 \times 10^7) = 5.65 \times 10^{-6}$ kg/Ns, and the effective overall efficiency is about $5.65/62.30 = 0.0907$. Thus, for a given flight speed, lowering the SFC is possible only by using a better overall thermodynamic efficiency. Further, the engine running on ground u_∞ is very small and η_p is small; consequently, the SFC is *finite*.

On the other hand, for a propeller engine, the SFC is defined in terms of shaft power:

$$\text{SFC} = \frac{\dot{m}_f}{\dot{m}_a w} = \frac{f}{w} = \frac{f}{\eta_{th} q} = \frac{f}{f \eta_{th} \Delta H_p} = \frac{1}{\eta_{th} \Delta H_p} \qquad (1.14)$$

For a typical piston engine, let's take SFC $= 71.0 \times 10^{-6}$ kg/kJ. Now, with $\Delta H_p = 42,707$ kJ/kg for gasoline, $\eta_{th} = 1/(71.0 \times 10^{-6} \times 42,707) = 0.33$.

1.4 Fuel Combustion with Air

We consider here only the class of propellant systems in which the fuel burns with oxygen in the air. In a general case, let the fuel be a hydrocarbon of the type $C_m H_n O_p$. Thus, for a complete combustion with *stoichiometric* air, we can write

$$C_m H_n O_p + (m + n/4 - p/2)O_2 + 0.79(m + n/4 - p/2)/0.21 N_2$$
$$\rightarrow m CO_2 + (n/2)H_2O + 0.79(m + n/4 - p/2)/0.21 N_2$$

Thus, the minimum oxygen molecule and air requirements for the complete combustion of one mole of fuel are

$$O_{2\,min} = m + n/4 - p/2 \qquad (1.15a)$$

$$Air_{min} = (m + n/4 - p/2)/0.21 = O_{min}/0.21 \qquad (1.15b)$$

If we define an air–fuel *equivalence ratio* λ as the ratio of actual oxygen molecules to the minimum oxygen requirement, the above chemical reaction equation (for $\lambda > 1$) is

$$C_m H_n O_p + \lambda O_{2,min} + \lambda O_{2,min} \frac{0.79}{0.21} N_2$$
$$\rightarrow m CO_2 + (n/2)H_2O + (\lambda - 1)O_{2,min} + \lambda O_{2,min} \frac{0.79}{0.21} N_2$$

Instead of writing the chemical reaction equation in terms of moles, we may write the same in terms of mass, in which case we get (for $\lambda > 1$)

$$(12m + n + 16p)\text{kg}_{fuel} + 32\lambda O_{2,min}\text{kg}_{O_2} + 28\lambda O_{2,min} \frac{0.79}{0.21} \text{kg}_{N_2}$$
$$\rightarrow 44m\text{kg}_{CO_2} + (9n)\text{kg}_{H_2O} + 32(\lambda - 1)O_{2,min}\text{kg}_{O_2} + 28\lambda O_{2,min} \frac{0.79}{0.21} \text{kg}_{N_2}$$

Thus, the *fuel-to-air mass ratio* in kg-fuel/kg-air is

$$f = \frac{\dot{m}_f}{\dot{m}_a} = \left[\frac{12m + n + 16p}{\lambda\{32 + 28\frac{0.79}{0.21}\}\{m + n/4 - p/2\}} \right]$$

$$= \left[\frac{12m + n + 16p}{\lambda 137.333\{m + n/4 - p/2\}} \right] \tag{1.16}$$

Now, if the *heat of reaction* per kg of fuel is ΔH_p (in kJ/kg-fuel), then from *energy balance* we may write

$$(12m + n + 16p)\Delta H_p = (T_{\text{flame}} - T_{\text{ref}}) \sum \left[C''_{pj} x''_j \right] - (T_{\text{init}} - T_{\text{ref}}) \sum \left[C'_{pj} x'_j \right] \tag{1.17}$$

where (C_{pj}) is the *average* molar specific heat of the jth species at constant pressure (since for most of the combustion chambers, except in Otto engines, the heat input is a near-constant-pressure process), x_j is the mole fraction of various species, T_{ref} is the reference temperature from which the averaging of C_p is done, T_{init} is the initial temperature of the fuel–air mixture before combustion, and T_{flame} is the *adiabatic flame temperature* (maximum combustion temperature without losses). In addition, $()'$ and $()''$ refer to those before combustion and after combustion, respectively. It is convenient to describe C_p with the help of a polynomial, $C_p = a + bT + cT^2$, and the average (C_p) can be computed with the help of the relations

$$[C_p] = \frac{1}{T - T_{\text{ref}}} \int_{T_{\text{ref}}}^{T} C_p \, dT \tag{1.18}$$

Table 1.1 contains the coefficients of the *polynomial equation* for the specific heat at constant pressure for various gases, when T is in K and T_{ref} or the equivalent is 0°C, which can be used for evaluating (1.18). Note that for diatomic homopolar molecules, such as H_2, N_2, and O_2, the values of the coefficients are almost equal to each other, whereas the values for heteropolar diatomic molecules, such as CO, OH, and NO, are slightly different.

Table 1.2 gives the fuel data for some of the fuels, which either are in use presently or are potential fuel for the future. It also gives the heat content of each fuel. For the heat of reaction per kg of fuel, we have to differentiate between two situations. In the first case, the water produced by the chemical reaction remains in the vapor form, and the heat of reaction is the lower of the two (*lower heat of reaction*). In the second instance, when the water is in liquid form, we have the *higher heat of reaction*. For our purposes, we consider only the lower heat of reaction. It is interesting to note that the heat content per unit mass for gasoline is only 42,707 kJ/kg, whereas for hydrogen gas is about three times higher; for liquid hydrogen, the value will be still higher. However, when it is considered per

Table 1.1 Coefficients for polynomial equation for specific heat,$C_p = a + bT + cT^2$ (kJ/kmole-K) for perfect gas (valid for 0–3,000°C)

Gas	a	b	c
H_2	27.30	3.620e-3	−1.495e-7
N_2	27.35	6.382e-3	−1.035e-6
O_2	27.84	7.491e-3	−1.158e-6
CO_2	31.98	2.447e-2	−4.699e-6
H_2O	28.99	1.494e-2	−1.998e-6
CO	27.40	6.760e-3	−1.146e-6
OH	28.50	3.025e-3	−5.183e-8
NO	28.38	6.482e-3	−1.124e-6
N_2O	35.41	2.111e-2	4.169e-6

Table 1.2 Some fuel data

Fuel	Heat content (kJ/kg)	Heat content (kJ/kmole)	No. of atoms C	H	O	Fuel molar mass
Gasoline	42,707	1.09331e7	17	36	0	53
$C_2H_5OH(l)$	26,964	1.24036e6	2	6	1	31
$CH_3OH(l)$ (95% pure)	25,290	8.09263e5	1	4	1	32
$H_2(g)$ (298 K)	119,955	2.41830e5	0	2	0	2

kmole of the fuel, the value for hydrogen is the lowest. Because of that, a fire with hydrogen will not give a higher flame temperature than with other fuel: Gasoline is the biggest culprit. The same conclusion occurred following a fire experiment with a complete aircraft model conducted by NASA. However, liquid hydrogen, with its much lower density than gasoline, will require a much larger volume of the fuel.

Table 1.3 gives *postcombustion data* for selected fuels listed in Table 1.2 burned in air for different equivalence ratios. We see that for stoichiometric combustion (equivalence ratio = 1) with air, the air-to-fuel mass ratio is about 20, or the mass of fuel is only about 5% of that of air. It is thus often possible to ignore the mass of fuel during the cycle analysis of a thermodynamic process for an engine.

Table 1.4 gives the data for JP fuels, where the subscript "c" denotes critical conditions, ρ_l is the density of the liquid fuel, and σ and ε/k_B are the molecular diameter and collisional potential data, respectively. JP-4, JP-5, and JP-8 are types of aviation kerosene, which are broadly a blend of kerosene and gasoline. JP-4 is used mainly by NATO air forces; it is highly volatile, with a vapor pressure of about 0.17 bar at 37°C and a *flash point* of about 25°C. On the other hand, because of the likelihood of postcrash fires and other accidents, commercial airlines use the kerosene-based fuel JP-8, with a typical vapor pressure of 0.007 bar at 37°C and a flash point at 52°C. Unique problems associated with fuels used in shipboard gas turbines require the use of a third fuel, called JP-5, which has a still-higher flash point of 65°C and a lower vapor pressure of 0.003 bar at 37°C. For the above fuel data and initial (precombustion) temperature of 25°C, the flame temperature and

Table 1.3 Postcombustion data for fuel with air

Fuel	Equiv. ratio	T_{flame} (K)	Molar mass (mixture)	Mole fraction				A/F (kg/kg)
				x_{CO2}	x_{H2O}	x_{N2}	x_{O2}	
Gasoline(l)	1.0	2,102	27.55	0.10	0.11	0.79	0.00	20.03
	1.5	1,578	28.65	0.07	0.07	0.79	0.07	30.04
	2.0	1,291	28.70	0.05	0.05	0.79	0.10	40.06
C_2H_5OH(l)	1.0	1,875	28.09	0.09	0.14	0.77	0.00	13.43
	1.5	1,422	28.34	0.06	0.09	0.78	0.07	20.55
	2.0	1,170	28.46	0.05	0.07	0.78	0.10	26.87
CH_3OH(l)	1.0	2,088	27.68	0.08	0.16	0.76	0.00	10.73
	1.5	1,584	28.05	0.05	0.11	0.77	0.07	16.09
	2.0	1,124	28.36	0.03	0.03	0.78	0.12	26.87
H_2(g)	1.0	1,767	25.90	0.00	0.21	0.79	0.00	68.47
	1.5	1,326	26.88	0.00	0.14	0.79	0.07	103.0
	2.0	1,091	27.37	0.00	0.10	0.79	0.12	137.3

Table 1.4 Data for JP fuels

Fuel name	Chem. formula	Equiv. formula	Molar mass	T_c (K)	p_c (bar)	ρ_l (kg/m³)	Flash point (C)	Boil. point (C)	σ (ang.)	ε/k_B (K)
JP-4	$C_{8.5}H_{16.9}$	C_8H_{16}	119.03	599.6	31.00	776	–	207	6.241	461.7
JP-5	$C_{11.9}H_{22.2}$	$C_{12}H_{23}$	165.18	671.6	19.96	816	60	247	6.840	517.1
JP-7	$C_{12.1}H_{24.4}$	$C_{12}H_{24}$	169.79	678.1	18.64	792	60	235	6.977	522.1
JP-8	$C_{10.9}H_{20.9}$	$C_{10}H_{21}$	151.87	660.6	21.53	807	38	252	6.681	508.7
JP-10	$C_{10}H_{16}$	$C_{10}H_{16}$	136.13	605.9	25.04	939	55	–	6.124	466.5

other data, given in Table 1.4, are calculated as follows. First, the composition after the combustion has completed is calculated. Then the flame temperature is estimated, and the average specific heat is calculated from (1.18), the coefficients given in Table 1.1, and the previously calculated composition. The flame temperature is then obtained from (1.17).

From the results given in Table 1.3, it is evident that even for a stoichiometric air–fuel mixture ratio, the quantity of fuel is about 5% by weight of air, and hence it is possible to idealize the combustion process by taking air as an ideal gas with a constant specific heat into which heat is added externally.

Hence, from the energy balance, we can write

$$\dot{m}_f \Delta H_p = \dot{m}_a c_p \Delta T$$

and thus,

$$f = \frac{\dot{m}_f}{\dot{m}_a} = \frac{c_p \Delta T}{\Delta H_p} \tag{1.19}$$

For air, taking $c_p = 1.005$ kJ/kg, the initial temperature $T_{init} = 298.15$ K, gasoline lower heat of reaction $\Delta H_p = 4.2707 \times 10^4$ kJ/kg, and fuel/air mass ratio $f = 1/20.03 \approx 0.05$, we can now evaluate the flame temperature to be 2,125 K, which is slightly above the value given in Table 1.3. We could better match the value in Table 1.3 if we took the actual average specific heat value into account.

While the mole fraction given in Table 1.3 assumes complete combustion and the existence of no other products such as the oxides of nitrogen or OH radicals, these can, of course, be estimated by more involved methods of chemical thermo-dynamics. Since in the temperature range and pressure range of operation for the combustion chamber under consideration, disassociation is not likely to play a significant role, the above method of calculation may be accurate enough for all practical purposes. It is more difficult to estimate the role of unburned hydrocarbon and smoke (carbon particles) in the exhaust gas from the combustion chamber.

1.4.1 Future Fuels

The 1974 study by the Club of Rome about the worldwide fuel supply predicted that all fossil liquid fuels would be used in about 30 years. While it may not have turned out to be accurate almost 40 years later, there are considerable concerns about the worldwide fuel supply in the near future.

In this connection, hydrogen economy was suggested as an alternate fuel, primarily because hydrogen fuel is ecofriendly. In a flame of pure hydrogen gas burning in air, hydrogen reacts with oxygen to form water and heat; it does not produce other chemical byproducts, except for a small quantity of oxides of nitrogen. When burning in air, the temperature is roughly 2,000°C, almost exactly the same with the petroleum-based fuels burning in air. It can provide power for aircraft applications, among other uses, and also for *fuel cell* applications, which can power an electric motor.

The current leading technology for producing hydrogen in large quantities is the steam reforming of methane gas $(C_2H_5 + 2H_2O \rightarrow CO_2 + 4H_2)$. In addition, hydrogen can be produced by electrolyzing water using cheap electric power involving solar or wind sources.

At the gas pressure between 350 and 700 bar at which hydrogen is typically stored, hydrogen requires more storage volume than the volume of petroleum-based fuels that produces the equivalent energy, but the weight of this hydrogen is nearly one third that of the petroleum-based fuel. Because of this fact, in aircrafts the petroleum-based fuel is stored in the wings (also to increase the mass moment of inertia of wings for structural reasons), but hydrogen has to be stored in fuselage. With regard to unwanted explosions, hydrogen fuel is at least as safe as aviation gasoline, as was demonstrated a few years back in a NASA-sponsored experiment by burning actually a hydrogen-fueled aircraft.

There have been some recent developments with nonpetroleum-based jet fuels in the pursuit of *alternative fuels* that can power jets and address rising fuel costs. The *Fischer–Tropsch* process is an important chemical reaction in which a synthetic gas—a mixture of carbon monoxide and hydrogen—is converted into liquid hydrocarbons of various forms for use as a lubricant and fuel.

A DC-8 aircraft, based at Dryden at Edwards Air Force Base in California, is the test vehicle in which the researchers are testing 100% *synthetic fuels* and 50/50 blends of synthetic and regular jet fuels.

In August 2010 at the Farnborough Airshow, environmental executives from Boeing, Airbus, and IATA discussed the development of alternative sustainable jet fuels on a commercial scale, with the target of a 30% adoption of biofuels by 2030. After being scaled up to a commercial-scale production, Solazyme's technology, which uses *algae* to convert biomass into oil using photosynthesis, could supply around 50–100 million gallons per year of jet fuel at $60–80/barrel.

Between February 2008 and June 2011, 11 airlines flew their various aircrafts with a mixed blend of biofuels, and between June 2011 and October 2011, five airlines flew Boeing and Airbus commercial aircrafts using biofuel. In October 2011, the UK's Thompson Airways flew a Boeing 757-200 aircraft using biofuel from used cooking oil. By using biofuels, an 85% reduction in greenhouse gases is estimated, and the biofuels do not contain sulfur compounds and thus do not emit sulfur dioxide.

Ethanol (C_2H_5OH) produced through fermentation from a number of raw materials and ethanol-blended gasoline have been used in cars for a number of years, but not so much for aircrafts. In Brazil, where ethanol is produced in large quantities and is used in blended form for cars, the price of ethanol has come down to the level of usual gasoline and lower.

Finally, *electric aircraft* have been used as model aircraft for many years. They run on electric motors rather than on the internal combustion engines, with electricity coming from fuel cells, ultracapacitors, solar cells, power beaming, and/or batteries. They are also currently being tested.

1.5   Specific Impulse and Range

The *specific impulse* is defined as the ratio of thrust per unit mass fuel flow rate and is given by the equation

$$I = \frac{F}{\dot{m}_f} \tag{1.20}$$

Thus, for a jet engine, except the fanjet engine,

$$I = \frac{\dot{m}_a(u_e - u_\infty)}{\dot{m}_f} = \frac{u_e - u_\infty}{f}, \frac{m}{s} \tag{1.21a}$$

generally independent of attitude $(f = \dot{m}_f/\dot{m}_a)$, and for a fanjet engine,

$$I = \frac{\dot{m}_H(u_H - u_\infty) + \dot{m}_c(u_c - u_\infty)}{\dot{m}_f}, \frac{m}{s} \qquad (1.21b)$$

With the *bypass ratio* $b = \dot{m}_c/\dot{m}_H$ and *fuel–air ratio* $f = \dot{m}_f/\dot{m}_H$, the *specific impulse for a fanjet engine* becomes

$$I = \frac{\dot{m}_H}{\dot{m}_f}[(u_H - u_\infty) + b(u_c - u_\infty)] = \frac{[(u_H - u_\infty) + b(u_c - u_\infty)]}{f} \qquad (1.21c)$$

Now for an aircraft to have level flight, the lift L must balance against the weight of the aircraft Mg, and the drag D must be equal to the thrust F. Therefore,

$$L = Mg; \quad F = D = L\left(\frac{D}{L}\right) = Mg\left(\frac{D}{L}\right)$$

Since

$$\dot{m}_f = -\frac{dM}{dt} = \frac{F}{I} = \frac{Mg}{I}\left(\frac{D}{L}\right)$$

we get by integration

$$M = M_{init}\exp^{\frac{g}{I}\left(\frac{D}{L}\right)t} \qquad (1.22)$$

where M_{init} is the initial mass of the aircraft (at time $t = 0$). Now if (D/L) is kept constant at minimum during cruise and also if I is constant, then we get the *maximum flight time* t_{max} from the relation

$$t_{max} = \frac{I}{g}\left(\frac{L}{D}\right)_{max} \ln\left(\frac{M_{init}}{M_{final}}\right) \qquad (1.23a)$$

and the *flight range* is

$$\text{Range} = u_\infty t_{max} = \frac{I u_\infty}{g}\left(\frac{L}{D}\right)_{max} \ln\left(\frac{M_{init}}{M_{final}}\right) \qquad (1.23b)$$

where M_{final} is the *final mass of the aircraft*.

One can, of course, estimate the specific impulse of the engine from the flight parameters. As a constant consumption of fuel

$$\dot{m}_f = \frac{F}{I} = \frac{1}{t_{max}}\left(\frac{M_{final}}{M_{init}}\right)$$

one gets an implicit relation to determine I.

While (1.23b) does not give any indication of the flight altitude, evidently for a given weight and flight speed, the lift has to be independent of the flight altitude, that is, independent of the ambient pressure. On the other hand, the drag depends on the ambient air density and pressure. Thus, the ratio L/D must increase with the altitude. At a lower altitude, (L/D), $(L/D)_{max}$, and the range will be smaller for cruise flights. It may also be noted that for a given flight speed, both the drag D and the engine thrust F are proportional to the flight altitude ambient pressure, p_∞^0.

1.6 Some Selected Aircraft Engines and Future Trends

We start our review of selected engines with the Rolls-Royce DART, which has been one of the most successful turboprop engines, a sketch of which appears in Fig. 1.5a, and the cyclic process involved is shown in Fig. 1.5b. There is a temperature-entropy (T, s) chart. The engine has, among others, three models: DART 510, 526, and 545, and consists of two-stage radial compressors, seven combustor cans and a two-stage (for DART 510) or three-stage (for DART 526/545) axial turbine. Since the turbine in a turboprop engine has to drive the propeller, there is very little expansion in the nozzle and state 4 in the (T, s) chart is very near state 5. Some of the performance parameters for various modules of the DART engine are given later in the Appendix, a summary of which is

Mass flow rate of air, $\dot{m}_a = 9.1$–10.0 kg/s
Overall compression ratio, $\pi_c = 5.5$–5.75
Combustion chamber temperature, $t_3 = 860$–$890°C$
rpm $= 14,500$–$15,000$ min^{-1}
Shaft power $= 1,200$–$2,200$ kW
SFC $= (106$–$115) \times 10^{-6}$ kg/kJ

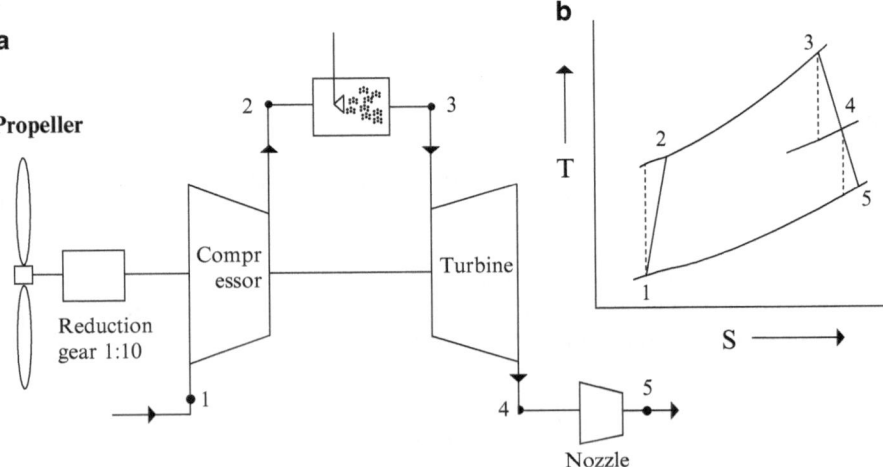

Fig. 1.5 (a) Schematic sketch of Rolls-Royce DART engine; (b) corresponding cyclic process

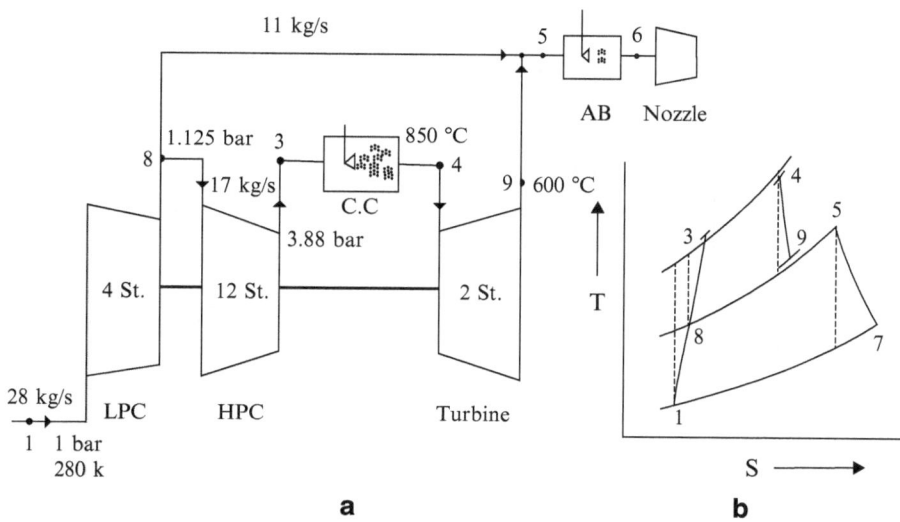

Fig. 1.6 (**a**) Schematic sketch of Rateau's A-69 engine; (**b**) associated cyclic process

The next engine being considered is the low-bypass turbojet engine A-65 of the French company Rateau, whose initial production was started in 1947. A schematic sketch of the engine is shown in Fig. 1.6a, and the associated thermodynamic process is shown in Fig. 1.6b.

The engine consists of a four-stage axial fan (low-pressure compressor) with the fan compression ratio $\pi_f = 1.125$ and a 12-stage axial high-pressure compressor with the overall compressive ratio $\pi_c = 3.88$, followed by nine reverse-flow combustor cans and a two-stage axial turbine. The rpm of the engine was 8,000 $\min^{-1}$. The bypass ratio was $b = 11/17 = 0.649$, which is quite low; the exhaust jet speed was normally 280 m/s and was 320 m/s with afterburner, with the corresponding thrust being 14 and 16 kN, respectively.

The low-bypass engine, in which a part of the air is bypassed around the combustion chamber and the turbine and mixed after the turbine, has an improved efficiency against the straight-jet engine because of a lower nozzle exit speed. However, increasing the bypass ratio in such engines means a lower and lower turbine exit pressure, and one can hardly have a bypass ratio greater than 1. Increasing the bypass ratio much beyond 1 became practical with the high-bypass fans, which is designated further as the *fanjet*. The main difference between the (*low-*) bypass jet and the fanjet is that in former the hot and cold gases mix inside the jet pipe, whereas in the latter, both are expelled. In the former, there are important installation advantages as the single pipe is less complicated, and the silencer and the thrust reverses need act only in one gas stream.

Modifying regular straight-jet engines, General Electric (U.S.) developed in 1953 a fanjet engine CJ805-21. In this the main 17-stage axial compressor was left untouched to compress 72.5 kg/s air with a compression ratio $\pi_c = 12$ driven by

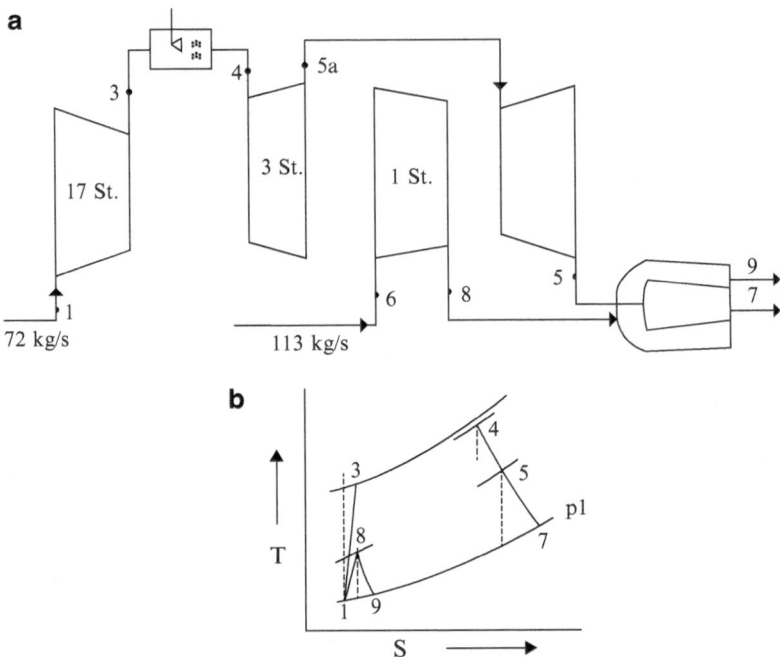

Fig. 1.7 (**a**) Schematic sketch of GE CJ-805-21; and (**b**) corresponding thermodynamic process in a (T, s) chart

a three-stage axial turbine. The fourth-stage axial turbine was used to drive a fan stage placed at the top of the turbine stage to compress 113 kg/s with a fan compression ratio $\pi_f = 1.6$. The arrangement of the engine and the equivalent thermodynamic cycle on a temperature-entropy (T, s) chart are shown in Fig. 1.7a, b. While such an aft fan, designed as an extension of earlier straight-jet engines, is much simpler in comparison to later-developed front-fan engines, the former blades, made out of the same material as the expensive turbine blades, may be quite expensive. On the other hand, the air passing through the fan of the front-fan engines is compressed before going to the high-pressure system to increase the overall pressure ratio.

A variation of GE's fanjet engine, a front-fan engine (Fig. 1.7a), came a few years later, for example, with Pratt & Whitney's P&WJT3D engine for Boeing 707 aircraft coming into production in 1960; Pratt & Whitney's P&WJT9D engine for Boeing 747 first produced in 1967 and subsequently for DC-10 in 1972 and Airbus A300B4 in 1973; Pratt & Whitney's P&WJT8D first produced in 1970 for Boeing 727 and later for Boeing 757, Aerospatialle Caravelle, DC-9-20, etc.; and finally, Rolls-Royce RB211 for L-1011 Tristar, Boeing 757, and Airbus A300 and A310. A schematic sketch of such engines is given in Fig. 1.8, whereas the thermodynamic process is essentially the same as that given in Fig. 1.7b.

There is a great difference in magnitude between the bypass flow turboprop and turbofan. In the turboprop the mass flow rate through propeller may be 50

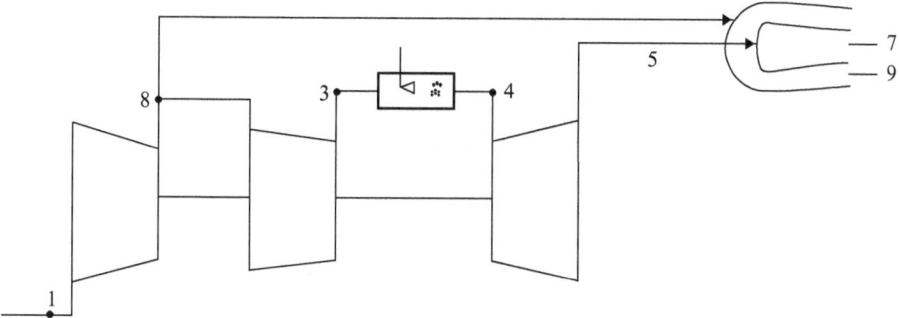

Fig. 1.8 Schematic sketch of a fanjet engine

or more times than the turbofan jet pipe mass flow, whereas the bypass ratio (maximum) is about 10.

Some other modern-day engines that should also be mentioned at this stage are the following:

1. Pratt & Whitney's P&WJT3D (1960) was built for Boeing 707 aircraft. It was a two-spool engine in which the inner spool running at $n_1 = 6,185$ min^{-1} consisted of one fan stage, seven L.P. compressor stages, and one L.P. turbine stage, and the outer spool running at $n_2 = 9,655$ min^{-1} consisted of seven H.P. compressor stages and one H.P. turbine stage. The engine was 3.46 m long, with a 1.35-m maximum diameter and a weight of 1,892 kg. The engine had a cannular combustion chamber. Several different models were built, with the thrust ranging between 7,720 and 9,500 kg, a mass flow rate between 204 and 227 kg/s, a bypass ratio between 1.24 and 1.37, a fan pressure ratio between 1.71 and 1.90, an overall compression ratio between 13.0 and 16.10, and a take-off specific fuel consumption (SFC) between 0.52 and 0.605 kg/(kgf-h).
2. Pratt & Whitney's JT8D was first built in 1970 for Boeing 727, but it was used subsequently for Boeing 737, Aerospatialle Caravelle, MacDonnell Douglas DC-9, etc. Approximately 1,000 units of these engine types were built per year, and until 1982, around 9,700 units were built. It again has two spools, the inner spool consisting of one fan stage with titanium blades, six axial LP compressors, and a seven-stage axial LP turbine, and an outer stage consisting of a seven-stage HP axial turbine and a one-stage axial HP turbine. There are nine cannular low-emission burners. It gives 9,400 kg of thrust. Other data for the engine include an overall compression ratio of 18.6, an SFC of 21.32 mg/Ns, a mass flow rate of 217 kg/s, a maximum diameter of 1.25 m, and an engine length of 3.911 m.
3. Rolls-Royce's RB211 engine, built for Lockheed L-1011 TriStar, Boeing 757, Airbus A300 and A310 aircrafts, certification given first in 1973, is a three-spool engine. The innermost spool has one fan and one turbine stage, the intermediate spool has seven compressor stages and one turbine stage, and the outermost spool consists of six compressor stages and one turbine stage. The engine length is 3 m and the diameter is 2.18 m. Further engine data are an overall compression

Fig. 1.9 General Electric's GE-90 engine (Courtesy of GE)

ratio of 25:1, a bypass ratio of 5, thrust is 18,000 kgf at 34.2°C, and the mass flow
rate is 657.7 kg/s.
4. The P&W JT9D was first built in 1967 and was flown first with Boeing 747 in
   February 1979, with DC10-40 in February 1972, and with A300B4 in 1978.
   Several different models were built, but the general arrangement of this two-
   spool engine was as follows:

Inner spool: LP 1 fan + 3 compressors + 4 turbine, $n_1 = 3{,}415$ min^{-1}
Outer spool: HP 11 compressors + 2 turbines, $n_2 = 7{,}650$ min^{-1}
Idle speed (64%) $= 4{,}950$ min^{-1}

The fan stage was with titanium blades, while the rest were made of a titanium
alloy or a high-nickel alloy. The engine had a diameter of 2.43 m and a length of
3.26 m. Some of the temperatures in model 3A are: in engine inlet 27°C, fan exit
640°C, low pressure compressor (LPC) exit 112°C, high pressure compressor
(HPC) exit 505°C, high pressure Turbine, (HPT) inlet 1,220°C (HPT inlet for latest
model: 1,370°C), low pressure turbine (LPT) inlet 780°C, and exhaust 494°C.
Furthermore, the characteristic speeds in the same model are inlet 1,500 m/s, LPC
exit 170 m/s, HPC exit 130 m/s, jet exhaust 440 m/s, and fan exhaust 271 m/s.
5. In the 1990s, two fanjet engines, Pratt & Whitney's PW4000 (Fig. 1.8) and
   General Electric's GE-90 (Fig. 1.9), were developed. The latter was first certified
   in 1995 with a thrust of 380 kN/kg (force). About 15 years later, GE90-115B had
   a certified thrust of 510 kN and was found in several models of Boeing 777
   aircrafts. GE-90's 3.07-m fan can provide a 9:1 bypass ratio and a high-pressure
   compressor with a 23:1 pressure ratio. The turbine disks, blades, and vanes were
   made out of monocrystal alloys.

Pratt & Whitney's PW4000 (Fig. 1.10), with a thrust in the range of 250–280
kN, received its certification in 1986, and it is supposed to grow to an additional
10% thrust later.

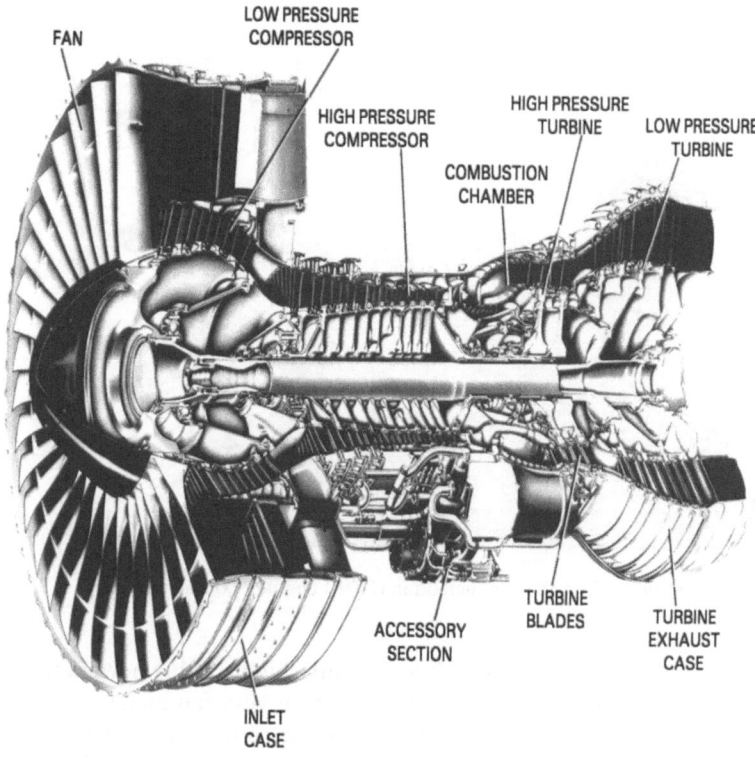

Fig. 1.10   Pratt & Whitney's PW4000 Engine (Courtesy of Pratt & Whitney)

Looking at the trends in fan engine design, it is evident that as the previous millennium came to an end, the overall pressure ratio went to about 45, the bypass ratio to about 9, and the turbine inlet temperature to about 1,500°C. It has also been observed that for a given thrust, if the velocity in the outstream is higher than that of the core, a noise reduction of the order of 10 pNdB may occur. Further, the fuel flow decreases 0.7% for each 1% increase in fan efficiency. Around 1976, the average operating time of an aircraft engine was 14–25 h. In recent years, no engine has entered service without between one or two orders of magnitude more life.

6. A new development in the 1990s was the development of *unducted fan* (*propfan*) engines with counter-rotating propellers with a large solidity (Fig. 1.11). An unducted fan is a two-spool engine with a low-pressure counter-rotating turbine and has an effective bypass ratio of 25:1. The turbine blades rotate at the same speed but in opposite directions, so that the effect is that of flow output in the axial direction and there is no need for a gear box. To alleviate the effects of high relative Mach numbers, the blades are thin and "swept back" at large radii. A 4-m-diameter propfan may be able to generate 160 kN of thrust.

7. In the new millennium, some very interesting engines have been seen. Among the GE engines for commercial aircrafts to be mentioned is the GEnX family of engines, with a thrust range between 24,000 and 33,000 kgf, found in Boeing

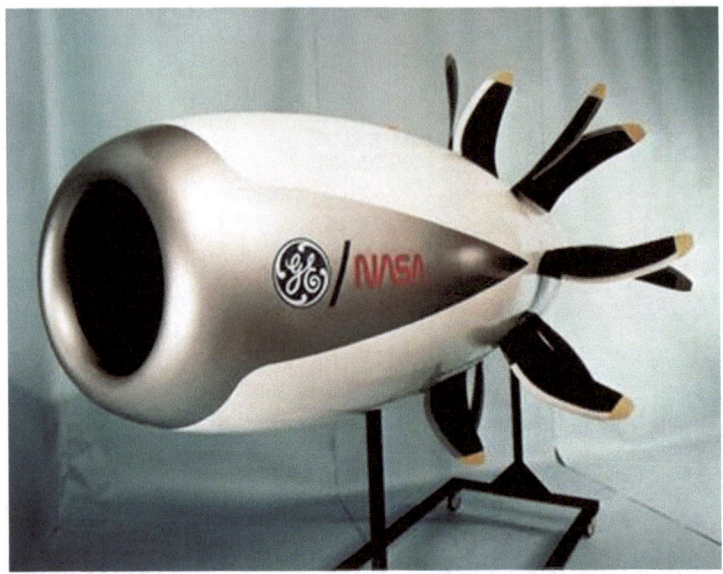

Fig. 1.11 Picture of a GE/NASA unducted fan (UDF) engine (Courtesy of GE)

787 Dreamliners and Boeing 747-8. The first flight of this new engine was on June 16, 2010. It is the world's only jet engine with fan blades and a fan case made of camber-fiber composites. It has one fan, four LPC, 10 HPC, an annular combustor, two HPT, and seven LPT, with a compression ratio of 23:1.

A joint venture between GE and Pratt & Whitney is the GE7000; with a take-off thrust 311 kN and cruising at 11.5-km altitude at M = 0.85, a cruise thrust of 56 kN is developed. It is earmarked for double-decker Airbus A380, with one fan, five LPC, nine HPC, a single annular combustor, two HPT, and six LPT. The physical dimensions of the engine are as follows: length = 4.75 m, diameter = 3.16 m, and fan diameter = 2.96 m. A further development of the same engine is the GP7277, with a take-off thrust of 343 kN and a cruise thrust of 61 kN. Another joint venture of GE and P&W for use in double-decker Airbus A380, it had its first flight on Emirates on August 1, 2008.

Pratt & Whitney's PW5000 (internal designation F119) developed an afterburning turbofan for supersonic flights that even without the use of an after-burner delivers 160 kN thrust. It is a twin-spool, augmented turbofan of length 5.16 m, dry weight 1,775 kg, and with an annular combustor, axial flow-centered rotating turbines with thrust-to-weight ratio of 9:1. A further Pratt & Whitney engine F135 is an afterburning turbofan developed for F-35 Lightning II single-engine strike fighters. The first production engine was scheduled for 2009.

UK's Rolls-Royce Company, which is the world's second-largest aero-engine manufacturing company after GE, developed an engine Trent 900 from Rolls-Royce RB211 in 1990 for Airbus A380 and Boeing 787 (Dreamliner). It is a three-spool engine, with a counter-rotating high spool. There are several variants, with a maximum thrust going up to 360 kN and including a 2.95-m-diameter swept back fan. Another engine from Rolls-Royce, V2533-A5, has a thrust of 146.80 kN, a bypass ratio of 4.5:1, a compression ratio of 34.2:1, a fan diameter of 1.613 m, a total length of 3.2 m, and a weight of 2,359 kg; production started in 1996.

1.7 Exercises

1. The equation for the pressure differential due to an air column is $dp = -\rho g dH$, where H is the height. The equation can be integrated easily for a constant air temperature and constant gravity constant, and one gets an expression for the pressure (or density) ratio at a height divided by these values at zero altitude (*barometric pressure formula*):

$$\frac{p}{p_{H=0}} = \frac{\rho}{\rho_{H=0}} = \exp^{-0.114 H_{km}}$$

   If an aircraft flies at subsonic-level flights at different altitudes by keeping the flight speed the same, how do the following variables for the aircraft and the engine change: lift, L; drag, D; mass flow rate of air and fuel, $\dot{m}_a$ and $\dot{m}_f$; fuel–air ratio, $\dot{m}_f/\dot{m}_a$; engine thrust, F; and specific fuel consumption, SFC?

2. For aircraft fuels given in Table 1.4, estimate for stoichiometric mixtures the fuel–air ratio, composition after burning, and adiabatic flame temperature.

3. Consider the engine data given in the database contained in this book or elsewhere, and examine the engine specification values for different type of engines, weight-to-thrust ratio, etc.

4. Estimate the fuel–air stoichiometric mass ratio f_{stoic} (kg_{fuel}/kg_{air}) and corresponding temperature in combustion chamber of all the fuels given in Table 1.3.

5. Given that a commercial airliner with a dry mass of the aircraft 800 t ($= 8 \times 10^5$ kg) and $L/D = 1.15$ has a range of 10,000 km using 200 t ($= 2 \times 10^5$ kg) of aviation fuel, estimate the range of the aircraft burning the same volume of the liquid or gaseous hydrogen. The fuel properties are as follows:

Fuel	Heat content (kJ/kg)	Fuel's molar mass	f_{stoic}	Liquid density	I (m/s)
Gasoline	42,707	53	0.4990	800.0	1,560
$H_{2,gas}$ at 298 K	119,955	2	0.0147	70.8	53,318

6. Enumerate the advantages and disadvantages of nonpetroleum-based fuels vis-à-vis petroleum-based aviation fuels.

7. Readers are encouraged to study various aircraft topics, including those related to fuels, on the Internet.

Chapter 2
Thermodynamic Ideal Cycle Analysis

In studying an ideal cycle relevant to aircraft propulsion, one can have a good insight into the relevant important parameters that give the performance parameters of a given system. For such a study, the following assumptions are usually taken:

1. Compression and expansion processes are isotropic.
2. The working fluid is a perfect gas with a constant specific heat and a constant adiabatic exponent.
3. The fuel mass added to air is neglected, and heat is added as if it is from an external source.

Under the above assumptions, we will first study piston engines followed by jet engines, including gas turbines.

2.1 Propeller-Driven Engines

Among the propeller-driven engines, we discuss first the piston engines and later the turboprops. A schematic sketch of a piston engine is shown in Fig. 2.1, in which TDC and BDC refer to the top dead center and bottom dead center, respectively, as per the convention to describe extreme positions of a vertically operating piston. Further, V_c is the *clearance volume*, V_d is the *displacement volume*, and the sum of the two, $V_t = V_c + V_d$, is the *total volume*. The diameter of the cylinder is D, known as the *bore*; the length of the cylinder that is traversed by the piston is the stroke length, or *stroke*. Thus, the displacement volume is the product of the stroke length and cross section of the cylinder.

In the operation of piston engines, we have to differentiate between *two-stroke* and *four-stroke* engines. In the four-stroke engine, air is allowed inside the cylinder by opening the valve and withdrawing the piston, creating a vacuum inside the cylinder. Then the valve is closed and air is compressed by the piston's pushing forward. The fuel is ignited with the resultant withdrawal of the piston, giving the

T. Bose, *Airbreathing Propulsion: An Introduction*,
Springer Aerospace Technology, DOI 10.1007/978-1-4614-3532-7_2,
© Springer Science+Business Media, LLC 2012

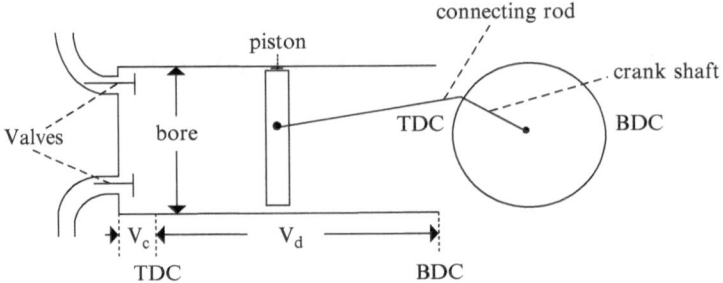

Fig. 2.1 Schematic sketch of a piston engine

effective work (*power stroke*), and, finally, the valve opens and the exhaust gas is pushed out. Thus, in a four-stroke engine, there are four different strokes in two complete cycles of motion of the crank shaft for each power stroke. For ideal cycle analysis, however, it has been found more expedient to consider the two-stroke engine, in which there are no separate sections and exhaust strokes; they are combined as well as possible in the design, near the bottom dead center, which ideally is a constant-volume process. Incidentally, in piston engines, a *volume compression ratio* is defined as $\varepsilon = V_t/V_c$, the ratio of the total volume to the clearance volume, whereas for gas turbines, usually the compression ratio means the ratio of pressure across the compression. The volume compression ratio for aircraft piston engines with Otto cycles is between 5.9 and 10.5 and with fiesel cycles is between 15 and 22; the aircraft diesel engine in the databank with this book has a volume compression ratio of 27.0. The pressure in a diesel engine rises to about 40.0 bar, while in a gasoline engine, after the compression stroke, is between 8 and 14 bar.

With simple algebraic manipulation, we can get the following relations among the three volumes (see Fig. 2.2):

$$V_c = V_t/\varepsilon = V_2; \ V_d = V_t\frac{\varepsilon - 1}{\varepsilon}; \ V_t = V_1$$

We now separately consider the ideal *Otto cycle* and the ideal *diesel cycle. Ideal cycle* means the isentropic process is without any heat exchange with the wall, the *isobar* process is completely at constant temperature, and the *isochore* (constant-volume) process involves instantaneous change in the pressure at a constant volume. The gas behaves like an ideal gas, and the ideal change of the isentropic process is through the isentropic coefficient for ideal air, $\gamma = 1.4$.

2.1.1   Ideal Otto Cycle

In an Otto cycle, a fuel–air mixture is introduced during the changing of the cylinder, and near the top dead center (TDC), the fuel–air mixture is ignited

Fig. 2.2 Thermodynamic process in Otto cycle of a piston engine

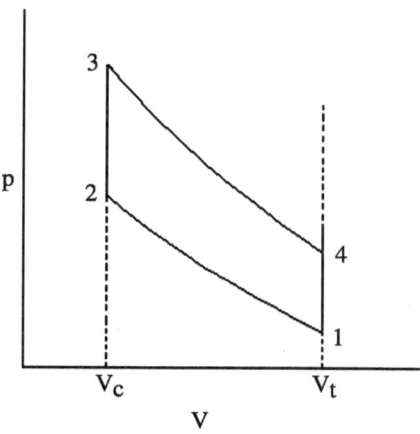

electrically with the help of a spark-ignition system. Ideally, the combustion is instantaneous and can be considered a constant-volume process. It is therefore also called a *spark-ignition (SI)* engine. Hence, the thermodynamic cycle consists of a four-part process, shown schematically in Fig. 2.2, as follows: Air is brought into the cylinder at state 1 (*intake stroke*); it is compressed isentropically (with constant entropy) to state 2 (*compression stroke*); combustion takes place at a constant volume to reach state 3; it is expanded isentropically to state 4 (*power stroke*); the exhaust gas is expelled (*exhaust stroke*), and simultaneously a fresh change in the fuel–air mixture is introduced at a constant volume (*isochore*). This is when all four processes are completed within one cycle in two half-cycles (*two-stroke engines*), as in the present ideal case, or in four half-cycles; the last two are separate processes for the suction of air and expelling the hot gas (*four-stroke engine*).

In addition to the *compression ratio* $\varepsilon = V_t/V_c$, let the other operational parameter be $\Theta_3 = T_3/T_1$, which depends on the *fuel–air ratio*. Now, for the two *isentropic processes*,

$$p_1 V_t^\gamma = p_2 V_t^\gamma \tag{2.1a}$$

and

$$p_4 V_t^\gamma = p_3 V_c^\gamma \tag{2.1b}$$

and we get

$$\frac{p_2}{p_1} = \frac{p_3}{p_4} = \left(\frac{V_t}{V_c}\right)^\gamma = \varepsilon^\gamma = \left(\frac{T_2}{T_1}\right)^{\gamma/(\gamma-1)} = \left(\frac{T_3}{T_4}\right)^{\gamma/(\gamma-1)} \tag{2.2}$$

Thus,

$$\frac{T_2}{T_1} = \frac{T_3}{T_4} = \varepsilon^{(\gamma-1)} \tag{2.3}$$

and

$$\frac{p_3}{p_2} = \frac{p_4}{p_1} = \frac{T_3}{T_2} = \left(\frac{T_3}{T_1}\right)\left(\frac{T_1}{T_2}\right) = \frac{\Theta}{\varepsilon^{(\gamma-1)}} \tag{2.4}$$

Similarly,

$$\frac{T_4}{T_1} = \left(\frac{T_4}{T_3}\right)\left(\frac{T_3}{T_1}\right) = \frac{\Theta}{\varepsilon^{(\gamma-1)}} \tag{2.5}$$

Now, heat added to the process at a constant volume is $q_a = q_{23} = c_v(T_3 - T_2)$, and heat rejected at a constant volume is $q_r = q_{41} = c_v(T_1 - T_4)$. Therefore,

$$\frac{q_a}{c_v T_1} = \frac{T_3}{T_1} - \frac{T_2}{T_1} = \Theta - \varepsilon^{\gamma-1} \tag{2.6a}$$

and

$$\left|\frac{q_r}{c_v T_1}\right| = \frac{T_4}{T_1} = -1 = \frac{\Theta - \varepsilon^{(\gamma-1)}}{\varepsilon^{\gamma-1}} \tag{2.6b}$$

Furthermore, $q_{12} = q_{34}$. Now the compression work is $w_{12} = c_v(T_2 - T_1)$, and the expansion work is $w_{34} = c_v(T_4 - T_3)$. For the other two-part processes, $w_{23} = w_{41} = 0$ since both are constant-volume processes. Thus,

$$\left|\frac{w_{12}}{c_v T_1}\right| = \frac{T_2}{T_1} - 1 = \varepsilon^{(\gamma-1)} - 1 \tag{2.7a}$$

and

$$\left|\frac{w_{12}}{c_v T_1}\right| = \left(\frac{T_3}{T_4}\right)/T_1 = \Theta \frac{\left(\varepsilon^{(\gamma-1)} - 1\right)}{\varepsilon^{(\gamma-1)}} \tag{2.7b}$$

Now, in a cyclic process, the overall work gained must be the same as the overall heat exchanged, and thus,

$$\frac{w}{c_v T_1} = \left|\frac{w_{12}}{c_v T_1}\right| - \left|\frac{w_{12}}{c_v T_1}\right| = \frac{\left(\varepsilon^{(\gamma-1)} - 1\right)\left(\Theta - \varepsilon^{(\gamma-1)}\right)}{\varepsilon^{(\gamma-1)}} \tag{2.8a}$$

$$\frac{q}{c_v T_1} = \left|\frac{q_a}{c_v T_1}\right| - \frac{q_r}{c_v T_1} = \frac{\left(\Theta - \varepsilon^{(\gamma-1)}\right)\left(\varepsilon^{(\gamma-1)} - 1\right)}{\varepsilon^{(\gamma-1)}} \tag{2.8b}$$

Obviously, in view of the *first law of thermodynamics*, the *overall work* and *overall heat* must be the same. However, the thermodynamic efficiency is defined as the overall work divided by the added heat. Thus, the *thermodynamic efficiency* is

$$\eta_{\text{th}} = \frac{w}{q_a} = \frac{(\varepsilon^{\gamma-1} - 1)}{\varepsilon^{\gamma-1}} \tag{2.8}$$

and the two limiting values of the thermodynamic efficiency are

$$\varepsilon \to 1 : \eta_{\text{th}} \to 0 \quad \text{and} \quad \varepsilon \to \infty : \eta_{\text{th}} \to 1$$

However, the compression ratio for Otto cycles with gasoline as fuel for a typical aircraft engine is within the range 6.3–10.5, and any higher value is restricted due to autoignition.

Although Θ as a *temperature ratio* is introduced here as an independent parameter, it depends on the fuel–air ratio, the maximum inverse value of which can be taken from Table 1.3. Analogous to (1.19), we may write

$$\dot{m}_f \Delta H_v = \dot{m}_a c_v (T_3 - T_2) \tag{2.9a}$$

and hence,

$$f = \frac{\dot{m}_f}{\dot{m}_a} = \frac{c_v T_1}{\Delta H_v} (\Theta - \varepsilon^{\gamma-1}) \tag{2.9b}$$

In the above equation, ΔH_v is the heat of reaction, the *lower heat of reaction*, at a constant volume. Noting further that the relation between the heat of reaction for constant-pressure combustion and for constant-volume combustion is given by the relation

$$\Delta H_p = \Delta H_v + v \Delta p$$

it follows that

$$\Delta H_v = \Delta H_p - \frac{V_c}{M_1 \varepsilon} \left(\frac{p_3}{p_1} - \frac{p_2}{p_1} \right)$$

where M_1 is the mass of air at point 1.

Noting further that

$$M_1 = \frac{p_1 V_t}{RT}; \frac{p_2}{p_1} = \varepsilon^{\gamma}; \frac{p_3}{p_1} = \Theta \varepsilon$$

we get

$$\frac{\Delta H_v}{c_v T_1} = \frac{\Delta H_p}{c_v T_1} - (\gamma - 1)(\Theta - \varepsilon^{\gamma-1})$$

By rearrangement, we therefore get

$$\Theta - \varepsilon^{\gamma-1} = \gamma f \frac{\Delta H_p}{c_p T_1 [1 + f(\gamma - 1)]} \approx \frac{\Delta H_p}{c_p T_1}$$

It is worth mentioning again that in the above, $M_1 = p_1 V_t / (R T_1)$ is the mass of the medium (air) in the cylinder. Further, the *mean indicated pressure*, which is an important design parameter, is defined as the work done in a cycle, divided by the displacement volume. Thus, $p_m = w M_1 / V_d$, from which it follows that

$$\frac{p_m}{p_1} = (\gamma - 1)\frac{\varepsilon}{\varepsilon - 1}(\Theta - \varepsilon^{\gamma-1})\left(\frac{\varepsilon^{\gamma-1} - 1}{\varepsilon^{\gamma-1}}\right) \qquad (2.10)$$

Now, for piston engines, the power developed is given by the relation

$$P_E = p_m V_d \left(\frac{n}{60}\right)\left(\frac{2}{N}\right) \qquad (2.11)$$

where n is the rpm, and $N = 2$ for a two-stroke and $N = 4$ for a four-stroke engine. In (2.11), if p_m is in N/m^2 and V_d is in m^3, then P_E is in watts. For a practical aircraft piston engine, $p_m = 7.8$–18.7 bar, but generally it is 12–13.5 bar.

We may recall that the *specific fuel consumption* is given by the relation

$$\text{SFC} = \frac{\dot{m}_f}{\dot{m}_a w} = \frac{f}{c_v T_1} \cdot \frac{c_v T_1}{w} = \frac{\varepsilon^{\gamma-1}}{[\Delta H_v (\varepsilon^{\gamma-1} - 1)]} = \frac{1}{(\Delta H_v \eta_{th})}$$

The above definition of the specific fuel consumption is dependent on the heat content of the fuel, and hence is dimensional. Thus, we define a *nondimensional specific fuel consumption*

$$\text{SFC}^* = \text{SFC}.\Delta H_v = 1/\eta_{th}$$

which is inversely proportional to the thermodynamic efficiency, which depends on the compression ratio only.

We will now make an estimate of the compression ratio based on the performance parameters of the engine. We take gasoline as the fuel, for which, according to Tables 1.2 and 1.3, $\Delta H_p = 4.2707 \times 10^4$ kJ/kg and *stoichiometric*

Table 2.1 Some nondimensional results for different compression ratios in Otto cycles

ε	4	5	6	8	10	15
T_2/T_2	1.741	1.903	2.048	2.297	2.511	2.954
Θ	11.710	11.880	12.020	12.270	12.490	12.930
$w/(c_v T_1)$	4.247	4.736	5.104	5.634	6.005	6.600
p_m/p_1	14.150	14.800	15.310	16.090	16.680	17.680
η_{th}	0.426	0.475	0.510	0.564	0.602	0.661
SFC*	1.954	1.852	1.771	1.710	1.662	1.621
$\eta_{th,actual}$	0.310	0.360	0.370	0.420	0.460	0.510

fuel–air ratio $f = 1/20.03 \approx 0.05$. Assuming further that the specific heat of air is $c_p = 1.005$ kJ/(kgK), ambient temperature $T_1 = 298.15$ K, and *specific heat ratio* $\gamma = 1.4$, we get for the *temperature ratio*

$$\Theta = \varepsilon^{0.4} + \left[\frac{1.4 \times 0.05 \times 4.2707 \times 10^4}{1.005 \times 298.15} \right] = \varepsilon^{0.4} + 9.976914$$

Some of these results are computed and given in Table 2.1.

It is evident that for an Otto motor to have better thermodynamic efficiency and better specific work, the compression ratio must be increased. This is not possible because of the autoignition of the fuel–air mixture if the temperature after the compression, T_2, is too high. It is therefore necessary to introduce the fuel just around the top dead center, but in order to keep the already high pressure within limits, fuel is introduced at a controlled rate. This leads us to a compression-ignition engine, better known as a diesel engine.

2.1.2 Ideal Diesel Cycle

In a diesel cycle, fuel is introduced at a controlled rate to produce a constant-pressure combustion, while the piston is withdrawn to a cylinder volume V_3, and under ideal conditions, further expansion takes place isentropically. The process is shown schematically in Fig. 2.3.

Let's reintroduce first the three nondimensional ratios: the *volume compression ratio*, $\varepsilon = V_t/V_c$, the *combustion volume ratio*, $\Omega = V_3/V_c$, and the *temperature ratio*, $\Theta = T_3/T_1$ Obviously, $\Omega < \varepsilon$, and the value of Ω depends on both Θ and ε. Since

$$\frac{T_3}{T_2} = \frac{V_3}{V_c} = \Omega = \left(\frac{T_3}{T_1} \right) \left(\frac{T_1}{T_2} \right) = \frac{\Theta}{\varepsilon^{\gamma-1}}$$

Fig. 2.3 Schematic sketch
of a compression
ignition engine

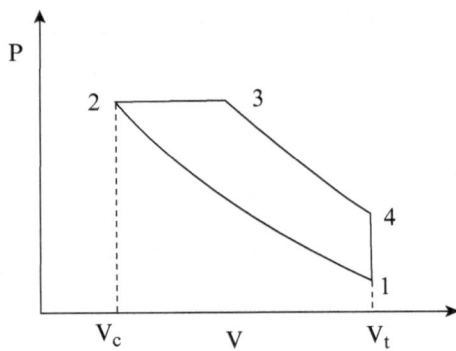

we get

$$\Omega = \frac{\Theta}{\varepsilon^{\gamma - 1}} \tag{2.12}$$

Further, since $p_2 = p_3$ and $p_1 V_t^\gamma = p_2 V_c^\gamma$, we get

$$\frac{p_2}{p_1} = \frac{p_3}{p_1} = \varepsilon^\gamma \tag{2.13a}$$

Now,

$$\frac{p_3}{p_4} = \left(\frac{V_t}{V_c}\right)^\gamma = \left[\left(\frac{V_t}{V_c}\right)\left(\frac{V_c}{V_c}\right)\right]^\gamma = \left(\frac{\varepsilon}{\Omega}\right)^\gamma$$

and thus,

$$\frac{p_4}{p_1} = \frac{p_3}{p_1} \cdot \frac{p_4}{p_3} = \Omega^\gamma \tag{2.13b}$$

Further,

$$\frac{T_2}{T_1} = \left(\frac{p_2}{p_1}\right)^{\frac{\gamma-1}{\gamma}} \tag{2.13c}$$

and

$$\frac{T_4}{T_1} = \left(\frac{T_4}{T_3}\right)\left(\frac{T_3}{T_1}\right) = \left(\frac{T_3}{T_1}\right)\left(\frac{p_4}{p_3}\right)^{\frac{\gamma-1}{\gamma}} \Theta \left(\frac{\Omega}{\varepsilon}\right)^{\gamma-1} \Omega^\gamma \frac{p_4}{p_1} \tag{2.13d}$$

Thus, the specific work done (per unit mass) for various part processes are

$$\left|\frac{w_{12}}{c_v T_1}\right| = \left(\frac{T_2}{T_1}\right) - 1 = \varepsilon^{\gamma-1} - 1 \tag{2.14a}$$

$$\left|\frac{w_{34}}{c_v T_1}\right| = \frac{T_3 - T_4}{T_1} = (\Theta - \Omega^\gamma) \tag{2.14b}$$

$$w_{41} = 0 \tag{2.14c}$$

Further, since $w_{23} = p_2 - (V_3 - V_2)/M_1$ and $M_1 = p_1 V_t/(RT_1)$ is the mean of air inside the cylinder, we get

$$\left|\frac{w_{23}}{c_v T_1}\right| = (\gamma - 1)(p_2 - p_1)\frac{(V_3 - V_c)}{V_t} = (\gamma - 1)(\Omega - 1)\varepsilon^{\gamma - 1}$$

$$= (\gamma - 1)\left(\Theta - \varepsilon^{\gamma - 1}\right) \tag{2.14d}$$

On the other hand, the *heat added* is

$$q_a = c_p(T_3 - T_2)$$

and the *heat rejected* is

$$q_r = c_v(T_4 - T_1)$$

from which, after some manipulation, we get

$$\left|\frac{q_a}{c_v T_1}\right| = \gamma\left(\Theta - \varepsilon^{\gamma - 1}\right) \tag{2.15a}$$

and

$$\left|\frac{q_r}{c_v T_1}\right| = \Omega^\gamma - 1 \tag{2.15b}$$

From the *principle of conservation of energy*, the total heat exchanged must be equal to the sum of work of all part processes; as such, the overall specific work is

$$\frac{w}{c_v T_1} = \gamma\left(\Theta - \varepsilon^{\gamma - 1}\right) - (\Omega^\gamma - 1) \tag{2.16}$$

Thus, the thermodynamic cycle efficiency is given by the relation

$$\eta_{th} = \frac{w}{q_a} = 1 - \left[\frac{\Omega^\gamma - 1}{\gamma\left(\Theta - \varepsilon^{\gamma - 1}\right)}\right] = 1 - \left[\frac{\Omega^\gamma - 1}{\gamma\varepsilon^{\gamma - 1}(\Omega - 1)}\right] \tag{2.17a}$$

Since $1 \leq \Omega \leq \varepsilon$ the two limiting cases are

$$\Omega \to 1 : \eta_{th} \to \frac{\varepsilon^{\gamma - 1} - 1}{\varepsilon^{\gamma - 1}} \tag{2.17b}$$

which is the same as (2.8b) for the Otto cycle, and

$$\Omega \to \varepsilon : \eta_{\text{th}} \to 1 - \frac{\varepsilon^{\gamma-1} - 1}{\varepsilon^{\gamma-1}} - (\Omega - 1) \tag{2.17c}$$

Once again, as in the Otto cycle, the overall heat q depends actually on the fuel–air ratio. From (1.19), we have

$$\dot{m}_{\text{f}} \Delta H_{\text{p}} = \dot{m}_{\text{a}} c_{\text{p}}(T_3 - T_2) = \gamma \dot{m}_{\text{a}} c_{\text{v}}(T_3 - T_2)$$

and hence the *fuel–air ratio*,

$$f = \frac{\dot{m}_{\text{f}}}{\dot{m}_{\text{a}}} = \frac{c_{\text{p}} T_1}{\Delta H_{\text{p}}} \frac{T_3 - T_2}{T_1} = \frac{c_{\text{p}} T_1}{\Delta H_{\text{p}}} \left(\Theta - \varepsilon^{\gamma-1} \right) \tag{2.18}$$

Thus, although Θ, as in the Otto cycle, depends on the fuel–air ratio f, unlike the Otto cycle, it also depends on Ω and ε.

Further, as in (2.10), the *mean indicated pressure*, p_{m}, is given by the relation

$$p_{\text{m}} = w.M_1/V_{\text{d}}$$

where M_1 is the mass of the medium (air) in the cylinder, which reduces the relation for the mean indicated pressure to

$$\frac{p_{\text{m}}}{p_1} = (\gamma - 1) \frac{\varepsilon}{\varepsilon - 1} \left[\gamma \left(\Theta - \varepsilon^{\gamma-1} \right) - (\Omega^{\gamma} - 1) \right] \tag{2.19}$$

For the power developed by the engine, (2.11) developed for the Otto cycle can be used. For the specific fuel consumption, we write $\text{SFC} = f/\Omega$, which reduces to

$$\text{SFC} = \frac{f}{(\eta_{\text{th}} q_{\text{a}})} = \frac{1}{\eta_{\text{th}} \Delta H_{\text{p}}} \tag{2.20}$$

which can be reduced further into a nondimensional form, $\text{SFC}^* = \text{SFC}.\Delta H_{\text{p}}$, which is again inversely proportional to the thermodynamic efficiency.

Now again, as for the Otto cycles, we take $f = 0$, $\Delta H_{\text{p}} = 4.2707 \times 10^4$ kJ/kg, $c_{\text{p}} = 1.005$ kJ/kg/K. and $T_1 = 298.15$ K. Thus, from (2.18), we have

$$\Theta - \varepsilon^{\gamma-1} = f \Delta H_{\text{p}}(c_{\text{p}} T_1) = 7.126358.$$

Further results are computed from (2.12), (2.16), (2.17a), and (2.19) and are given in Table 2.2. By comparing it with Table 2.1, we can see that the diesel cycle has a much better thermodynamic efficiency. In addition, because of the limitation in the value of the compression ratio for the Otto cycle, the diesel cycle has a much

Table 2.2 Some performance parameters for the diesel cycle

ε	4	5	6	8	10	15	20	35
Θ	8.867	9.030	9.174	9.420	9.638	10.080	10.440	11.020
Ω	5.093	4.743	4.480	4.102	3.837	3.412	3.150	2.828
p_m/p_1	4.033	6.672	8.443	10.750	12.240	14.060	15.770	17.800
$w/(c_vT_1)$	1.210	2.135	2.814	3.363	4.406	5.402	5.992	6.690
η_{th}	0.125	0.214	0.282	0.377	0.442	0.541	0.600	0.670
$\eta_{th,actual}$	–	–	0.200	0.275	0.330	0.420	0.470	0.500

higher compression ratio. For example, there are a reported compression ratio of 27.0 and a mean indicated pressure of 59.3 bar. These results are, of course, for an ideal cycle without any losses. In real cycles, there are losses due to incomplete entry of the combustion volume, incomplete combustion (especially for diesel cycles, where one can observe lots of smoke through the exhaust pipe, especially in cold engines), and heat transfer from the hot gas to the colder wall. Finally, the data for piston engines have been given in Appendix to this book, which may be studied in order to have an overall view of manufactured engines.

2.1.3   Force and Moment Analysis for Piston Engines

Because piston engines are reciprocating engines, the forces and moments acting on the piston and other components of the engine are not uniform. To develop engine power, these forces and moments have to be taken into account. For this purpose, we show the lengths and angles schematically in Fig. 2.4.

The following geometrical relations are obvious:

$$x = l(1 - \cos \varphi) + \frac{d}{2}(1 - \cos \theta) \tag{2.21a}$$

where θ is the *crank angle,* l is the *connecting rod length,* and d is the *stroke length.* Since

$$X = \frac{d}{2} + l - x$$

it follows that

$$\frac{X}{l} = 1 + \frac{d}{2l} - \frac{x}{l} \tag{2.21b}$$

Fig. 2.4 Lengths and angles
in a single-cylinder piston
engine

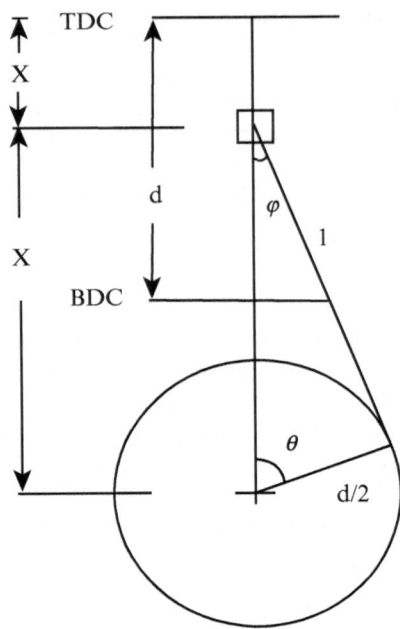

Now,

$$l \sin \varphi = \frac{d}{2} \sin \theta \qquad\qquad (2.21c)$$

and so it follows again that

$$\sin \varphi = \frac{d}{2l} \sin \theta$$

$$\cos \varphi = \sqrt{1 - \sin^2 \varphi} = \sqrt{1 - \left(\frac{d}{2l}\right) \sin^2 \theta}$$

and

$$\tan \varphi = \frac{\frac{d}{2l} \sin \theta}{\sqrt{1 - \left(\frac{d}{2l}\right)^2 \sin^2 \theta}}$$

Therefore,

$$\frac{x}{l} = \left[1 - \sqrt{1 - \left(\frac{d}{2l}\right)^2 \sin^2 \theta}\right] + \frac{d}{2l}(1 - \cos \theta) \qquad (2.21d)$$

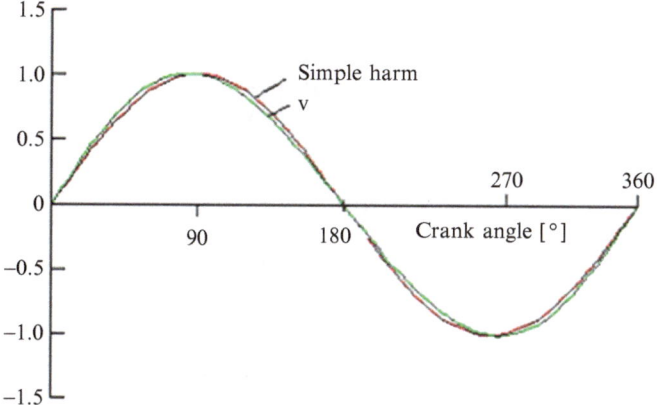

Fig. 2.5 Comparison of piston velocity for ($l/d = 3$) with simple harmonic motion

and

$$\frac{X}{l} = \frac{d}{2l} \cos\theta + \sqrt{1 - \left(\frac{d}{2l}\right)^2 \sin^2\theta} \qquad (2.21e)$$

Noting that $\dot\theta = \pi n/30$ is constant, we differentiate (2.21d) to get the expression for velocity:

$$v = \frac{dx}{dt} = l\frac{d(x/l)}{d\theta}.\dot\theta; \dot\theta = \frac{\pi n}{30}; \frac{d(x/l)}{d\theta} = \frac{d}{2l}\sin\theta\left[1 - \frac{d}{2l}\frac{\cos\theta}{\sqrt{1 - \left(\frac{d}{2l}\sin\theta\right)^2}}\right]$$

and hence,

$$v = \frac{\pi n d}{60}\sin\theta\left[1 - \frac{d}{2l}\frac{\cos\theta}{\sqrt{1 - \left(\frac{d}{2l}\sin\theta\right)^2}}\right] \qquad (2.22a)$$

Here the two limiting cases, described as top dead center (TDC) and bottom dead center (BDC) since these are extreme points of historically vertical machines, are

$$\theta = 0°(\text{TDC}) : x = 0, \ X = 1 + d/2, \ v = 0$$
$$\theta\pi(\text{BDC}) : x = d, \ X = 1 - d/2, \ v = 0 \qquad (2.22b)$$

and in between, the absolute piston velocity is maximum when the crank angle is 90°. The relative values of the piston's velocity, v, are plotted in Fig. 2.5, by taking the full equation (2.22a) for (l/d) =3, and also if only the first term in brackets is

Fig. 2.6 Forces acting
on cylinder wall, piston,
and connecting rod

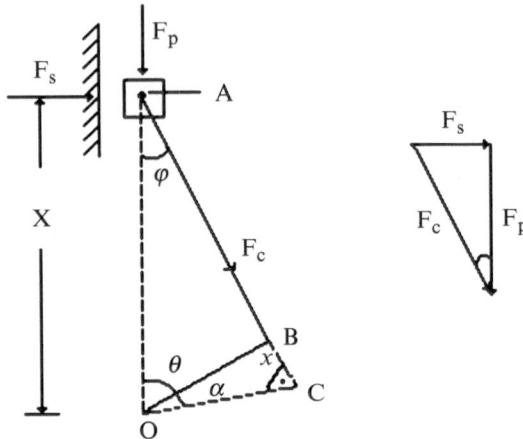

taken into consideration (simple harmonic motion). It is now possible to calculate
the inertial and total forces on the piston; for the latter, it is necessary to have the
data on the cylinder pressure, p, as a function of the crank angle.

Now, for the inertial and pressure force on the piston, F_{inertia} and F_{p}, respectively,

$$F_{\text{inertia}} = -M_{\text{p}}\dot{v} = -M_{\text{p}}\left(\frac{\pi^2 n^2 d}{1800}\right)\left[\cos\theta + \frac{d}{2l}\cos^2\theta\right]$$

$$F_{\text{p}} = p\frac{\pi}{4}D^2$$

and a sum of the two is the total force on the piston. In above expressions, M_{p} is the
mass of the piston and D is the diameter of the piston. These forces, and two other
forces—the side force on the cylinder, F_{s}, and the force on the connecting rod, F_{c}—
are shown schematically in Fig. 2.6. From the condition of equilibrium of forces on
the crank pin of the piston, we may write

$$F_{\text{c}} = F_{\text{s}} + F_{\text{p}}; \ F_{\text{c}} = F_{\text{p}}/\cos\theta$$

Normally, the pressure force has to be evaluated separately, following which one
may evaluate the torque moment. However, it can be safe to assume that at the time
of combustion, the pressure force is opposite the inertial force.

The *torque moment* is evaluated with the help of the equation

$$M_{\text{t}} = F_{\text{c}}x\sin\varphi = F_{\text{p}}X\tan\varphi$$

$$= F_{\text{p}}l\left[1 - \frac{d^2}{30l^2}\sin^2\theta + \frac{d}{2l}\cos\theta\right]\frac{d}{2l}\frac{\sin\theta}{\sqrt{1 - \left(\frac{d}{2l}\right)^2\sin^2\theta}} \qquad (2.22c)$$

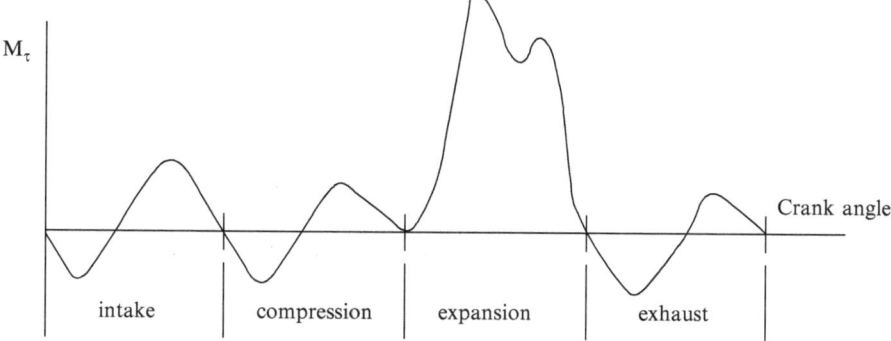

Fig. 2.7 Torque moment of the piston engine

It is obvious that for multicylinder engines the total torque moment has to be evaluated carefully by adding the moment of individual pistons, in order for the engine to run smoothly. The result of calculating (2.22c) is shown in Fig. 2.7, which shows that there are positive and negative torque moments in each part of the cycle, except in the expansion process (*power stroke*), where the positive torque moment has a large value.

2.1.4 An Interactive Code for Performance of Otto/Diesel Engines

The following program analyzes the performance of Otto and Diesel engines.

```
      PROGRAM OTTDIS
C  PERFORMANCE ROUTINE FOR OTTO/DIESEL CYCLE
15 FORMAT(1X,' OUTPUT UNIT NO.:',$)
16 FORMAT(1X,' DATE=',9A1)
17 FORMAT(1X,'PERFORMANCE ROUTINE FOR OTTO/DIESEL CYCLE')
      WRITE(7,15)
      READ(5,*)NOUT
      WRITE(NOUT,17)
      WRITE(NOUT,7)
10 WRITE(7,1)
1  FORMAT(1X,' READ EQUIV. RATIO,COMP. RATIO(2F6.2)=',$)
2  FORMAT(2F6.2)
      READ(5,2)EQUIV,CR
      IF(EQUIV.GE.1.0.AND.CR.GT.1.0)GOTO5
      WRITE(7,6)
6  FORMAT(1X,'PARAMETRIC ERROR : TRY AGAIN ')
7  FORMAT(1X,'EQUIV.RATIO,COMP.RATIO,T2/T1,(T3/T1,ETA,WNET,PME)OTTO '
     1,/,1X,'(T3/T1,OMEGA,ETA,WNET,PME) DIESEL')
```

```
      GOTO10
 5   CONTINUE
      XA=CR**0.4
      XB=13.267*(CR-1.0)/(EQUIV*CR)
      TR=XA+XB
      TRD=XA+(XB/1.4)
      ETA=(XA-1.0)/XA
      WNET=(TR-XA)*(XA-1.0)/XA
      PME=WNET*2.5*CR/(CR-1.0)
      OM=TRD/XA
      ETAD=1.0-((OM**0.4-1.0)/((TRD-XA)*1.4))
      WNETD=((TRD-XA)*1.4)-(OM**0.4-1.0)
      PMED=WNETD*2.5*CR/(CR-1.0)
      WRITE(NOUT,20)EQUIV,CR,XA,TR,ETA,WNET,PME,TRD,OM,ETAD,
        WNETD,PMED
 20  FORMAT(1X,1P12E10.3)
      GOTO10
      END
```

2.1.5   Ideal Turboprop Cycles

The engine being considered in this section is only partially a jet engine, but its main power delivery is through a rotating shaft to a propeller. As such, it can be considered in between a piston engine and the "pure" jet engines, like the ramjets, turbojets, and turbofans. In all of these jets, it is assumed, for the evaluation of an ideal cycle, that the working gas medium does not change its composition due to burning of the fuel (heat addition, for all practical purposes, is considered externally added at constant pressure), the specific heat of which is assumed to be constant. Further, all compressions and expansions are assumed to be isentropic.

Figure 2.8a is a schematic of a turboprop engine, and Fig. 2.8b is the corresponding schematic sketch of the ideal turboprop cycle. External air is assumed to enter through the inlet (state 1); it is compressed in the compressor isentropically to state 2; heat is added at constant pressure to state 3; and it is expanded isentropically in the turbine or nozzle to the ambient pressure (state 4). Normally, maximum expansion of the gas takes place in the turbine, which drives the compressor and (through a gear box) the propeller, but a small expansion (in pressure) is required in the nozzle for a smooth flow out of the engine. In this process, the nozzle develops a small thrust, the major thrust being through the propeller. Although in principle modern-day turboprops such as propfans can run for aircrafts flying at moderately high subsonic speeds, in the present case it is assumed that the flight speed and all other speeds inside the engine are small enough so that the stagnation state and the static state are practically the same.

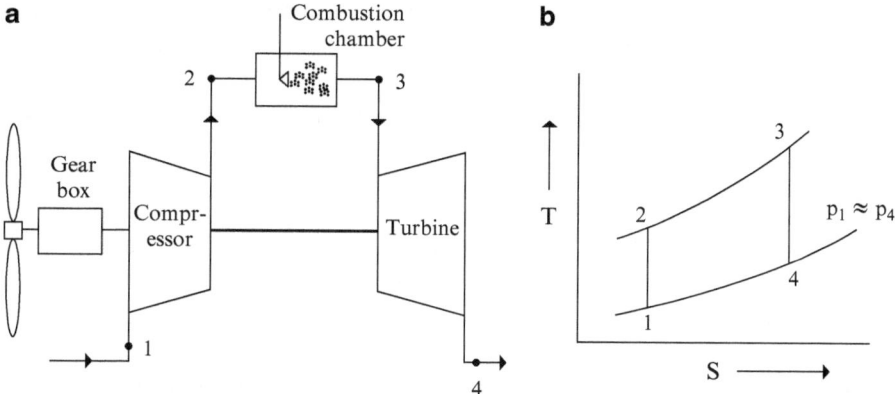

Fig. 2.8 Schematic sketch of a turboprop engine

As initial performance parameters, let's consider $\pi_c = p_2/p_1 = p_3/p_4$ as the *compression ratio* and $\Theta = T_3/T_1$ as the *temperature ratio* of the compressor given. Here we talk about the pressure compression ratio, as is usual in jet engines, instead of the volume compression ratio, as is usual for petrol engines. Obviously,

$$\pi_c = \Theta^{\gamma/(\gamma-1)}$$

Now

$$T_2/T_1 = T_3/T_4 = \pi_c^{(\gamma-1)/\gamma}$$

and thus,

$$\frac{T_4}{T_1} = \frac{\Theta}{\pi_c^{(\gamma-1)/\gamma}} \tag{2.23}$$

The technical work per unit mass gas in a flowing system for isentropic change of state is the difference of enthalpy; as such, for compression in the compressor and expansion in the turbine, these are

$$\left|\frac{w_{12}}{c_p T_1}\right| = \pi_c^{(\gamma-1)/\gamma} - 1 \tag{2.24a}$$

and

$$\left|\frac{w_{34}}{c_p T_1}\right| = \Theta \frac{\left(\pi_c^{(\gamma-1)/\gamma} - 1\right)}{\pi_c^{(\gamma-1)/\gamma}} \tag{2.24b}$$

On the other hand, heat added (process 2-3) and heat rejected (4-1) per unit mass are given by the expressions

$$\left|\frac{q_a}{c_p T_1}\right| = \left(\Theta - \pi_c^{(\gamma-1)/\gamma}\right) \tag{2.25a}$$

and

$$\left|\frac{q_r}{c_p T_1}\right| = \frac{\left(\Theta - \pi_c^{(\gamma-1)/\gamma}\right)}{\pi_c^{(\gamma-1)/\gamma}} \tag{2.25b}$$

Note that in the overall cycle analysis for an ideal case, the overall work is the same, whether it is computed from the difference of the two heats (heat added – heat rejected) or from consideration of all part work in various engine components (work developed in the turbine – work needed in the compressor). This overall work

$$\frac{|w|}{c_p T_1} = \frac{|w_{34}| - |w_{12}|}{c_p T_1} = \frac{|q|}{c_p T_1} = \frac{|q_a| - |q_r|}{c_p T_1}$$

$$= \left(\Theta - \pi_c^{(\gamma-1)/\gamma}\right)\left(\frac{\pi_c^{(\gamma-1)/\gamma} - 1}{\pi_c^{(\gamma-1)/\gamma}}\right) \tag{2.26a}$$

goes to drive the propeller. Now the thermodynamic efficiency is given by the ratio of overall work to heat added, and thus,

$$\eta_{th} = \frac{|w|}{|q_a|} = \left(\frac{\pi_c^{(\gamma-1)/\gamma} - 1}{\pi_c^{(\gamma-1)/\gamma}}\right) \tag{2.26b}$$

Now two special limiting cases for the given Θ are considered. The first one is when $\pi_c \to 1$, for which $|q_a| = |q_r| = \Theta - 1, w = 0, \eta_{th} \to 0$. In the other limiting case, let $\pi_c = \Theta^{\gamma/(\gamma-1)}$. Then $|q_a| = |q_r| = 1, w = 0$, and $\eta_{th} = (\Theta - 1)/\Theta$, which is the maximum efficiency of a Carnot cycle. Between these two limiting cases, w is maximum (optimum π_c). To find this, let $x = \pi_c^{(\gamma-1)/\gamma}$, and the maximum value of w is obtained by differentiating (2.26a) with respect to x and setting it equal to 0. The result is

$$\left(\pi_c^{(\gamma-1)/\gamma}\right)_{opt} = \sqrt{\Theta} \tag{2.27a}$$

and

$$\left[\frac{w}{c_p T_1}\right]_{opt} = \left(\sqrt{\Theta} - 1\right)^2 \tag{2.27b}$$

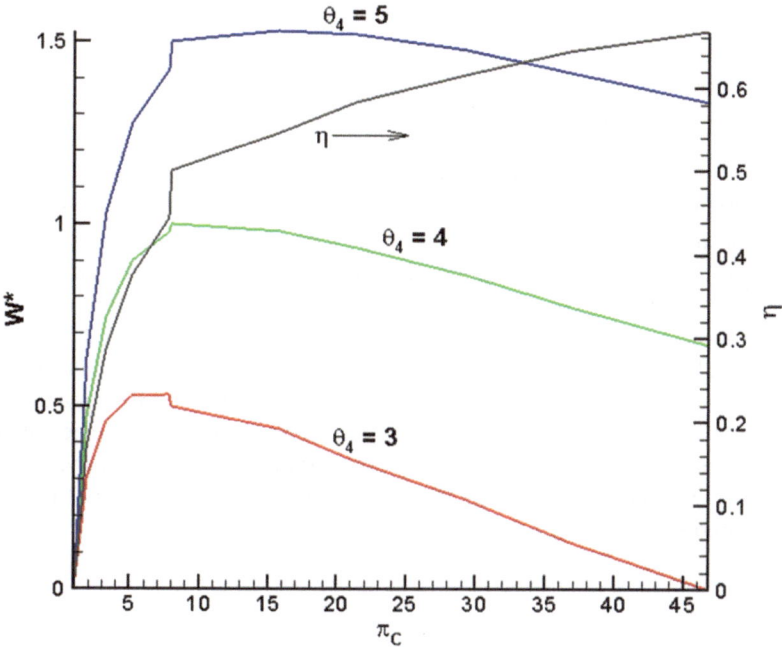

Fig. 2.9 Different values (as described in the text) vs. π_c

Table 2.3 $[w/(c_p T_1)]_{opt}$ and Corresponding $\pi_{c,opt}$ vs. Θ

Θ	3	4	5	6
π_c	46.760	128.000	279.510	529.090
$\pi_{c,opt}$	6.838	11.313	16.718	23.002
$[w/(c_p T_1)]_{opt}$	0.536	1.000	1.528	2.101

The results of calculating η_{th} and $w/(c_p T_1)$, the latter for different values of Θ, are presented vs. π_c in Fig. 2.9. It is seen that η_{th} increases monotonically with π_c, whereas $w/(c_p T_1)$ changes with π_c with a maximum (optimum) for different values of Θ. The optimum values of $[w/(c_p T_1)]_{opt}$ and the corresponding $\pi_{c,opt}$ are given in the Table 2.3.

As per the analysis of these results, it is evident that higher the compression ratio, the higher is the thermodynamic efficiency. However, increasing the compression ratio too much beyond the optimum compression ratio for a given temperature ratio results in a lower work output.

The fuel–air ratio is now given by (1.19a), and we get

$$f = \frac{c_p T_1}{\Delta H_p}\left(\frac{T_3}{T_1} - \frac{T_2}{T_1}\right) = \frac{c_p T_1}{\Delta H_p}\left(\Theta - \pi_c^{(\gamma-1)/\gamma}\right) \tag{2.28a}$$

The nondimensional *specific fuel consumption* is

$$\text{SFC} = \frac{f}{(\eta_{th} q_a)} = \frac{1}{(\eta_{th} \Delta H_p)} \rightarrow \text{SFC}^* = \text{SFC}.\Delta H_p = \frac{1}{\eta_{th}} \qquad (2.28b)$$

and is a measure of the thermodynamic efficiency.

It is also quite evident that by increasing the compression ratio as much as possible up to the limit of $\pi_c > \pi_{c,opt} > \pi_{c,max} = \Theta^{\gamma/(\gamma-1)}$, the thermodynamic efficiency can increase up to the maximum of the Carnot cycle efficiency, $\eta_{Carnot} = 1 - (1/\Theta) = 1 - (T_1/T_3)$, only, the specific work output is zero under the limit, and the maximum specific output is obtained under the condition of (2.27b). For this maximum, all three—that is, the heat addition, the heat rejection, and the work output—are zero, and no fuel can be introduced, but the thermodynamic efficiency is nonzero. On the other hand, for the minimum compression ratio, the heat added = heat rejected is nonzero, but both the work output and the thermodynamic efficiency are zero. For a proper understanding of turboprop engines, the data on manufactured engines given in Appendix may be studied.

2.2 Jet Engines

So far we have discussed engines that operate at comparatively low flight speeds, so that the static and stagnation states of air can be considered to be the same. Jet engines, including ramjets, straight turbojets, bypass jets, and fanjets not only operate at higher flight speeds than the propeller-driven engines, but they also have to develop high jet speeds at the exit nozzle. Later we will show that for the extraction or introduction of work in a turbine or a compressor, the change in *stagnation states* has to be considered, rather than the change of *static states*. Therefore, when analyzing jet engines, the usual concepts of the stagnation pressure and the stagnation temperature are introduced. In addition, here we again make the assumption of ideal gas (c_p = constant) and ideal thermodynamic cycle (compression and expansion as isentropic, and heat addition and heat rejection at constant pressure). Further, no change in the composition of gas (air) is assumed in the combustion chamber. From the equation of energy, we have

$$c_p T^o = c_p T + u^2/2 \qquad (2.29a)$$

where T^o is the stagnation temperature, T is static temperature, u is (one-dimensional) gas speed, and c_p is the specific heat at constant pressure. With $c_p/R = \gamma/(\gamma - 1)$, where R is the *gas constant*, and from the definition of Mach number $M = u/\sqrt{\gamma RT}$, we write

$$\frac{T^o}{T} = 1 + \frac{(\gamma - 1)}{2} M^2 \qquad (2.29b)$$

$$\frac{p^o}{p} = \left(\frac{T^o}{T}\right)^{\frac{\gamma}{\gamma-1}} \tag{2.29c}$$

We now introduce the definition of nondimensional stagnation temperature and stagnation pressure,

$$\Theta = \frac{T^o}{T_\infty} \tag{2.30a}$$

and

$$\delta = \frac{p^o}{p_\infty} \tag{2.30b}$$

where the subscript "∞" refers to the ambient state of air, and those of stagnation temperature and stagnation pressure ratio across components as

$$\tau = \frac{T^{o\prime\prime}}{T^\prime}, \tag{2.30c}$$

$$\pi = \frac{p^{o\prime\prime}}{p^{o\prime}} \tag{2.30d}$$

Therefore, at the inlet,

$$\Theta_\infty = \frac{T^o_\infty}{T_\infty} = 1 + \frac{(\gamma-1)}{2}M^2_\infty \tag{2.31a}$$

$$\delta_\infty = \frac{p^o_\infty}{p_\infty} = \Theta^{\gamma/(\gamma-1)}_\infty \tag{2.31b}$$

Now we'll start with the simplest of the jet engines, that is, the *ramjet*, which does not have any moving parts but operates only at high flight speeds.

2.2.1   *Ideal Ramjet Cycle*

Figure 2.10a shows schematically a ramjet engine and Fig. 2.10b depicts the *ideal ramjet cycle process* in a (*T*, *s*) diagram. It has an inlet, which actually has a much more complicated shape than the simple divergent diffusor shown in the figure to operate efficiently at the Mach number of operation, and it must suitably be altered at flight design and off-design Mach numbers. Fuel is introduced in the combustion

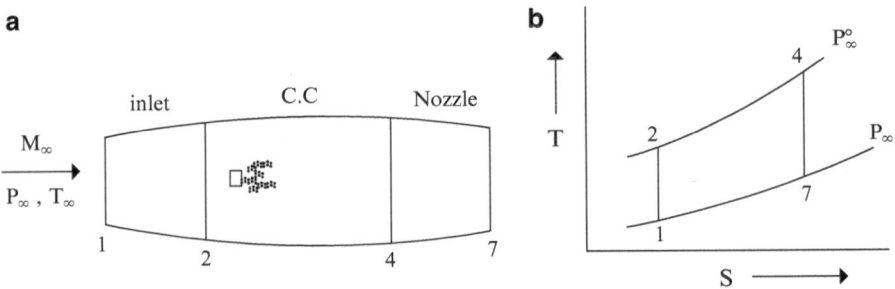

Fig. 2.10 Sketch of (**a**) a ramjet engine, and (**b**) a thermodynamic cycle process in a (T, s) chart

chamber at state 2, a complete combustion is assumed at state 4, and the gas is expelled through the nozzle (normally, it is a convergent–divergent nozzle for a supersonic exit) to reach ambient pressure at state 7.

Let's consider the role of the basic parameters for the ramjet: the approaching flow Mach number M_∞ and $\Theta_4 = T_4^o / T_\infty^o$.

Now,

$$\Theta_4 = \Theta_1 = \Theta_2 = \frac{T_\infty^o}{T_\infty} = \frac{T_1^o}{T_\infty} = \frac{T_2^o}{T_\infty} = 1 + \frac{(\gamma - 1)}{2} M_\infty^2 \qquad (2.32a)$$

and

$$\delta_4 = \delta_1 = \delta_2 = \frac{p_\infty^o}{p_\infty} = \frac{p_1^o}{p_\infty} = \frac{p_2^o}{p_\infty} = \frac{p_4^o}{p_\infty} = \frac{p_4^o}{p_\infty} = \Theta_\infty^{\gamma/(\gamma-1)} \qquad (2.32b)$$

Obviously, $(\Theta_4 - \Theta_\infty)$ depends on the fuel–air ratio and the efficiency with which the fuel is burned in the combustion chamber, and it will be maximum at a stoichiometric fuel–air ratio. A discussion on these has been given later in this section. One could therefore perhaps think that especially for ramjets with no moving parts, it should be possible to have a large Θ_4 for a large Θ_∞. It needs to be pointed out, however, that the static combustion chamber is limited by the temperature at which substantial dissociation may take place in carbon dioxide and oxygen molecules, which may be around 3,500 K at a gas pressure of 1 bar and may be substantially lower at lower pressures (higher altitude). Since the combustion chamber pressure is directly proportional to ambient pressure at a constant flight Mach number, but otherwise depends on about the sixth power of the flight Mach number, the dissociation temperature's dependence on the ambient pressure may not be very critical. However, a high Mach number is also associated with a very high-shock-pressure loss in the inlet region, especially if the nozzle is not operating under a fully isentropic change of state. It is therefore necessary to judiciously analyze the combination of flight Mach number, ambient temperature, and ambient pressure in an integrated way along the design of the inlet region.

Since

$$\frac{T_4^o}{T_\gamma} = \left(\frac{p_4^o}{p_\gamma}\right)^{(\gamma-1)/\gamma} = \Theta_\infty$$

we get

$$\frac{T_\gamma}{T_\infty} = \left(\frac{T_4^o}{T_\infty}\right)\left(\frac{T_\gamma}{T_4^o}\right) = \frac{\Theta_4}{\Theta_\infty} \tag{2.32c}$$

The *jet exit speed* is, therefore,

$$u_7 = \sqrt{2c_p(T_4 - T_7)} = \sqrt{2c_p T_\infty\left(\Theta_4 - \frac{T_7}{T_\infty}\right)} \tag{2.33a}$$

Further,

$$u_\infty = M_\infty\sqrt{\gamma R T_\infty} = M_\infty\sqrt{(\gamma-1)c_p T_\infty} \tag{2.33b}$$

and the ratio of the two speeds is

$$\frac{u_7}{u_\infty} = \frac{1}{M_\infty}\sqrt{\frac{2}{(\gamma-1)}\frac{\Theta_4(\Theta_\infty - 1)}{\Theta_\infty}} \tag{2.33c}$$

Note that for $u_\infty \to 0, M_\infty \to 0, \Theta_\infty \to 1$, and $u_7 \to 0$, but the ratio (u_7/u_∞) $\to \sqrt{\Theta_4}$ is a finite quantity.

The heat added in the combustion chamber and the heat rejected are

$$|q_a| = c_p\left(T_4^o - T_2^o\right) \quad \text{and} \quad |q_r| = c_p(T_7 - T_\infty)$$

from which we write

$$\frac{|q_a|}{c_p T_\infty} = \Theta_4 - \Theta_\infty \tag{2.34a}$$

and

$$\frac{|q_r|}{c_p T_\infty} = \frac{T_7}{T_\infty} - 1 = \frac{\Theta_4 - \Theta_\infty}{\Theta_\infty} \tag{2.34b}$$

Thus, the *overall specific work* in the system is

$$\frac{|w|}{c_p T_\infty} = \frac{|q_a| - |q_1|}{c_p T_\infty} = \frac{(\Theta_4 - \Theta_\infty)(\Theta_\infty - 1)}{\Theta_\infty} \tag{2.34c}$$

and the *thermodynamic efficiency* is

$$\eta_{th} = \frac{|w|}{|q_a|} = \frac{(\Theta_\infty - 1)}{\Theta_\infty} \tag{2.35a}$$

Note further from (2.33a) and (2.33b) that

$$\frac{1}{2}\left(u_7^2 - u_\infty^2\right) = c_p T_\infty = \frac{(\Theta_4 - \Theta_\infty)(\Theta_\infty - 1)}{\Theta_\infty} = |w| \tag{2.34d}$$

and hence the effective work gained out of the ramjet is equal to the increase in the kinetic energy. This result is important, since for the first time in a jet machine it is shown that work output is not with the help of a mechanical engine, but through the change in kinetic energy.

For $M_{inf} \to 0, \Theta_\infty \to 1$, both w and η_{th} tend to zero. On the other hand, for $\Theta_4 = \Theta_\infty$ (no fuel introduced), $w = 0$, but η_{th} is the maximum value of the Carnot cycle efficiency. Between these two extreme cases, we can examine the condition for maximum specific work by differentiating (2.33b) with respect to Θ_∞ and set it equal to zero to get the result that $\lfloor w/(c_p T_\infty) \rfloor$ will be maximum, and $\Theta_\infty = \sqrt{\Theta_4}$. Substituting back into (2.34c), we therefore get the maximum work

$$[w/(c_p T_\infty)]_{max} = \left(\sqrt{\Theta_4} - 1\right)^2 \tag{2.36}$$

Now, the *propulsive efficiency* is given by the relation

$$\eta_p = \frac{2}{\left[1 + \left(\frac{u_7}{u_\infty}\right)\right]} = \frac{2}{\left[1 + \sqrt{\frac{\Theta_4}{\Theta_\infty}}\right]} \tag{2.37a}$$

Since $\Theta_4 > \Theta_\infty$, $\eta_p < 1$. On the other hand, for $\Theta_\infty > 1$, $\eta_p = 2/\lfloor 1 + \sqrt{\Theta_4} \rfloor$ and $\Theta_4 = \Theta_\infty$, $\eta_p = 1$. It is evident that for $M_\infty \to 0, \Theta_\infty \to 1$, and the jet speed goes to zero, but η_p is finite.

Since the thrust is given by the expression

$$F = \dot{m}_a(u_7 - u_\infty) = \dot{m}_a u_\infty \left(\frac{u_7}{u_\infty} - 1\right)$$

we can write for the *specific thrust*, after some manipulation,

$$\frac{F}{\dot{m}_a u_\infty} = \sqrt{\frac{2}{\gamma - 1}(\Theta_\infty - 1)}\left[\sqrt{\frac{\Theta_4}{\Theta_\infty}} - 1\right] = M_\infty \left[\sqrt{\frac{2\Theta_4}{2 + (\gamma - 1)M_\infty^2}} - 1\right]$$

Fig. 2.11 Specific thrust
vs. flight Mach number

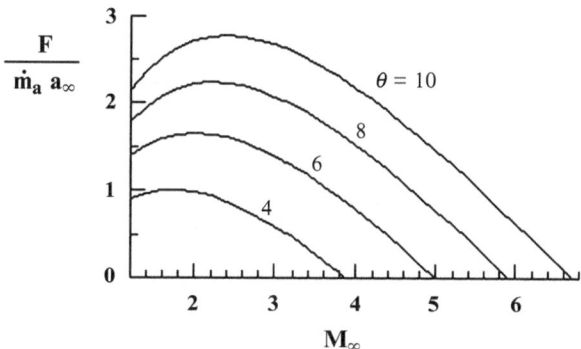

From the above equation, it can be seen that the specific thrust is equal to zero if $M_\infty \to 0$, $\Theta_\infty \to 1$, or $\Theta_4 = \Theta_\infty$. While the first case refers to zero flow, where the thrust has to be zero anyway, the second condition refers to the case that by just gas dynamic compression the combustion chamber temperature is reached and no additional fuel can be added. In between is the condition for maximum specific thrust. By differentiating the above equation with respect to M_∞ and setting it equal to zero, we find the specific thrust optimum (maximum) if $\Theta_\infty = \Theta_4^{1/3}$ and the maximum specific thrust becomes

$$\left(\frac{F}{\dot{m}_a u_\infty}\right)_{\mathrm{max}} = \sqrt{\frac{2}{(\gamma - 1)}\left(\Theta_4^{1/3} - 1\right)^3} \tag{2.37b}$$

We should mention, however, that since the mass flow rate is directly proportional to the flight speed (or flight Mach number), determining the flight Mach number at which the thrust is maximum requires a slightly different modification of the specific thrust equation. In this case, the dependent variable is $F/(p_{\mathrm{int}} A_{\mathrm{inlet}})$, where A_{inlet} is the inlet diffuser entry cross section. Both of these are plotted in Figs. 2.11 and 2.12 as a function of the flight Mach number, respectively. Figure 2.13 shows the results of calculating the specific work as a function of the flight Mach number.

We therefore have two values of Θ_∞ for optimization: One is for the specific thrust if $\Theta_\infty = \Theta_4^{1/3}$, and the other is for maximum specific work if $\Theta_\infty = \sqrt{\Theta_4}$. The two corresponding optimum Mach numbers are designated as $M_{1\infty,\mathrm{opt}}$ and $M_{2\infty,opt}$, respectively. These are given in Table 2.4.

Now, we calculate the fuel–air ratio from (1.19) to get

$$f = \frac{\dot{m}_f}{\dot{m}_a} = \frac{c_p \Delta T}{\Delta H_p} = \frac{c_p T_\infty}{\Delta H_p}(\Theta_4 - \Theta_\infty) \tag{2.38a}$$

which decreases continuously from $\Theta_\infty = 1$ (that is, $M_\infty \to 0$) to $\Theta_\infty = \Theta_4$. On the other hand, the specific thrust first increases and then decreases. Usually, the

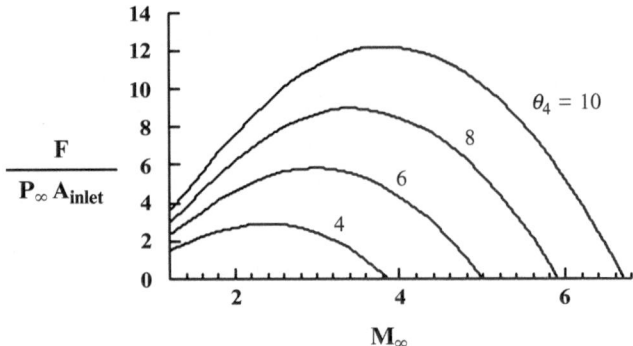

Fig. 2.12 Thrust per unit inlet area vs. flight Mach number

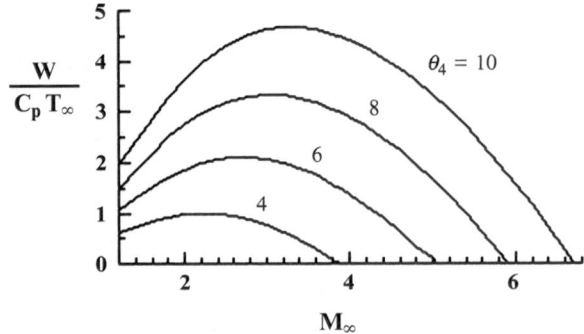

Fig. 2.13 Specific work vs. flight Mach number

Table 2.4 Two corresponding optimum Mach numbers for different Θ_4

Θ_4	3	4	5	6	10
$M_{1\infty,opt}$	1.913	2.236	2.486	2.692	3.288
$\left\lfloor w_{opt}/(c_p T_\infty) \right\rfloor$	0.536	1.000	1.527	2.101	4.675
$M_{2\infty,opt}$	1.487	1.713	1.884	2.021	2.402
$\left(\frac{F}{\dot{m}_a a_\infty} \right)_{opt}$	0.658	1.007	1.338	1.652	2.773

specific fuel consumption [kg/J] is computed by dividing the fuel flow rate $\dot{m}_f$ [kg/s] by the work produced per kg of air [J/kg]. However, for aircraft engine applications, it is computed by dividing the fuel mass flow rate by the thrust [N], and hence the specific fuel consumption [kg/(N.s) $\equiv$ s/m] is defined as

Fig. 2.14 Overall efficiency of ramjet vs. flight Mach number and Θ_4 as parameter

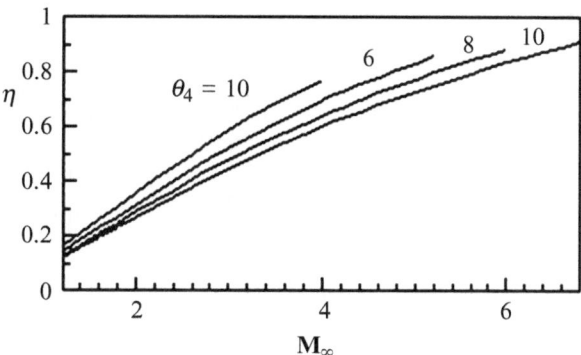

Fig. 2.15 Nondimensional SFC of ramjet vs. flight Mach number and Θ_4 as parameter

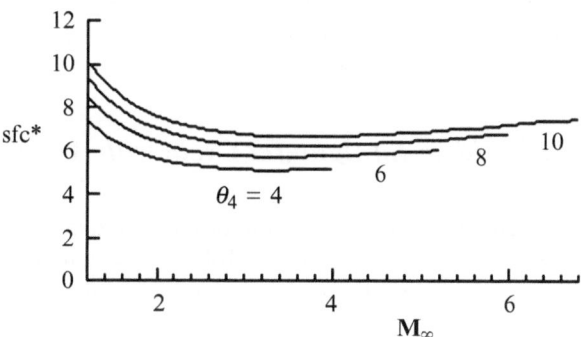

Thus, the, which is the fuel mass flow rate per unit thrust, is a somewhat complicated relation as follows:

$$\text{SFC} = \frac{\dot{m}_f}{F} = \frac{f}{M_\infty a_\infty \left(\frac{u_7}{u_\infty}\right)} = \frac{f}{a_\infty \left[\frac{2\Theta_4}{(\gamma-1)} \frac{(\Theta_\infty-1)}{\Theta_\infty} - M_\infty\right]} \qquad (2.38b)$$

For $M_\infty = 0$, both $f\,(\Theta_4 = \Theta_\infty$: if there is no airflow, then there is also no fuel flow either) and $F = 0$, and SFC is indeterminate. From (2.38b) we can now define a *nondimensional specific fuel consumption*:

$$\text{SFC}^* = \text{SFC}.a_\infty = \frac{f}{a_\infty\left[\sqrt{\frac{2\Theta_4}{(\gamma-1)} \frac{(\Theta_\infty-1)}{\Theta_\infty}} - M_\infty\right]} \qquad (2.38c)$$

The optimum values are given in Table 2.4. Further various parameters have been computed as a function of M_∞ and Θ_4 (all calculations have been done for $\gamma = 1.4$), as shown in Figs. 2.14 and 2.15. While the thermodynamic efficiency for a

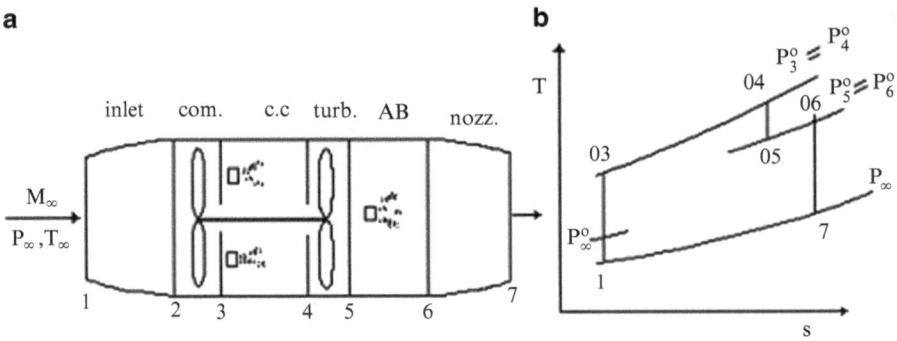

Fig. 2.16 Schematic sketch of an ideal straight-jet engine with (T, s) chart

ramjet is a function of the flight Mach number alone (the higher the flight Mach number, the better the thermodynamic efficiency), the overall efficiency depends on both M_∞ and Θ_4. Similarly, SFC* depends on both.

2.2.2 Ideal Straight-Jet Cycle with Afterburner

A schematic sketch of the straight-jet engine is shown again in Fig. 2.16a, and the ideal process is shown in Fig. 2.16b. Once again, it has an inlet, which actually may look quite different because of a supersonic inlet flow, compression in a compressor, expansion in a turbine, a combustion chamber, an afterburner, and a nozzle. For example, a supersonic inlet may have a built-in multishock system designed to have a minimum loss at the inlet, and a supersonic exit nozzle of convergent–divergent type. Similarly, both the compressor and turbine will have multiple stages with many blades in each stage, the design of which will depend on the axial or radial flow of the gas. Again, the combustion chamber and the afterburner are designed to have the best possible combustion characteristics with minimal pressure loss. In this system, air at state (p_∞, T_∞) is compressed in the inlet diffuser to state 02 (state: $0\infty = 01 = 02$); it is compressed further with the help of the compressor to state 03; fuel is introduced and burned at constant (stagnation) pressure to state 04; it is expanded through a turbine to state 05 (for without the afterburner operation, $05 = 06 = 07$, and for the case of the afterburner operation to reach state $06 = 07$); and, finally, the exhaust gas is expanded in the nozzle to reach state 07 at the ambient pressure p_∞. Thus, the state 04 is the turbine inlet state.

Let's assume the parameters given are $\pi_c = p_3^o/p_2^o$ as the compression ratio, $\Theta_4 = T_4^o/T_\infty$ as the temperature ratio, M_∞ as the approaching flow Mach number, and

$$\Delta\Theta_{AB} = (\Theta_6 - \Theta_5) = \frac{(T_6^o - T_5^o)}{T_\infty}$$

as the stagnation temperature increase in the afterburner (which is, of course, zero if the afterburner is not in operation). Thus, even without the afterburner, we have an

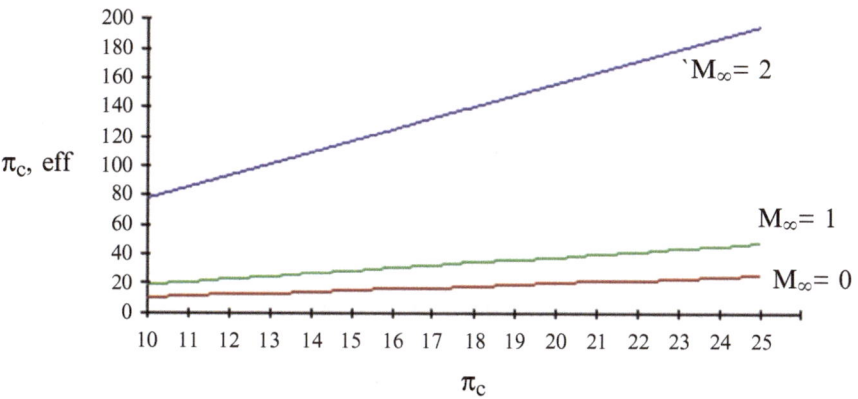

Fig. 2.17 Effective compression ratio vs. compression ratio with Mach number as parameter

increase in operational parameters by 1 over these for ramjets, although, as we will show, by defining an effective compression ratio $\pi_{c,\text{eff}}$, we would reduce the number of operational parameters to 2 again (without afterburner), excluding M_∞.

Now,

$$\Theta_\infty = \Theta_1 = \Theta_2 = \frac{T^o_\infty}{T_\infty} = \frac{T^o_1}{T_\infty} = \frac{T^o_2}{T_\infty} = 1 + \frac{\gamma - 1}{2} M^2_\infty \qquad (2.39a)$$

and

$$\delta_\infty = \delta_1 = \delta_2 = \frac{p^o_\infty}{p_\infty} = \frac{p^o_1}{p_\infty} = \frac{p^o_2}{p_\infty} = \Theta^{\gamma/(\gamma-1)}_\infty \qquad (2.39b)$$

Further,

$$\delta_3 = \delta_4 = \frac{p^o_3}{p_\infty} = \frac{p^o_4}{p_\infty} = \pi_c \delta_\infty = \pi_c \Theta^{\gamma/(\gamma-1)}_\infty \qquad (2.39c)$$

which we call the *effective compression pressure ratio*, $\pi_{c,\text{eff}}$, and

$$\Theta_3 = \frac{T^o_3}{T_\infty} = \pi_c \delta_\infty = \delta^{(\gamma-1)/\gamma} = \Theta_\infty \pi^{(\gamma-1)/\gamma}_c \qquad (2.39d)$$

which we call the *effective compression temperature ratio*, $\tau_{c,\text{eff}}$ (for the ideal case only, $p^o_3 = p^o_4$, since here there would not be any pressure loss being considered due to friction and heat addition in the combustion chamber, but in the actual case, $p^o_3 \geq p^o_4$). It is strongly dependent on the approaching flow Mach number and compression pressure ratio and may indicate the limitation in both for a given temperature ratio. Hence, the relationship is plotted in Fig. 2.17.

Further, the temperature ratios across the compressor and combustion chamber are

$$\tau_c = \frac{T_3^o}{T_2^o} = \frac{\Theta_3}{\Theta_2} = \frac{\Theta_3}{\Theta_\infty} = \pi_c^{(\gamma-1)/\gamma} \geq 1 \qquad (2.40a)$$

and

$$\tau_b = \frac{T_4^o}{T_3^o} = \frac{\Theta_4}{\Theta_3} = \frac{\Theta_4}{\tau_c \Theta_\infty} \qquad (2.40b)$$

and the *effective compression pressure ratio* and *effective temperature ratio* across the compressor are, respectively,

$$\pi_{c,eff} = \pi_c \delta_\infty = \delta_3; \qquad (2.40c)$$

$$\tau_{c,eff} = \tau_c \Theta_\infty = \Theta_3 \qquad (2.40d)$$

Since the turbine and the compressor sit on the same shaft, work required by the compressor is supplied by the turbine, and thus,

$$T_3^o - T_2^o = T_4^o - T_5^o$$

from which it follows that

$$\Theta_3 - \Theta_2 = \Theta_\infty(\tau_c - 1) = \Theta_4 - \Theta_5 \qquad (2.40e)$$

Therefore,

$$\Theta_3 = \frac{T_5^o}{T_\infty} = \Theta_4 - \Theta_\infty(\tau_c - 1) = \Theta_4 + \Theta_\infty - \tau_{c,eff} \qquad (2.40f)$$

and the (stagnation) temperature ratio across the turbine is

$$\tau_t = \frac{\Theta_5}{\Theta_4} = 1 - \Theta_\infty \frac{(\tau_c - 1)}{\Theta_4} \leq 1 \qquad (2.40g)$$

since $\tau_c \geq 1$.

At the lower limit, $\tau_c = \pi_c = 1$, there is no compression in the compressor and no expansion in the turbine ($\pi_t = 1, \tau_t = 1$). On the other hand, it is possible that the (given) turbine inlet temperature is reached by compression alone and that there is no injection of the fuel ($\Theta_4 = \Theta_3 = \Theta_\infty \tau_c$).

We will now first analyze the rest of process *without the afterburner*, for which $\Theta_5 = \Theta_6 = \Theta_7$.

Now, since

$$\frac{T_5^o}{T_7} = \left(\frac{p_5^o}{p_\infty}\right)^{\frac{\gamma-1}{\gamma}} = \delta_5^{\frac{\gamma-1}{\gamma}} = \tau_t \delta_4^{\frac{\gamma-1}{\gamma}} = \tau_t \delta_3^{\frac{\gamma-1}{\gamma}} = \tau_t \Theta_3$$

$$= \tau_t \tau_c \Theta_\infty = \left[1 - \Theta_\infty \frac{(\tau_c - 1)}{\Theta_4}\right] \tau_c \Theta_\infty = \tau_c \Theta_\infty \frac{\Theta_5}{\Theta_4}$$

we get

$$\frac{T_7}{T_\infty} = \frac{T_5^o}{T_\infty} \cdot \frac{T_7}{T_5^o} = \frac{\Theta_5 \Theta_4}{\tau_c \Theta_\infty \Theta_5} = \frac{\Theta_4}{\tau_c \Theta_\infty} = \frac{\Theta_4}{\tau_{c,\mathrm{eff}}} \tag{2.40h}$$

Heat added to and *rejected* by the system are

$$|q_a| = c_p\left(T_4^o - T_4^o\right) = c_p T_\infty(\Theta_4 - \Theta_3) = c_p T_\infty(\Theta_4 - \tau_{c,\mathrm{eff}})$$
$$= c_p T_\infty(\Theta_5 - \Theta_\infty \tau_c - \Theta_\infty - \tau_c \Theta_\infty) = c_p T_\infty$$

and

$$|q_r| = c_p(T_7 - T_\infty) = c_p T_\infty\left(\frac{T_7}{T_\infty} - 1\right) = c_p T_\infty \frac{(\Theta_4 - \tau_{c,\mathrm{eff}})}{\tau_{c,\mathrm{eff}}} \tag{2.41b}$$

Therefore, the nondimensional work and thermodynamic efficiency are

$$\frac{w}{c_p T_\infty} = \frac{(|q_a| - |q_r|)}{c_p T_\infty} = (\Theta_4 - \tau_{c,\mathrm{eff}}) \frac{(\tau_{c,\mathrm{eff}} - 1)}{\tau_{c,\mathrm{eff}}} \tag{2.41c}$$

and

$$\eta_{\mathrm{th}} = \frac{w}{q_q} = \frac{(\tau_{c,\mathrm{eff}} - 1)}{\tau_{c,\mathrm{eff}}} = \frac{\pi_{c,\mathrm{eff}}^{\frac{\gamma-1}{\gamma}} - 1}{\pi_{c,\mathrm{eff}}^{\frac{\gamma-1}{\gamma}}} \tag{2.41d}$$

which for $\tau_{c,\mathrm{eff}} = 1$ (no compression in inlet or compressor), both w and η_{th} are zero, for $(\Theta_4 - \tau_{c,\mathrm{eff}}) = 0$ (no fuel injection), $w = 0$, but η_{th} attains the maximum possible efficiency of a Carnot cycle. It is interesting to note from (2.41d) that the two limiting values of the thermal efficiency are $\pi_{c,\mathrm{eff}} = 1 : \eta_{\mathrm{th}} = 0$ and $\pi_{c,\mathrm{eff}} \to \infty : \eta_{\mathrm{th}} \to 1$, and thus it would appear that the compression ratio must be as large as possible. However, that is not possible, since the maximum effective compression ratio is restricted by the maximum given temperature ratio with no fuel injection, where again the specific work goes to zero. Hence, the two limits of $\tau_{c,\mathrm{eff}}$ are $1 \le \tau_{c,\mathrm{eff}} \le \Theta_4$. At the second limit, $\tau_{c,\mathrm{eff}} = 1 : \eta_{\mathrm{th}} = (\Theta_4 - 1)/\Theta_4$ thermodynamic efficiency of the *Carnot cycle (Carnot cycle efficiency)*.

Between these two extreme cases, we can now find the condition for maximum specific work by differentiating (2.41c) with respect to $\tau_{c,\text{eff}}$ and setting it equal to zero to get

$$\left(\tau_{c,\text{eff}}\right)_{\text{opt}} = \sqrt{\Theta_4} \tag{2.42a}$$

and the optimum (maximum) nondimensional work is

$$\frac{w_{\text{opt}}}{c_p T_\infty} = \left(\sqrt{\Theta_4} - 1\right)^2 \tag{2.42b}$$

Note that (2.41c) and (2.41d) about work output and thermodynamic efficiency here are exactly the same as (2.26a) and (2.26b) for turboprops, except that the variables $\pi_{c,\text{eff}}$ and $\tau_{c,\text{eff}}$ here are replaced in the latter by π_c and τ_c, respectively.

Now we'll calculate the jet exhaust speed. From the principle of energy consumption,

$$c_p T_7^o = c_p T_7 + \frac{1}{2} u_7^2$$

we write

$$
\begin{aligned}
\frac{u_7}{u_\infty} &= \frac{1}{M_\infty} \frac{\sqrt{c_p(T_5^o - T_7)}}{\gamma R T_\infty} = \frac{1}{M_\infty} \sqrt{\frac{2}{\gamma - 1}\left(\Theta_5 - \frac{T_7}{T_\infty}\right)} \\
&= \frac{1}{M_\infty} \sqrt{\frac{2}{\gamma - 1}\left[\Theta_4 + \Theta_\infty - \tau_{c,\text{eff}} - \frac{\Theta_4}{\tau_{c,\text{eff}}}\right]} \\
&= \frac{1}{M_\infty} \sqrt{\frac{2}{\gamma - 1}\left[\Theta_4 \frac{(\tau_{c,\text{eff}} - 1)}{\tau_{c,\text{eff}}} - \Theta_\infty(\tau_c - 1)\right]} \\
&= \frac{1}{M_\infty} \sqrt{\frac{2}{\gamma - 1}\left[\Theta_4 \frac{(\Theta_\infty \tau_c - 1)}{\Theta_\infty \tau_c} - \Theta_\infty(\tau_c - 1)\right]}
\end{aligned}
\tag{2.43a}
$$

and we get an expression for the *propulsive efficiency*:

$$\eta_p = \frac{2}{\left[1 + \frac{u_7}{u_\infty}\right]} \tag{2.43b}$$

On the other hand, work is gained through the difference in the kinetic energy of the jet and the inlet flow, which are

$$\frac{u_7^2}{2 c_p T_\infty} = \Theta_5 - \frac{T_7}{T_\infty} = \Theta_4 + \Theta_\infty - \tau_{c,\text{eff}} - \frac{\Theta_4}{\tau_{c,\text{eff}}} \tag{2.43c}$$

and

$$\frac{u_\infty^2}{2c_p T_\infty} = \frac{\gamma - 1}{2} M_\infty^2 = \Theta_\infty - 1 \qquad (2.43d)$$

Subtracting one from the other, we get

$$\frac{u_7^2 - u_\infty^2}{2c_p T_\infty} = \Theta_4 + 1 - \tau_{c,\text{eff}} - \frac{\Theta_4}{\tau_{c,\text{eff}}} = \Theta_4 \frac{\tau_{c,\text{eff}} - 1}{\tau_{c,\text{eff}}} - (\tau_{c,\text{eff}} - 1)$$

$$= (\Theta_4 - \tau_{c,\text{eff}}) \frac{(\tau_{c,\text{eff}} - 1)}{\tau_{c,\text{eff}}} = \frac{w}{c_p T_\infty}$$

and we show again, just like for ramjets, that the overall work out of the thermodynamic cycle is due to the increase in the kinetic energy.

Now, the thrust to be developed is

$$F = \dot{m}_a (u_7 - u_\infty)$$

In the above expression, and contrary to that for the ramjet engine, the mass flow rate and u_∞ are not completely on the flight speed of the engine, since even at zero flight speed on ground, air is sucked into the engine due to the rotating gas turbine engine.

Now, the nondimensional specific thrust, with the help of (2.43a), is

$$\frac{F}{\dot{m}_a a_\infty} = M_\infty \left(\frac{u_7}{u_\infty} - 1 \right) = M_\infty \left[\frac{1}{M_\infty} \sqrt{\frac{2}{\gamma - 1} \Theta_4 \frac{\Theta_\infty \tau_c - 1}{\Theta_\infty \tau_c} - \Theta_\infty (\tau_c - 1) - 1} \right]$$

$$= \sqrt{\frac{2}{\gamma - 1} \Theta_4 \left(\frac{\Theta_\infty \tau_c - 1}{\Theta_\infty \tau_c} - \Theta_\infty (\tau_c - 1) \right)} - M_\infty$$

$$(2.44a)$$

Equation 2.44a is a somewhat complicated function of $(M_\infty, \pi_c, \Theta_4)$. However from (2.44a), it can be seen that the specific thrust is maximum at zero Mach number $(\Theta_\infty = 1)$ and decreases linearly with increasing flight Mach number. This is because although the mass flow rate increases with the flight Mach number, the difference in the two velocities continuously decreases with the increasing flight Mach number. Thus, thrust is maximal at zero Mach number, as follows:

$$\left(\frac{F}{\dot{m}_a a_\infty} \right)_{M_\infty = 0} = \sqrt{\frac{2}{\gamma - 1} \Theta_4 \left(\frac{\tau_c - 1}{\tau_c} - (\tau_c - 1) \right)}$$

$$= \sqrt{\frac{2}{\gamma - 1} \left(\frac{w}{c_p T_\infty} \right)_{M_\infty = 0}} \qquad (2.44b)$$

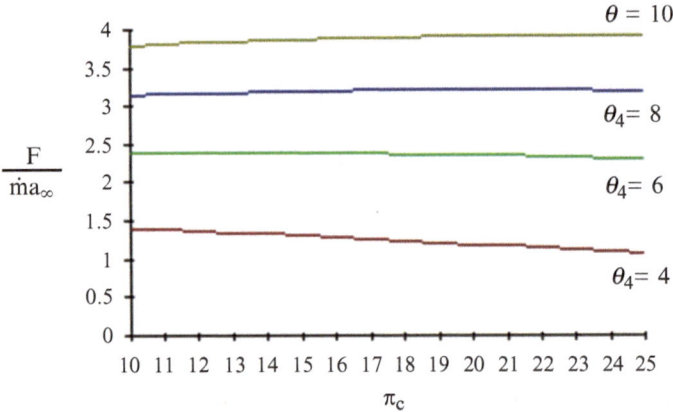

Fig. 2.18 Results of calculating (2.44a) for different Θ_4

Results of calculating (2.44a) at zero Mach number are given in Fig. 2.18.

Thus, the condition for maximum thrust for $M_\infty = 0$ is the same as that for maximum work, as we can verify by comparing with (2.41c).

Now, for the energy input in the combustion chamber,

$$\dot{m}_a c_p (T_4^o - T_3^o) = \dot{m}_f \Delta H_p = \dot{m}_a q_a = \frac{\dot{m}_a w}{\eta_{th}} \qquad (2.44c)$$

which gives the relation for the *fuel–air ratio* as

$$f = \frac{\dot{m}_f}{\dot{m}_a} = \frac{c_p T_\infty}{\Delta H_p}(\Theta_4 - \Theta_3) = \left(\frac{c_p T_\infty}{\Delta H_p}\right)(\Theta_4 - \tau_{c,\text{eff}}) \qquad (2.45a)$$

Let's discuss the specific fuel combustion further. Usually, the specific fuel consumption [kg/J] is computed by dividing the fuel flow rate $\dot{m}_f$ [kg/s] by the work produced per kg of air [J/kg]. However, for aircraft engine applications, it is computed by dividing the fuel flow rate by the thrust [N].

Hence, the specific fuel consumption is defined as

$$\text{SFC} = \frac{\dot{m}_f}{F} = \frac{f}{a_\infty \left[\sqrt{\frac{2}{\gamma-1}\left\{\Theta_4 \frac{(\Theta_\infty \tau_c - 1)}{\Theta_\infty \tau_c}\right\}} - M_\infty\right]} \qquad (2.45b)$$

and one can define a *nondimensional specific fuel consumption*,

$$\text{SFC}^* = \text{SFC}.a_\infty = \frac{f}{\left[\sqrt{\frac{2}{\gamma-1}\left\{\Theta_4 \frac{(\Theta_\infty \tau_c - 1)}{\Theta_\infty \tau_c}\right\}} - M_\infty\right]} \qquad (2.45c)$$

We have therefore shown that although the fuel–air ratio (without afterburner operation) is reduced linearly with τ_c, it is also proportional to Θ_4. On the other hand, the nondimensional specific fuel consumption is a complicated function of $(\Theta_4|\Theta_\infty)$, that is, M_∞, τ_c, and η_{th}, the last variable again being dependent on τ_c and Θ_∞.

We will now consider the situation with *afterburner*. Instead of (2.41a), the *heat added* is

$$|q_a| = c_p\left(T_4^o - T_3^o + T_6^o - T_5^o\right) = c_p T_\infty(\Theta_4 - \Theta_3 + \Theta_6 - \Theta_5)$$

Since from (2.40d) and (2.40f),

$$\Theta_3 = \tau_c \Theta_\infty \quad \text{and} \quad \Theta_5 = \Theta_4 - \Theta_\infty(\tau_c - 1)$$

we get

$$\begin{aligned}
|q_a| &= c_p T_\infty(\Theta_4 - \Theta_\infty \tau_c + \Theta_6 - \Theta_4 + \Theta_\infty \tau_c - \Theta_\infty) = c_p T_\infty(\Theta_6 - \Theta_\infty) \\
&= c_p T_\infty[(\Theta_6 - \Theta_5) + (\Theta_4 - \Theta_\infty - \tau_c)] \\
&= \frac{a_\infty^2}{(\gamma - 1)}[(\Theta_6 - \Theta_5) + (\Theta_4 - \Theta_\infty - \tau_c)]
\end{aligned}$$

$$(2.46a)$$

Since

$$\frac{T_7^o}{T_7} = \left(\frac{p_7^o}{p_7}\right)^{\frac{\gamma-1}{\gamma}} = \delta_5^{\frac{\gamma-1}{\gamma}} = \tau_t \delta_3^{\frac{\gamma-1}{\gamma}} = \tau_t \Theta_3 = \tau_c \Theta_5 \frac{\Theta_\infty}{\Theta_4} = \Theta_7 = \Theta_6 \quad \text{and} \quad \tau_t = \frac{\Theta_5}{\Theta_4}$$

we get

$$\begin{aligned}
\frac{T_7}{T_\infty} &= \frac{\Theta_6}{(\Theta_\infty \tau_t \tau_c)} = \frac{\Theta_6 - \Theta_5 + \Theta_5}{(\Theta_\infty \tau_t \tau_c)} = \Theta_4 \frac{\Theta_6 - \Theta_5 + \Theta_5}{(\Theta_\infty \tau_c \Theta_5)} \\
&= \frac{\Theta_4}{\Theta_\infty \tau_c} + \Theta_4 \frac{\Theta_6 - \Theta_5}{(\Theta_\infty \tau_c \Theta_5)}
\end{aligned}$$

$$(2.47)$$

and hence, the *reject heat* is

$$\begin{aligned}
|q_r| &= c_p T_\infty\left(\frac{T_7}{T_\infty} - 1\right) = c_p T_\infty\left[\frac{\Theta_4}{\Theta_\infty \tau_c} - 1 + \Theta_4 \frac{\Theta_6 - \Theta_5}{\Theta_\infty \tau_c \Theta_5}\right] \\
&= c_p T_\infty\left[\frac{\Theta_4 - \Theta_\infty \tau_c}{\Theta_\infty \tau_c} + \Theta_4 \frac{\Theta_6 - \Theta_5}{\Theta_\infty \tau_c \Theta_5}\right]
\end{aligned}$$

$$(2.46b)$$

Thus, the specific work gained is

$$\frac{w}{c_p T_\infty} = \frac{|q_a| - |q_r|}{c_p T_\infty} = (\Theta_6 - \Theta_5) + (\Theta_4 - \Theta_\infty \tau_c) - \frac{(\Theta_4 - \Theta_\infty \tau_c)}{\Theta_\infty \tau_c} - \frac{(\Theta_6 - \Theta_5)}{\Theta_\infty \tau_c \Theta_5}$$

$$= (\Theta_4 - \Theta_\infty \tau_c)\frac{(\Theta_\infty \tau_c - 1)}{\Theta_\infty \tau_c} + (\Theta_6 - \Theta_5)\frac{(\Theta_\infty \tau_c \Theta_5 - \Theta_4)}{\Theta_\infty \tau_c \Theta_5}$$

$$= (\Theta_4 - \Theta_\infty \tau_c)\frac{(\Theta_\infty \tau_c - 1)}{\Theta_\infty \tau_c} + (\Theta_6 - \Theta_5)\left(1 - \frac{\Theta_4}{\Theta_\infty \tau_c \Theta_5}\right)$$

$$= (\Theta_4 - \tau_{c,\text{eff}})\frac{(\tau_{c,\text{eff}} - 1)}{\tau_{c,\text{eff}}} + (\Theta_6 - \Theta_5)\left(1 - \frac{\Theta_4}{\tau_{c,\text{eff}} \Theta_5}\right)$$

$$(2.46c)$$

Now,

$$w = \frac{1}{2}\left(u_7^2 - u_\infty^2\right)\frac{w}{c_p T_\infty} = \frac{\gamma - 1}{2}M_\infty^2\left[\left(\frac{u_7}{u_\infty}\right)^2 - 1\right]$$

$$= (\Theta_4 - \tau_{c,\text{eff}})\frac{(\Theta_4 - \tau_{c,\text{eff}} - 1)}{\tau_{c,\text{eff}}} + (\Theta_6 - \Theta_5)\left(1 - \frac{\Theta_4}{\tau_{c,\text{eff}} \Theta_5}\right)$$

Hence,

$$\frac{u_7}{u_\infty} = \sqrt{1 + \frac{2}{(\gamma - 1)}M_\infty^2\left[(\Theta_4 - \tau_{c,\text{eff}})\frac{(\tau_{c,\text{eff}} - 1)}{\tau_{c,\text{eff}}} + (\Theta_6 - \Theta_5)\left(1 - \frac{\Theta_4}{\tau_{c,\text{eff}} \Theta_5}\right)\right]}$$

and the *nondimensional specific thrust* is

$$\frac{F}{\dot{m}_a a_\infty} = M_\infty\left(\frac{u_7}{u_\infty} - 1\right)$$

$$= M_\infty\left[\sqrt{1 + \frac{2}{(\gamma - 1)}M_\infty^2\left[(\Theta_4 - \tau_{c,\text{eff}})\frac{(\tau_{c,\text{eff}} - 1)}{\tau_{c,\text{eff}}} + (\Theta_6 - \Theta_5)\left(1 - \frac{\Theta_4}{\tau_{c,\text{eff}} \Theta_5}\right)\right]} - 1\right]$$

$$(2.46d)$$

Now, the maximum of (2.46d) is again similar to that given by (2.44a), and its maximum in comparison to (2.44b), which is without an afterburner, is shifted toward a high effective compression ratio.

Now, similar to for the case without an afterburner, (2.44c), we have

$$\dot{m}_f \Delta H_p = \dot{m}_a q_a = \dot{m}_a c_p T_\infty \left[(\Theta_6 - \Theta_5) + (\Theta_4 - \tau_{c,\text{eff}})\right]$$

Thus, the *fuel–air ratio* is

$$f = \frac{\dot{m}_f}{\dot{m}_a} = \frac{c_p T_\infty}{\Delta H_p}\left[(\Theta_6 - \Theta_5) + (\Theta_4 - \tau_{c,\text{eff}})\right]$$

and the thermodynamic efficiency is

$$\eta_{\text{th}} = \frac{\left\{\frac{(\Theta_4 - \Theta_\infty \tau_c)(\Theta_\infty \tau_c - 1)}{\Theta_\infty \tau_c} + (\Theta_6 - \Theta_5)\left[1 - \frac{\Theta_4}{\Theta_\infty \tau_c \Theta_5}\right]\right\}}{\left\{(\Theta_4 - \Theta_\infty \tau_c) + (\Theta_6 - \Theta_5)\right\}}$$

$$= \frac{\left\{\frac{(\Theta_4 - \tau_{c,\text{eff}})(\tau_{c,\text{eff}} - 1)}{\tau_{c,\text{eff}}} + (\Theta_6 - \Theta_5)\left[1 - \frac{\Theta_4}{\tau_{c,\text{eff}}\Theta_5}\right]\right\}}{\left\{(\Theta_4 - \tau_{c,\text{eff}}) + (\Theta_6 - \Theta_5)\right\}}$$

Now, the *thrust* is given by

$$F = M_\infty \dot{m}_a a_\infty$$
$$\left[\sqrt{1 + \frac{2}{(\gamma - 1)}M_\infty^2\left[(\Theta_4 - \tau_{c,\text{eff}})\frac{(\tau_{c,\text{eff}} - 1)}{\tau_{c,\text{eff}}} + (\Theta_6 - \Theta_5)\left(1 - \frac{\Theta_4}{\tau_{c,\text{eff}}\Theta_5}\right)\right]} - 1\right], N$$

Hence, the nondimensional specific fuel consumption is

$$\text{SFC} = \frac{\dot{m}_f}{F} \frac{f}{a_\infty\left[\sqrt{1 + \frac{2}{(\gamma-1)}M_\infty^2\left[(\Theta_4 - \tau_{c,\text{eff}})\frac{(\tau_{c,\text{eff}}-1)}{\tau_{c,\text{eff}}} + (\Theta_6 - \Theta_5)\left(1 - \frac{\Theta_4}{\tau_{c,\text{eff}}\Theta_5}\right)\right]} - 1\right]}$$

and the nondimensional specific fuel combustion is

$$\text{SFC}^* = \text{SFC}.a_\infty \frac{f}{\left[\sqrt{1 + \frac{2}{(\gamma-1)}M_\infty^2\left[(\Theta_4 - \tau_{c,\text{eff}})\frac{(\tau_{c,\text{eff}}-1)}{\tau_{c,\text{eff}}} + (\Theta_6 - \Theta_5)\left(1 - \frac{\Theta_4}{\tau_{c,\text{eff}}\Theta_5}\right)\right]} - 1\right]}$$

For the afterburner operation, f increases, but with increasing flight Mach number (increasing Θ_∞) it decreases, since for a given Θ_4 the scope for fuel flow rate decreases (before the combustion chamber, the air is heated), and the specific fuel consumption changes in a similar fashion.

It is seen that the thermodyamic efficiency is considerably reduced against that without an afterburner. As before, the optimum specific work is obtained by differentiating (2.46c) with respect to τ_c and equating it to zero to get

$$(\tau_c)_{\text{opt}} = \sqrt{\Theta_4 | \Theta_6 / \Theta_5} \tag{2.48a}$$

and the *optimum specific work* is

$$\frac{w_{opt}}{c_p T_\infty} = (\Theta_6 - \Theta_5)\left[1 - \sqrt{\frac{\Theta_4 \Theta_6}{\Theta_5}} - \Theta_6\right] + \sqrt{\frac{\Theta_5}{\Theta_6}\left(\Theta_4 \sqrt{\frac{\Theta_6}{\Theta_5}}\right)}\left(\sqrt{\Theta_4 \frac{\Theta_6}{\Theta_5}} - 1\right)$$

$$(2.48b)$$

From these equations, we can see that for a given combustion chamber tempera-ture ratio, π_c can be considerably increased at higher flight Mach numbers if the afterburner is operational; however, it has the penalty of a higher specific fuel consumption. An aircraft, like a supersonic transporter to operate at Mach >2, can therefore take off the ground and increase its speed with or without an afterburner. With increasing flight speed, the thrust will come to zero sooner if the afterburner does not work. Therefore, with the afterburner working, it can maintain its thrust level sufficiently high to reach the required level. At the time the afterburner is operating, the tailpipe's cross-sectional area has to be increased to accommodate a higher-volume flow rate; otherwise, pressure can be fed back upstream through the turbine, combustion chamber, and compressor to reduce the mass flow rate.

In order to show the effect from the afterburner working, results of $w/(c_p T_\infty)$, $F/(\dot{m}_a a_\infty)$, and SFC* are computed for $(\Theta_4 = 4)$ for two values each of M_∞ and $\Delta\Theta_{AB} = (\Theta_6 - \Theta_5)$ and are shown in Figs. 2.19, 2.20, and 2.21.

2.2.3   Ideal Bypass Jet Cycle with Afterburner

We have shown earlier that in order to get better propulsive efficiency, it is necessary to have the average jet speed lower and as close to the flight speed as possible. For this purpose, two different methods have been considered. In the first method, the incoming air after the low-pressure compressor is split into two parts: one going through the high-pressure compressor, the combustion chamber, and the turbine (internal bypass and mixing), and the other bypasses both the combustion chamber and the turbine. For the first, there are problems because in order to match the air stream pressure after the turbine, the bypass ratio is limited. But this has a comparatively small front area and can be useful for military engines. On the other hand, if a very high volume of bypassed air is exited right after the low-pressure compressor, then one can considerably increase the bypass ratio (fanjet), which is the prevalent engine for transonic flying commercial jets. For the first type, a schematic sketch of a bypass (internal mixing) jet with an afterburner is shown in Fig. 2.22a, and the corresponding ideal thermodynamic cyclic process appears in Fig. 2.22b. Here b is the bypass ratio, which is the ratio of bypassed (cold) air to the core (hot) air, $b = \dot{m}_C/\dot{m}_H$. Since $\dot{m}_C + \dot{m}_H = \dot{m}_a$, which is the total air mass flow ratio, it is evident that

$$\dot{m}_H = \frac{\dot{m}_a}{1 + b}$$

$$(2.50a)$$

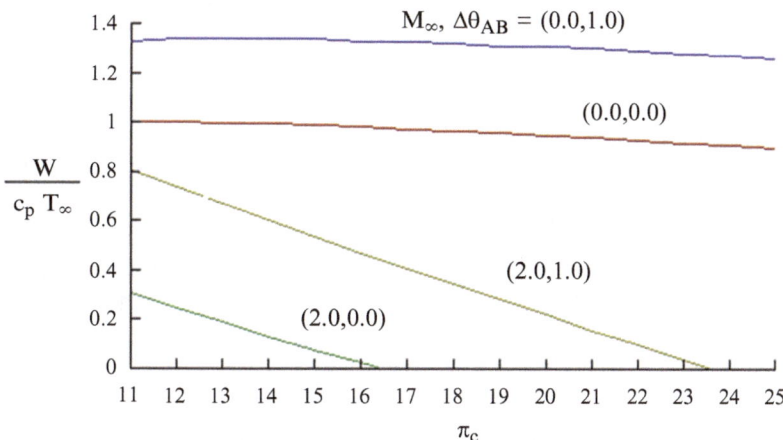

Fig. 2.19 Specific work vs. pressure ratio for afterburner operation

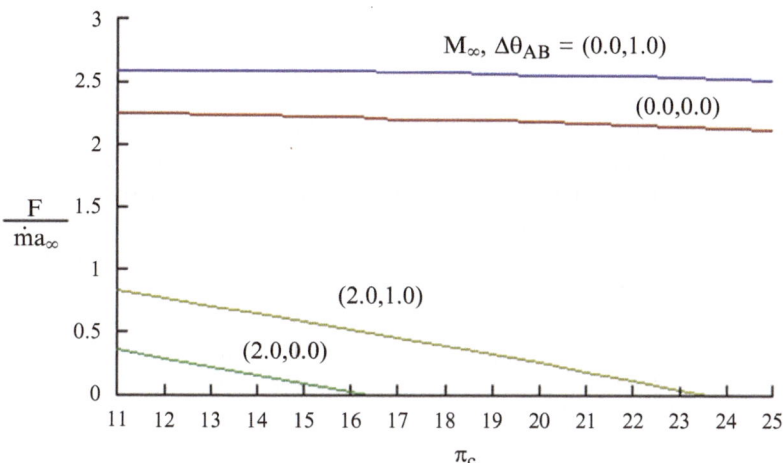

Fig. 2.20 Specific thrust vs. pressure ratio for afterburner operation

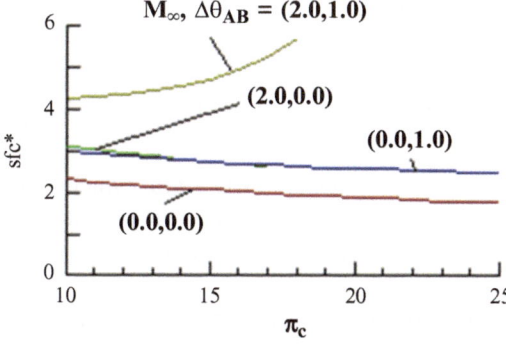

Fig. 2.21 Nondimensional SFC vs. pressure ratio for afterburner operation

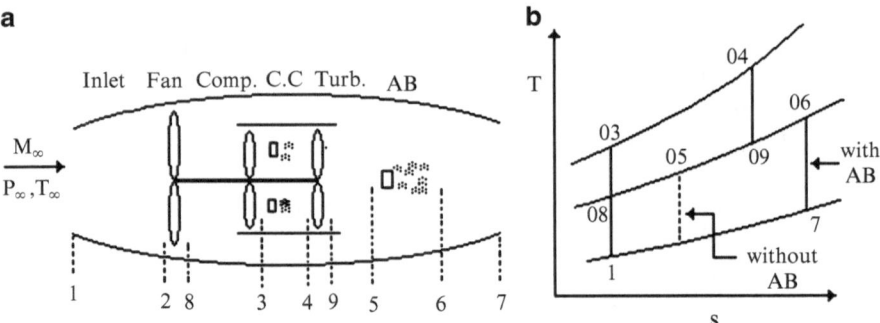

Fig. 2.22 (**a**) Schematic sketch of a bypass jet (with internal mixing); (**b**) corresponding sketch of the thermodynamic process in a (T, s) chart

and

$$\dot{m}_C = \frac{\dot{m}_a b}{1+b} \tag{2.50b}$$

For this case, the bypass ratio b is being introduced as a parameter in addition to the parameters already introduced in the previous case, that is, the approaching flow Mach number, the combustion chamber temperature ratio, and the afterburner stagnation temperature increases.

Thus, let's assume that the parameters given are $\pi_c = p_3^o/p_2^o$ as the *overall compressor ratio*, $\Theta_4 = T_4^o/T_\infty$ as the main *combustion chamber temperature ratio = turbine inlet temperature ratio*, M_∞ as the *approaching flow Mach number*, b as the *bypass ratio*, and $\Delta\Theta_{AB} = \left(T_6^o - T_5^o\right)/T_\infty$ as the *stagnation temperature increase in the afterburner*. Thus, we have an increase in the basic parameter by 1 over those for the straight-jet engine.

Now,

$$\Theta_\infty = \Theta_1 = \Theta_2 = \frac{T_\infty^o}{T_\infty} = \frac{T_1^o}{T_\infty} = \frac{T_2^o}{T_\infty} = 1 + \frac{\gamma-1}{2}M_\infty^2 \tag{2.51a}$$

and

$$\delta_\infty = \delta_1 = \delta_2 = \frac{p_\infty^o}{p_\infty} = \frac{p_1^o}{p_\infty} = \frac{p_2^o}{p_\infty} = \left(1 + \frac{\gamma-1}{2}M_\infty^2\right)^{\frac{\gamma}{\gamma-1}} \tag{2.51b}$$

Let $\pi_f = p_8^o/p_\infty^2$ be the *fan compression ratio*, which is not an independent parameter, but depends on b to match the fan exit pressure with the turbine exit pressure. Thus, the following relations are obvious:

$$\delta_8 = \delta_9 = \delta_5 = \delta_6 = p_8^o/p_\infty = p_9^o/p_\infty = p_5^o/p_\infty = p_6^o/p_\infty = \pi_f\delta_\infty$$

$$= \pi_f\Theta^{\gamma(\gamma-1)} = \pi_{f,eff} \tag{2.51c}$$

$$\delta_3 = \delta_4 = p_3^o/p_\infty = p_4^o/p_\infty = \pi_c \delta_\infty = \pi_c \Theta_\infty^{\gamma/(\gamma-1)} = \pi_{c,\text{eff}} \qquad (2.51d)$$

$$\Theta_8 = T_8^o/T_\infty = \pi_{c,\text{eff}}^{(\gamma-1)/\gamma} = \Theta_\infty \pi_f^{(\gamma-1)/\gamma} \qquad (2.51e)$$

and

$$\Theta_3 = T_3^o/T_\infty = \pi_{c,\text{eff}}^{(\gamma-1)/\gamma} = \Theta_\infty \pi_c^{(\gamma-1)/\gamma} \qquad (2.51f)$$

In above equation, $\pi_{c,\text{eff}}$ and $\pi_{f,\text{eff}}$ are the *effective overall compression ratio* and the *effective fan compression ratio*, respectively.

Noting that the turbine is driving the compressor and assuming a *single-spool engine*, we have

$$T_4^o - T_9^o = \left(T_3^o - T_2^o\right) + b\left(T_8^o - T_2^o\right)$$

from which it follows that

$$
\begin{aligned}
\Theta_9 &= \Theta_4 - [(\Theta_3 - \Theta_2) + b(\Theta_8 - \Theta_2)] \\
&= \Theta_4 - \Theta_\infty \left\{ \left(\pi_c^{(\gamma-1)/\gamma} - 1\right) + b\left(\pi_f^{(\gamma-1)/\gamma} - 1\right) \right\} \\
&= \Theta_4 - \left\{ \left(\pi_{c,\text{eff}}^{(\gamma-1)/\gamma} - 1\right) + \left(\pi_{f,\text{eff}}^{(\gamma-1)/\gamma} - 1\right) + (1+b)\Theta_\infty \right\} \qquad (2.52a)
\end{aligned}
$$

Furthermore,

$$\frac{T_4^o}{T_9^o} = \frac{\Theta_4}{\Theta_9} = \left(\frac{p_2^o}{p_9^o}\right)^{\frac{\gamma-1}{\gamma}} = \left[\frac{\pi_c \Theta_\infty}{\pi_f \Theta_\infty}\right]^{\frac{\gamma-1}{\gamma}} = \left(\frac{\pi_{c,\text{eff}}}{\pi_{f,\text{eff}}}\right)^{\frac{\gamma-1}{\gamma}} \qquad (2.52b)$$

From (2.52a) and (2.52b), we can therefore write

$$\Theta_9 = \Theta_4 \left(\frac{\pi_f}{\pi_c}\right)^{\frac{\gamma-1}{\gamma}} = \{\Theta_4 - \Theta_\infty\}\left[\left(\pi_c^{(\gamma-1)/\gamma} - 1\right) + b\left(\pi_f^{(\gamma-1)/\gamma}\right)\right]$$

from which it follows that

$$\left(\frac{\pi_f}{\pi_c}\right)^{\frac{\gamma-1}{\gamma}} = \frac{\{\Theta_4 - \Theta_\infty\}}{\Theta_4}\left[\left(\pi_c^{(\gamma-1)/\gamma} - 1\right) + b(\pi_f^{(\gamma-1)/\gamma} - 1\right] \qquad (2.53a)$$

Thus,

$$\frac{\pi_f}{\pi_c} = f(\Theta_4, \Theta_\infty, \pi_c, b, \gamma) \qquad (2.53b)$$

Table 2.5 Effect of Bypass Ratio on the Fan Pressure Ratio	b	0.4	0.8	1.2	1.6
	π_f/π_c	0.1720	0.0522	0.0291	0.0042

For $\gamma = 1.4$ and $M_\infty \to 0$, that is, $\Theta_\infty = 1$, we have still three independent parameters on which the ratio (π_f/π_c) depends. Obviously, the limit of $b \to 0$ for the ratio (π_f/π_c) has no meaning, since then the pressure behind the turbine is not related to the pressure behind the fan, although we get a finite ratio. However, (π_f/π_c) is computed for nominal values of $\gamma = 1.4, \Theta_4 = 4.0$ and $\pi_c = 16.0$, and the results are presented for $b = 0.4, 0.8, 1.2$, and 1.6 in Table 2.5.

The results therefore show that with the increasing bypass ratio, the fan pressure ratio decreases. This can be explained due to the fact that with the increasing bypass ratio, the turbine has to produce more work to allow compression of more and more air flow, and hence the turbine pressure must decrease, which, to balance the pressure at that point, must reduce the fan pressure.

From energy balance, we now have

$$bc_p(T_5^o - T_8^o) = c_p(T_9^o - T_5^o)$$

which results in the relation

$$b(\Theta_5 - \Theta_8) = \Theta_9 - \Theta_5$$

Hence,

$$\Theta_5(b+1) = \Theta_9 + \Theta_8 = \Theta_4 \left(\frac{\pi_f}{\pi_c}\right)^{\frac{\gamma-1}{\gamma}} + b\Theta_\infty \pi_f^{\frac{\gamma-1}{\gamma}} = \pi_f^{\frac{\gamma-1}{\gamma}} \left[b\Theta_\infty + \frac{\Theta_4}{\pi_c^{\frac{\gamma-1}{\gamma}}}\right]$$

Since

$$\pi_f^{\frac{\gamma-1}{\gamma}} \left[b\Theta_\infty + \frac{\Theta_4}{\pi_c^{\frac{\gamma-1}{\gamma}}}\right] = \Theta_4 - \Theta_\infty \left(\pi_c^{(\gamma-1)/\gamma} - 1 - b\right)$$

we get

$$\Theta_5 = \frac{\Theta_4 - \pi_{c,eff}^{(\gamma-1)/\gamma} + \Theta_\infty(1-b)}{1+b} \tag{2.54a}$$

It follows further that

$$\delta_5 = \frac{p_5^o}{p_\infty} = \left(\frac{p_4^o}{p_\infty}\right)\left(\frac{p_5^o}{p_4^o}\right) = \left(\frac{p_4^o}{p_\infty}\right)\left(\frac{p_9^o}{p_4^o}\right) = \left(\frac{p_3^o}{p_\infty}\right)\left(\frac{p_8^o}{p_3^o}\right)$$

$$= \delta_8 \frac{p_8^\infty}{p_\infty} = \pi_{c,eff}\left(\frac{p_9^o}{p_4^o}\right)^{\frac{\gamma}{\gamma-1}} = \pi_{c,eff}\left(\frac{\Theta_9}{\Theta_4}\right)^{\frac{\gamma}{\gamma-1}} \frac{\pi_f}{\pi_c} = \pi_{f,eff} \tag{2.54b}$$

We will now first study the case *without* an afterburner, that is, $T_5^o = T_6^o = T_7^o$. Thus,

$$u_7 = \sqrt{2c_p(T_5^o - T_7)}$$

from which it follows that

$$\frac{u_7}{u_\infty} = \frac{1}{M_\infty} \sqrt{\frac{2}{\gamma - 1} \left(\Theta_5 - \frac{T_7}{T_\infty} \right)}$$

Now since

$$\frac{T_7}{T_\infty} = \left(\frac{T_5^o}{T_\infty} \right) \left(\frac{T_7}{T_5^o} \right) = \Theta_5 \left(\frac{p_\infty}{p_5^o} \right)^{\frac{\gamma-1}{\gamma}} = \frac{\Theta_5}{\left(\Theta_\infty \pi_f^{(\gamma-1)/\gamma} \right)} = \frac{\Theta_5}{\left(\pi_{f,eff}^{(\gamma-1)/\gamma} \right)} = \frac{\Theta_5}{\tau_{f,eff}} \tag{2.55a}$$

it follows that

$$\left(\frac{u_7}{u_\infty} \right) = \frac{1}{M_\infty} \sqrt{\frac{2}{\gamma - 1} \left(1 - \pi_{f,eff}^{(\gamma-1)/\gamma} \right)} = \frac{1}{M_\infty} \sqrt{\frac{2}{\gamma - 1} \left(1 - \tau_{f,eff}^{(\gamma-1)/\gamma} \right)} \tag{2.55b}$$

Now

$$\frac{u_7^2}{2c_p T_\infty} = \frac{(T_5^o - T_7)}{T_\infty} = \Theta_5 - \frac{T_7}{T_\infty} = \Theta_5 \frac{(\tau_{f,eff} - 1)}{\tau_{f,eff}} \tag{2.55c}$$

and

$$\frac{u_\infty^2}{2c_p T_\infty} = M_\infty^2 \frac{(\gamma - 1)}{2} = \Theta_\infty - 1 \tag{2.55d}$$

From the difference in (2.55c) and (2.55d), we get the overall work (per kg of mixture) as

$$\frac{|w|}{c_p T_\infty} = \frac{u_7^2 - u_\infty^2}{2c_p T_\infty} = \Theta_\infty \frac{(\tau_{f,eff} - 1)}{\tau_{f,eff}} - (\Theta_\infty - 1) \tag{2.56a}$$

We get the same result from the overall balance of the heat input and output. First, the heat added (per unit mass of hot gas) is q_a, which, when written in nondimensional form, is

$$\frac{|q_a|}{c_p T_\infty} = \Theta_4 - \Theta_3 = \Theta_4 - \tau_{c,eff}$$

which, by taking (2.54a) into consideration, becomes

$$\frac{|q_a|}{c_p T_\infty} = (\Theta_5 - \Theta_\infty).(1 + b) \qquad (2.56b)$$

On the other hand, the heat added (per unit of total gas mixture) is

$$|\bar{q}_a| = \frac{|q_a|}{1 + b}$$

and thus,

$$\frac{|\bar{q}_a|}{c_p T_\infty} = (\Theta_5 - \Theta_\infty). \qquad (2.56c)$$

However, the rejected heat (per unit mass of total gas mixture) is

$$|q_r| = c_p(T_7 - T_\infty)$$

and, written in nondimensional form, it is

$$\frac{|\bar{q}_a|}{c_p T_\infty} = \left(\frac{T_7}{T_\infty} - 1\right) = \left(\frac{\Theta_5 - \tau_{f,\text{eff}}}{\tau_{f,\text{eff}}}\right) \qquad (2.56d)$$

From (2.56c) and (2.55d), we get the overall work (per kg of mixture),

$$\frac{|w|}{c_p T_\infty} = \frac{|\bar{q}_a|}{c_p T_\infty} - \frac{|\bar{q}_r|}{c_p T_\infty} = (\Theta_5 - \Theta_\infty) = \left(\frac{\Theta_5 - \tau_{f,\text{eff}}}{\tau_{f,\text{eff}}}\right)$$

$$= \Theta_5\left(\frac{\tau_{f,\text{eff}} - 1}{\tau_{f,\text{eff}}}\right) - (\Theta_\infty - 1)$$

which is the same as (2.56a). We therefore prove, once again, that the overall work in a low-bypass jet, as in the case of the ramjet and straight turbojet (with or without afterburner), is equal to the increase in the kinetic energy.

Now the *thermodynamic efficiency* is

$$\eta_{th} = \frac{|w|}{|\bar{q}_a|} = \left\{\Theta_5\left(\frac{\tau_{f,\text{eff}} - 1}{\tau_{f,\text{eff}}}\right) - \frac{(\Theta_\infty - 1)}{(\Theta_5 - \Theta_\infty)}\right\} = 1 - \frac{(\Theta - \tau_{f,\text{eff}})}{(\Theta_5 - \Theta_\infty)\tau_{f,\text{eff}}} \qquad (2.57a)$$

and the *propulsive efficiency* is

$$\eta_p = \frac{2}{1 + \frac{u_7}{u_\infty}} = \frac{2}{1 + \frac{1}{M_\infty}\sqrt{\frac{2\Theta_5}{\gamma - 1}\left(1 - \frac{1}{\tau_{f,\text{eff}}}\right)}} \qquad (2.57b)$$

which can, of course, give the *overall efficiency* as the product of the two. Now since the thrust is

$$F = \dot{m}_a(u_7 - u_\infty)$$

it can easily be shown that

$$\frac{F}{\dot{m}_a a_\infty} = M_\infty\left(\frac{u_7}{u_\infty} - 1\right) = \sqrt{\frac{2\Theta_5}{\gamma - 1}\left(1 - \frac{1}{\tau_{f,eff}}\right)} - M_\infty \qquad (2.57c)$$

which gives the surprising result that the specific thrust depends mainly on how much the temperature ratio at point 5 is across the far, which, of course, depends on various parameters, as shown in (2.55b). It is therefore evident, that a high specific thrust in a low-bypass jet requires a smaller $\tau_{f,eff}$ and a high Θ_5. This last one, according to (2.54a), depends on Θ_4, $\pi_{c,eff}$, and b.

We will now consider the situation with the afterburner operating. Let's specify the temperature increase in the afterburner, $\Delta\Theta_{AB} = \Theta_6 - \Theta_5$. Thus, the least heat added (per unit mass of mixed air) is

$$\frac{\bar{q}_a}{c_p T_\infty} = \Theta_5 - \Theta_\infty + \Theta_6 - \Theta_5 = \Theta_6 - \Theta_\infty \qquad (2.58a)$$

Now,

$$\frac{T_7}{T_\infty} = \left(\frac{T_6^o}{T_\infty}\right)\left(\frac{T_7}{T_6^o}\right) = \Theta_6\left(\frac{p_\infty}{p_6^o}\right)^{\frac{\gamma-1}{\gamma}} = \frac{\Theta_5}{\pi_{f,eff}^{(\gamma-1)/\gamma}} = \frac{\Theta_6}{\tau_{f,eff}} \qquad (2.58b)$$

and thus the reject heat in nondimensional form (per unit mass of air) is

$$\frac{|\bar{q}_r|}{c_p T_\infty} = \frac{T_7}{T_\infty} - 1 = \frac{(\Theta_6 - \tau_{f,eff})}{\tau_{f,eff}} \qquad (2.58c)$$

The *exhaust kinetic energy* in nondimensional form is now

$$\frac{u_7^2}{2c_p T_\infty} = \Theta_6 - \frac{T_{67}}{T_\infty} = \Theta_6\frac{(\tau_{f,eff} - 1)}{\tau_{f,eff}} \qquad (2.58d)$$

and the approaching flow kinetic energy in nondimensional form, before in (2.55d), is

$$\frac{u_\infty^2}{2c_p T_\infty} = \Theta_\infty - 1 \qquad (2.58e)$$

Thus, we can show that the overall work (per unit mass of mixed air) in nondimensional form is

$$\frac{w}{c_p T_\infty} = \frac{|\bar{q}_a| - |\bar{q}_r|}{c_p T_\infty} = (\Theta_6 - \Theta_\infty) - \frac{(\Theta_6 - \tau_{f,eff})}{\tau_{f,eff}} = \frac{(u_7^2 - u_\infty^2)}{2c_p T_\infty} \qquad (2.59a)$$

Now, the *thermodynamic efficiency* is

$$\eta_{th} = \frac{|w|}{|\bar{q}_a|} = 1 - \frac{(\Theta_6 - \tau_{f,eff})}{\tau_{f,eff}} (\Theta_6 - \Theta_\infty) \qquad (2.59b)$$

which replaces (2.57a) *without an afterburner*. From (2.58d) and (2.58e), the relation for the *speed ratio* is

$$\frac{u_7}{u_\infty} = \sqrt{\frac{\Theta_6}{(\Theta_\infty - 1)} \cdot \frac{(\tau_{f,eff} - 1)}{\tau_{f,eff}}} = \frac{1}{M_\infty} \sqrt{\frac{2}{\gamma - 1} \Theta_6 \left(1 - \frac{1}{\tau_{f,eff}}\right)} \qquad (2.58f)$$

and, thus, the *propulsion efficiency* is

$$\eta_p = \frac{2}{1 + \sqrt{\frac{\Theta_6}{(\Theta_\infty - 1)} \left(1 - \frac{1}{\tau_{f,eff}}\right)}} \qquad (2.59c)$$

which replaces (2.57b). The *specific thrust* relation with afterburner operating is now

$$\frac{F}{\dot{m}_a a_\infty} = M_\infty \left(\frac{u_7}{u_\infty} - 1\right) = \sqrt{\frac{2}{\gamma - 1} \Theta_6 \left(1 - \frac{1}{\tau_{f,eff}}\right)} - M_\infty \qquad (2.59d)$$

which replaces (2.57c).

Now, the fuel mass flow rate in the combustion chamber (without afterburner operating) is obtained (in term of unit mass of hot gas) by writing

$$\dot{m}_{fcc} = \dot{m}_H c_p T_\infty \left(\frac{\Theta_4 - \tau_{c,eff}}{\Delta H_p}\right)$$

which gives the *fuel–air ratio* in the combustion chamber as

$$f_{cc} = \frac{\dot{m}_{fcc}}{\dot{m}_H} = \frac{c_p T_\infty}{\Delta H_p} (\Theta_4 - \tau_{c,eff}) \qquad (2.60a)$$

The above fuel–air ratio in the combustion chamber must not be more than the maximum fuel–air ratio allowed for a particular fuel. Since, however, $\dot{m}_H = \dot{m}_a/(b+1)$, we get the fuel–air ratio after mixing the two cold and hot air streams as

$$\bar{f}_{cc} = \frac{\dot{m}_{fcc}}{\dot{m}_a} = \frac{f_{cc}}{(b+1)} = \frac{c_p T_\infty}{\Delta H_p} \frac{(\Theta_4 - \tau_{c,eff})}{(b+1)} \qquad (2.60b)$$

While the afterburner is in operation,

$$\dot{m}_{fAB} = \dot{m}_a c_p \frac{(T_6^o - T_5^o)}{\Delta H_p}$$

and thus, the fuel–air ratio added in the afterburner is

$$\overline{\Delta f}_{AB} = \frac{\dot{m}_{fAB}}{\dot{m}_a} \left[\frac{c_p T_\infty}{\Delta H_p} (\Theta_6 - \Theta_5) \right] \qquad (2.60c)$$

The total fuel flow rate is now

$$\dot{m}_f = \dot{m}_{fcc} + \dot{m}_{fAB} = \dot{m}_a \frac{c_p T_\infty}{\Delta H_p} \left[\frac{(\Theta_4 - \tau_{c,eff})}{(b+1)} + (\Theta_6 - \Theta_5) \right] \qquad (2.60d)$$

Thus, the specific fuel consumption (SFC) is given by the relation

$$\begin{aligned}
\mathrm{SFC} &= \frac{\dot{m}_f}{F} = \frac{\dot{m}_a}{\dot{m}_a(u_7 - u_\infty)} \cdot \frac{c_p T_\infty}{\Delta H_p} \left[\frac{(\Theta_4 - \tau_{c,eff})}{1+b} + (\Theta_6 - \Theta_5) \right] \\
&= \frac{1}{M_\infty} \cdot \frac{a_\infty}{(\gamma-1)\Delta H_p} \left[\frac{(\Theta_4 - \tau_{c,eff})}{1+b} + (\Theta_6 - \Theta_5) \right] \cdot \frac{1}{\left(\frac{u_7}{u_\infty} - 1 \right)} \\
&= \frac{a_\infty}{M_\infty(\gamma-1)\Delta H_p} \frac{\left[\frac{(\Theta_4 - \tau_{c,eff})}{1+b} + (\Theta_6 - \Theta_5) \right]}{\sqrt{\frac{2\Theta_6}{\gamma-1} \left\{ 1 - \frac{1}{\tau_{f,eff}} \right\}} - 1} \\
&= \frac{a_\infty}{(\gamma-1)\Delta H_p(1+b)} \frac{[(\Theta_4 - \Theta_3) + (\Theta_6 - \Theta_5)(1+b)]}{\sqrt{\frac{2\Theta_6}{\gamma-1} \left\{ 1 - \frac{1}{\tau_{f,eff}} \right\}} - M_\infty} \qquad (2.60e)
\end{aligned}$$

and the nondimensional specific fuel consumption is given by the relation

$$\mathrm{SFC}^* = \frac{\mathrm{SFC}.\Delta H_p}{a_\infty} = \frac{1}{(\gamma-1)(1+b)} \frac{[(\Theta_4 - \Theta_3) + (\Theta_6 - \Theta_5)(1+b)]}{\sqrt{\frac{2\Theta_6}{\gamma-1} \left\{ 1 - \frac{1}{\tau_{f,eff}} \right\}} - M_\infty} \qquad (2.60f)$$

Now, from (2.53b), we have

$$\frac{1}{\tau_{f,eff}} = \frac{\Theta_4 + b\tau_{c,eff}}{\tau_{c,eff}} \left[\Theta_4 + \Theta_\infty (1+b) - \tau_{c,eff} \right]$$

which can be substituted in (2.60f) to get an explicit expression for (SFC*) as a function of $\left(\Theta_4, b, \tau_{c,eff}, M_\infty \right)$ and an increase in the (nondimensional) stagnation temperature in the afterburner, $(\Theta_6 - \Theta_5)$.

2.2.4 Ideal Fanjet Cycle with Afterburner

While low-bypass jets, with or without an afterburner, are used for supersonic flights, giving it an adequate thrust with good performance, an increase in the bypass ratio is, of course, limited because of the minimum enthalpy difference required in the nozzle. In addition, in low-bypass jets, since the fan outlet pressure is limited to the nozzle inlet pressure, a high-bypass ratio in such engines means inevitably that the fan pressure ratio must go to 1 or very near 1. This second problem can be solved if the fan pressure ratio is made independent of the overall compression ratio by keeping the two streams separated after the fan stage. In addition, for the engine operating at moderate flight speeds, a high-bypass fan can ensure a high propulsive efficiency with adequate thrust to fly large-sized commercial aircrafts. A sketch of a fanjet engine that can operate at moderate transonic speeds and is capable of both high efficiency (low specific fuel consumption) and high thrust is shown in Fig. 2.23a, and the corresponding ideal thermodynamic process appears in Fig. 2.23b. Since in the fanjet, part of the air, which goes through the main combustion chamber and the turbine, has already been used to supply oxygen for the combustion of fuel, not much oxygen is left to burn in the afterburner. Therefore, this type of engine, used mainly for commercial aircrafts, has little scope for an afterburner. However, consistent with the previous analysis for straight jets and low-bypass jets, an afterburner has been included in Fig. 2.23a, b schematically.

We now have, contrary to the low-bypass jets, a change in all the basic parameters from Points 5 to 6. These parameters are now $\pi_c = p_3^o/p_2^o$ as the *overall compression ratio*, $\Theta_4 = T_4^o/T_\infty$ as the main *combustion chamber temperature ratio*, M_∞ as the *approaching flow Mach number*, b as the *bypass ratio*, $\Delta\Theta_{AB}$ as the *stagnation temperature increase in the afterburner*, and $\pi_f = p_8^o/p_7^o$ as the *fan compression ratio*.

Now,

$$\Theta_\infty = \Theta_1 = \Theta_2 = \frac{T_\infty^o}{T_\infty} = \frac{T_1^o}{T_\infty} = \frac{T_2^o}{T_\infty} = 1 + \frac{\gamma-1}{2} M_\infty^2 \qquad (2.61a)$$

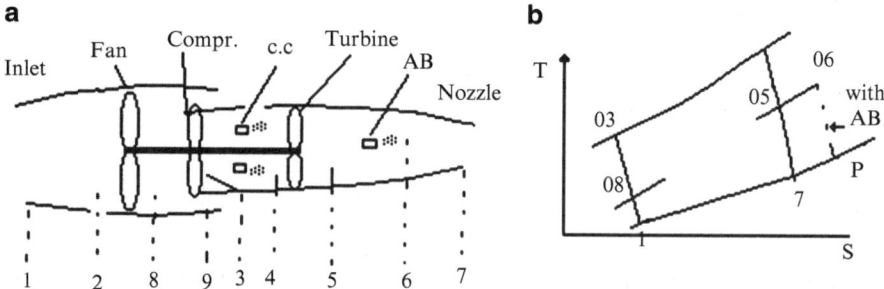

Fig. 2.23 (**a**) Schematic sketch of a fanjet engine with an afterburner; (**b**) the corresponding sketch of the thermodynamic process in a (T, s) chart

and

$$\delta_\infty = \delta_1 = \delta_2 = p_\infty^o = p_1^o = p_2^o = \left[1 + \frac{\gamma - 1}{2} M_\infty^2\right]^{\frac{\gamma}{\gamma - 1}} \tag{2.61b}$$

Further,

$$\Theta_8 = \frac{T_8^o}{T_\infty} = \left[\left(\frac{p_8^o}{p_2^o}\right)\left(\frac{p_2^o}{p_\infty}\right)\right]^{\frac{\gamma-1}{\gamma}} = \Theta_\infty . \pi_f^{(\gamma-1)/\gamma} = \pi_{f,\text{eff}}^{(\gamma-1)/\gamma} = \tau_{f,\text{eff}} = \Theta_\infty \tau_f \tag{2.61c}$$

and

$$\delta_8 = \frac{p_8^o}{p_\infty} = \delta_\infty \pi_f = \pi_{f,\text{eff}} \tag{2.61d}$$

and similarly,

$$\Theta_3 = \frac{T_3^o}{T_\infty} = \Theta_\infty . \pi_c^{(\gamma-1)/\gamma} = \pi_{c,\text{eff}}^{(\gamma-1)/\gamma} = \tau_{c,\text{eff}} = \Theta_\infty \tau_c \tag{2.61e}$$

and

$$\delta_3 = \frac{p_3^o}{p_\infty} = \frac{p_4^o}{p_\infty} = \delta_\infty \pi_c = \pi_\infty \pi_{c,\text{eff}} \tag{2.61f}$$

Now, since the turbine drives the fan and the main compressor, we write from the energy balance that

$$c_p\left[\{b(\{T_8^o - T_2^o\}) + (\{T_3^o - T_2^o\})\}\right] = c_p\left(T_4^o - T_5^o\right)$$

and further,

$$b\left(\tau_{\mathrm{f,eff}} - \Theta_\infty\right) + \left(\tau_{\mathrm{c,eff}} - \Theta_\infty\right) = \Theta_4 - \Theta_5$$

Thus,

$$\Theta_5 = \frac{T_5^o}{T_\infty} = \Theta_4 - \Theta_\infty[(\tau_{\mathrm{f}} - 1)b + (\tau_{\mathrm{c}} - 1)] \tag{2.61g}$$

Defining a turbine expansion pressure ratio, $\pi_{\mathrm{t}} = p_5^o/p_4^o$, and a turbine expansion temperature ratio,

$$\tau_{\mathrm{t}} = \frac{T_5^o}{T_4^o} = \pi_{\mathrm{t}}^{(\gamma-1)/\gamma}$$

we write

$$\tau_{\mathrm{t}} = \frac{\Theta_5}{\Theta_4} = 1 - \frac{\Theta_\infty}{\Theta_4}[(\tau_{\mathrm{f}} - 1)b + (\tau_{\mathrm{c}} - 1)] \tag{2.61h}$$

and

$$\delta_5 = \delta_6 = \frac{p_5^o}{p_\infty} = \frac{p_6^o}{p_\infty} = \delta_4\left(\frac{\Theta_5}{\Theta_4}\right)^{\frac{\lambda}{\gamma-1}} = \delta_4\tau_{\mathrm{t}}^{\gamma/(\gamma-1)} \tag{2.61i}$$

On the other hand, $\Theta_6 = T_6^o/T_\infty = \Theta_5 + \Delta\Theta_{\mathrm{AB}}$.
Now,

$$\begin{aligned}
\frac{T_6^o}{T_7} = \frac{T_7^o}{T_7} = \left(\frac{p_5^o}{p_\infty}\right)^{\left(\frac{\gamma-1}{\gamma}\right)} &= \delta_5^{(\gamma-1)/\gamma} = \left(\frac{\Theta_5}{\Theta_4}\right)\delta_4^{(\gamma-1)/\gamma} \\
&= \pi_{\mathrm{c,eff}}^{(\gamma-1)/\gamma}\left[1 - \frac{\Theta_\infty}{\Theta_4}\{(\tau_{\mathrm{f}} - 1)b + (\tau_{\mathrm{c}} - 1)\}\right] \\
&= \tau_{\mathrm{c}}\Theta_\infty\left[1 - \frac{\Theta_\infty}{\Theta_4}\{(\tau_{\mathrm{f}} - 1)b + (\tau_{\mathrm{c}} - 1)\}\right]
\end{aligned} \tag{2.61j}$$

and

$$\frac{T_7}{T_\infty} = \left(\frac{T_6^o}{T_\infty}\right)\left(\frac{T_7}{T_6^o}\right) = \Theta_6\frac{\Theta_4}{\left(\Theta_5\delta_4^{(\gamma-1)/\gamma}\right)} = \Theta_6\frac{\Theta_4}{\Theta_5\tau_{\mathrm{c,eff}}} \tag{2.61k}$$

Now, the kinetic energy of the exhaust gas in the main nozzle and the fan nozzle and that of the inlet gas are

$$\frac{u_7^2}{2c_{\mathrm{p}}T_\infty} = \frac{T_6^o}{T_\infty} - \frac{T_7}{T_\infty} = \Theta_6\left[1 - \frac{\Theta_4}{\Theta_5\tau_{\mathrm{c,eff}}}\right] \tag{2.62a}$$

$$\frac{u_9^2}{2c_p T_\infty} = \Theta_s - 1 = \tau_{f,\text{eff}} - 1 \tag{2.62b}$$

$$\frac{u_\infty^2}{2c_p T_\infty} = \Theta_\infty - 1 \tag{2.62c}$$

Therefore, the kinetic energy (nondimensional) of the gas per unit mass of total air is

$$\frac{\Delta E_{\text{kin}}}{c_p T_\infty} = \frac{1}{(1+b)}\left[\frac{u_7^2 - u_\infty^2}{2c_p T_\infty} + b\frac{\left(u_9^2 - u_\infty^2\right)}{2c_p T_\infty}\right]$$
$$= \frac{1}{1+b}\left[\Theta_6\left\{1 - \frac{\Theta_4}{\Theta_5 \tau_{c,\text{eff}}}\right\} + b(\tau_{f,\text{eff}} - 1) - (1+b)(\Theta_\infty - 1)\right] \tag{2.63}$$

Further, the heat added and rejected per unit mass of hot air are

$$\frac{|\overline{q}_a|}{c_p T_\infty} = (\Theta_4 - \Theta_3) + (\Theta_6 - \Theta_5) \tag{2.64a}$$

and

$$\frac{|q_r|}{c_p T_\infty} = \frac{T_7}{T_\infty} - 1 = \Theta_6 - \frac{\Theta_4}{\Theta_5 \tau_{c,\text{eff}}} - 1 \tag{2.64b}$$

The fuel–air ratio for hot gas is

$$f = \frac{\dot{m}_f}{\dot{m}_H} = \frac{c_p T_\infty}{\Delta H_p}\left[(\Theta_4 - \Theta_3) + (\Theta_6 - \Theta_5)\right] \tag{2.65}$$

in which the first part is for the combustion chamber and the second part is for the afterburner, and the sum of the two must be equal to or less than the maximum fuel–air ratio allowed for the particular fuel.

Now, the work done *per unit mass of total air* is obtained from (2.64a) and (2.64b) as

$$\frac{w}{c_p T_\infty} = \frac{1}{(1+b)}\left[\frac{|\overline{q}_a| - |q_r|}{c_p T_\infty}\right]$$
$$= \frac{1}{(1+b)}\left[(\Theta_4 - \Theta_3) + (\Theta_6 - \Theta_5) - \Theta_6\frac{\Theta_4}{\Theta_5 \tau_{c,\text{eff}}} + 1\right]$$
$$\frac{1}{(1+b)}\left[(1 + b\tau_{f,\text{eff}} - (1+b)\Theta_\infty) + \Theta_6\left(1 - \frac{\Theta_4}{\Theta_5 \tau_{c,\text{eff}}}\right)\right] \tag{2.66}$$

Adding and subtracting b within the brackets and performing some algebraic manipulation, it is easy for us to show that (2.66) is exactly the same as (2.63). We get also from (2.62a) to (2.62c) that

$$\frac{u_7}{u_\infty} = \frac{1}{M_\infty} \sqrt{\frac{2\Theta_6}{\gamma - 1} \left(1 - \frac{\Theta_4}{\Theta_5 \tau_{c,eff}} \right)} \qquad (2.67a)$$

and

$$\frac{u_9}{u_\infty} = \frac{1}{M_\infty} \sqrt{\frac{2}{\gamma - 1} (\tau_{f,eff} - 1)} \qquad (2.67b)$$

Further,

$$\frac{F}{\dot{m}_a a_\infty} = \frac{M_\infty}{1 + b} \left[b \left(\frac{u_9}{u_\infty} - 1 \right) + \left(\frac{u_7}{u_\infty} - 1 \right) \right]$$

$$= \frac{1}{1 + b} \left[b \sqrt{\frac{2}{\gamma - 1} (\tau_{f,eff} - 1)} + \sqrt{\frac{2\Theta_6}{\gamma - 1} \left(1 - \frac{\Theta_4}{\Theta_5 \tau_{c,eff}} \right)} - (1 + b) M_\infty \right]$$

$$(2.68a)$$

The *thermodynamic efficiency* is $\eta_{th} = \overline{w}(1 + b)/q_a$, and with the help of (2.64a) and (2.66), we get

$$\eta_{th} = \frac{1}{(\Theta_4 - \Theta_3) + (\Theta_6 - \Theta_5)}$$

$$\times \left[1 + b\tau_{f,eff} - (1 + b)\Theta_\infty + \Theta_6 \left\{ 1 - \frac{\Theta_4}{\Theta_5 \tau_{c,eff}} \right\} \right] \qquad (2.68b)$$

Similarly, for the *propulsive efficiency*, we can write

$$\eta_p = \frac{2 \left[\left(\frac{u_7}{u_\infty} - 1 \right) \right] + b \left(\frac{u_9}{u_\infty} - 1 \right)}{\left[\left(\frac{u_7}{u_\infty} \right)^2 - 1 \right] + b \left[\left(\frac{u_9}{u_\infty} \right)^2 - 1 \right]} \qquad (2.68c)$$

where for the speed ratio we can substitute an expression from (2.67a) and (2.67b). For the specific fuel consumption,

$$SFC = \frac{\dot{m}_f}{F} = \frac{c_p T_\infty}{\Delta H_p u_\infty} \frac{[(\Theta_4 - \Theta_3) + (\Theta_6 - \Theta_5)]}{\left[b \left(\frac{u_9}{u_\infty} - 1 \right) + \left(\frac{u_7}{u_\infty} - 1 \right) \right]} \qquad (2.68d)$$

Table 2.6 Results of SFC* (Θ_4, π_c, π_f, b)

Θ_4	π_c	π_f	b	Θ_3	Θ_f	SFC*
4	12	1.2	3	2.034	3.615	0.0543
4	12	1.2	4	2.034	3.561	0.0347
4	12	1.2	5	2.034	3.508	0.0241
4	12	1.4	3	2.034	3.472	0.0543
4	12	1.4	4	2.034	3.371	0.0347
4	12	1.4	5	2.034	3.270	0.0241
4	16	1.2	3	2.208	3.585	0.0466
4	16	1.2	4	2.208	3.532	0.0297
4	16	1.2	5	2.208	3.479	0.0206
4	16	1.4	3	2.208	3.443	0.0466
4	16	1.4	4	2.208	3.342	0.0297
4	16	1.4	5	2.208	3.241	0.0206
4	20	1.2	3	2.353	3.456	0.0410
4	20	1.2	4	2.353	3.509	0.0260
4	20	1.2	5	2.353	3.455	0.0180
4	20	1.4	3	2.353	3.420	0.0410
4	20	1.4	4	2.353	3.319	0.0260
4	20	1.4	5	2.353	3.218	0.0180
5	12	1.2	3	2.034	4.615	0.0819
5	12	1.2	4	2.034	4.561	0.0523
5	12	1.2	5	2.034	4.508	0.0363
5	12	1.4	3	2.034	4.472	0.0819
5	12	1.4	4	2.034	4.371	0.0523
5	12	1.4	5	2.034	4.270	0.0363
5	16	1.2	3	2.208	4.585	0.0726
5	16	1.2	4	2.208	4.532	0.0463
5	16	1.2	5	2.208	4.479	0.0320
5	16	1.4	3	2.208	4.443	0.0726
5	16	1.4	4	2.208	4.342	0.0463
5	16	1.4	5	2.208	4.241	0.0320
5	20	1.2	3	2.353	4.562	0.0659
5	20	1.2	4	2.353	4.509	0.0419
5	20	1.2	5	2.353	4.455	0.0289
5	20	1.4	3	2.353	4.402	0.0659
5	20	1.4	4	2.353	4.319	0.0419
5	20	1.4	5	2.353	4.218	0.0289

from which we can write for the nondimensional specific fuel consumption

$$
\begin{aligned}
\text{SFC}^* = \text{SFC} \cdot \frac{\Delta H_p}{a_\infty} &= \frac{1}{(\gamma - 1)M_\infty} \frac{[(\Theta_4 - \Theta_3) + (\Theta_6 - \Theta_5)]}{\left[b\left(\frac{u_9}{u_\infty} - 1\right) + \left(\frac{u_7}{u_\infty} - 1\right)\right]} \\
&= \frac{1}{1 + b} \frac{[(\Theta_4 - \Theta_3) + (\Theta_6 - \Theta_5)]}{\left[b\sqrt{\frac{2}{\gamma - 1}\left(\tau_{f,\text{eff}} - 1\right)} + \sqrt{\frac{2}{\gamma - 1}\left(1 - \frac{\Theta_4}{\Theta_5 \tau_{c,\text{eff}}}\right)} - (1 + b)M_\infty\right]}
\end{aligned} \tag{2.68e}
$$

To anayze the results, we set $\Theta_\infty = 1$ or $M_\infty = 1$. Further, we let $\Delta\Theta_{AB} = 0$. We now evaluate the results for SFC* with four variables, namely, (Θ_4, π_c, π_f, b), and present them in Table 2.6. For these, $\Theta_4 = 4$ and 5, $\pi_c = 12$, 16, and 20, $\pi_f = 1.2$ and 1.4, and $b = 3$, 4, and 5.

By analyzing the results, we see that SFC* does not depend on π_f, but on the other three. It increases substantially with Θ_4 and decreases with an increase in π_c and the bypass ratio b. However, for every Θ_4, there is an optimum π_c, which can be investigated by the reader to get better specific fuel consumption. It is of historical interest to note that when in Boeing 707s, the old straight JT3 engine was replaced by the equivalent PW3D fanjet engine, there was an immediate 15% fuel savings. Needless to say, the trend in design of modern fanjets is to have a high turbine inlet temperature, high compressor pressure ratio, and high bypass ratio, which, again, the reader can verify by examining the most recent engines.

2.2.5 An Interactive Computer Program for Ideal Jet Engine Analysis (PAGIC)

An interactive computer program has been created to run on personal computers to study ideal jet engines' thermodynamic cycle. It is called PAGIC, and a listing is iven below. The program was tested with Professional Fortran compiler, but it should work with other FORTRAN compilers also. It works in an interactive manner by asking for relevant data and giving self-explanatory results. Before running the code, an unformatted binary data file containing six single-precision real data called AGRDT2 has to be created, which will be called by this code under Unit 7.

```
C      Interactive gasturbine performance program for ideal case (PAGIC)
       Program PAGIC
       CHARACTER IANS
       CHARACTER*10 AGTRD2
       DIMENSION VI(6),VF(9,2)
       G=1.4
       Write(6,*)' Interactive Gasturbine Program for Ideal Gas'
10     WRITE(6,*)' Type 1/2/3/4 for RJ/SJ/BJ/FJ: '
       READ(5,*)NSCOPE
       IF(NSCOPE.LT.1.OR.NSCOPE.GT.4)GOTO10
       IF(NSCOPE.EQ.1)WRITE(6,*)' Gasturbine type: Ramjet '
       IF(NSCOPE.EQ.2)WRITE(6,*)' Gasturbine type:
       Straightjet '
       IF(NSCOPE.EQ.3)WRITE(6,*)' Gasturbine type: Bypassjet '
       IF(NSCOPE.EQ.4)WRITE(6,*)' Gasturbine type: Fanjet '
       NNEW=1
       WRITE(6,*)' Is it o.k.? '
       READ(5,1001)IANS
       IF(IANS.NE.'y'.and.IANS.NE.'Y')GOTO10
14     OPEN(7,FILE='AGTRD2',STATUS='OLD',ACCESS='SEQUENTIAL',
         FORM=1'UNFORMATTED')
       READ(7)(VI(I),I=1,6)
```

```
11    CONTINUE
      IF(NSCOPE.EQ.1)WRITE(6,2001)(VI(I),I=1,2)
      IF(NSCOPE.EQ.2)WRITE(6,2002)(VI(I),I=1,4)
      IF(NSCOPE.EQ.3)WRITE(6,2003)(VI(I),I=1,5)
      IF(NSCOPE.EQ.4)WRITE(6,2004)(VI(I),I=1,6)
      IF(NNEW.EQ.1)WRITE(6,*)'any change? '
      IF(NNEW.EQ.1)READ(5,1001)IANS
      IF(NNEW.EQ.0)IANS='Y'
      IF(IANS.NE.'Y'.and.IANS.NE.'y')GOTO12
      NNEW=1
      WRITE(6,*)' Parameter no.,value: '
      READ(5,*)I,VAL
      VI(I)=VAL
      GOTO11
12    NTEST=0
      IF(NSCOPE.NE.1.AND.(VI(2).LE.1.OR.VI(3).LE.1.))NTEST=1
      IF(NSCOPE.EQ.4.AND.VI(6).LE.1.)NTEST=1
      IF(NTEST.EQ.1)WRITE(6,*)' input parameter error. Reenter '
      IF(NTEST.EQ.1)GOTO11
      REWIND(7)
      WRITE(7)(VI(I),I1,6)
      CLOSE(7)
      IF(NSCOPE.NE.1)GOTO13
      VI(3)=1.
      VI(4)=0.
      VI(5)=0.
      VI(6)=0.
13    CONTINUE
      EA=(G-1.)/G
      EB=1./EA
      TTRI=((G-1.)*.5*VI(1)**2)+1.
      TPRI=TTRI**EB
      DO16I=1,2
      VF(I,1)=TTRI
16    VF(I,2)=TPRI
      VF(3,1)=TTRI*VI(3)**EA
      IF(VF(3,1).GT.VI(2))GOTO50
      VF(3,2)=VI(3)*VF(1,2)
      VF(4,1)=VI(2)
      VF(4,2)=VF(3,2)
      PIC=VI(3)
      XMU=0.
      IF(NSCOPE.GT.2)XMU=VI(5)/(VI(5)+1.)
      GOTO(25,25,26,27),NSCOPE
```

```
25    VF(5,1)=VF(4,1)-(TTRI*(PIC**EA-1.))
      XA=VF(5,1)
      GOTO30
26    PIF=(((((VF(4,1)-VF(3,1))*(1.-XMU))+TTRI)/(VF(4,1)-(XMU*(VF(4,1)
     1 -VF(3,1))))))**EB*PIC
      VI(6)=PIF
      VF(8,1)=TTRI*PIF**EA
      VF(9,1)=VF(4,1)-(TTRI*((PIC**EA-1.)+(XMU/(1.-XMU)*(PIF**EA
     1 -1.))))
      VF(8,2)=PIF*TTRI**EB
      VF(9,2)=VF(3,2)*(VF(9,1)/VF(4,1))**EB
      VF(5,1)=(XMU*VF(8,1))+(1.-XMU)*(VF(4,1)-(TTRI*((PIC**EA-1.)+
     1 (XMU/(1.-XMU)*(PIF**EA-1.)))))
      XA=VF(9,1)
      GOTO30
27    PIF=VI(6)
      VF(8,1)=TTRI*PIF**EA
      VF(8,2)=PIF*TTRI**EB
      VF(9,1)=VF(8,1)
      VF(9,2)=VF(8,2)
      VF(5,1)=VF(4,1)-(TTRI*((PIC**EA-1.)+XMU/(1.-XMU)*  (PIF**EA-1.)))
      XA=VF(5,1)
30    VF(5,2)=PIC*(TTRI*XA/VF(4,1))**EB
      VF(6,2)=VF(5,2)
      VF(7,2)=VF(6,2)
      AFBURN=0.
      IF(NSCOPE.GT.1)AFBURN=VI(4)
      VF(6,1)=VF(5,1)+AFBURN
      VF(7,1)=VF(6,1)
c     U7R=uinf/u7;T7R=T7/Tinf;T7RI=Tinf/T7
      T7R=VF(7,1)/VF(7,2)**EA
      T7RI=1./T7R
      XM7=SQRT(2./(G-1.)*((VF(7,1)*T7RI)-1.))
      U7R=VI(1)/XM7*SQRT(T7RI)
      QIN=VF(4,1)-VF(3,1)
      IF(NSCOPE.LT.3)QIN=QIN+AFBURN
      IF(NSCOPE.EQ.3)QIN=(QIN*(1.-XMU))+AFBURN
      IF(NSCOPE.EQ.4)QIN=(QIN+AFBURN)*(1.-XMU)
      QREJ=T7R-1.
      IF(NSCOPE.EQ.4)QREJ=QREJ*(1.-XMU)
      WJET=QIN-QREJ
      ETATH=WJET/QIN
       IF(NSCOPE.EQ.4)GOTO32
       ETAP=2.*U7R/(U7R+1.)
```

```
c       FPA=F/(pinf*A7)
        FPA=G*XM7**2*(1.-U7R)
        GOTO35
c       U9R=uinf/u9,UI9R=u9/u7=U7R/U9R
32      XM9=SQRT(2./(G-1.)*(VF(9,1)-1.))
        U9R=VI(1)/XM9
        UI9R=XM9/XM7*SQRT(T7RI)
        ETAP=2.*U7R*((1.-XMU)+(XMU*U9R)-U7R)/(1.-XMU+(XMU*U9R**2)
         -U7R 1 **2)
        FPA=G*XM7**2*(1.-U7R+(VI(5)*UI9R*(1.-U9R)))
35      A7AFR=SQRT(1.+(AFBURN/VF(5,1)))
c       FS=F/(mair*ainf);XMF=mfuel*fuelheat/(mair*ainf**2)
        FS=SQRT(2.*VF(6,1)/(G-1.)*(1.-(1./VF(6,2)**EA)))-VI(1)
        IF(NSCOPE.EQ.4)FS=(FS+(VI(5)*SQRT(2.*VF(8,1)/(G-1.)*(1.-(1./VF
       1 (8,2)**EA)))-VI(1)))/(VI(5)+1.)
        IF(NSCOPE.NE.3)XMF=(VI(2)-VF(3,1)+AFBURN)/(G-1.)
        IF(NSCOPE.EQ.3)XMF=(((VI(2)-VF(3,1))/(VI(5)+1.))+AFBURN)/(G-1.)
        IF(NSCOPE.EQ.4)XMF=XMF/(VI(5)+1.)
        SFCS=XMF/FS
        ETAT=ETAP*ETATH
        WRITE(6,*)'Performance of an Ideal Aircraft Gasturbine'
        IF(NSCOPE.EQ.1)WRITE(6,*)'Gasturbine type: Ramjet '
        IF(NSCOPE.EQ.2)WRITE(6,*)'Gasturbine type:
        Straightjet '
        IF(NSCOPE.EQ.3)WRITE(6,*)'Gasturbine type:
        Bypassjet '
        IF(NSCOPE.EQ.4)WRITE(6,*)'Gasturbine type: Fanjet '
        WRITE(6,1026)
        IF(NSCOPE.EQ.1)WRITE(6,2001)(VI(I),I=1,2)
        IF(NSCOPE.EQ.2)WRITE(6,2002)(VI(I),I=1,4)
        IF(NSCOPE.EQ.3)WRITE(6,2003)(VI(I),I=1,5)
        IF(NSCOPE.EQ.4)WRITE(6,2004)(VI(I),I=1,6)
        WRITE(6,1022)T7R,QIN,QREJ,WJET,FS,SFCS
        WRITE(6,1023)ETATH,ETAP,ETAT,XM7
        IF(NSCOPE.NE.4)WRITE(6,1030)U7R,FPA
        IF(NSCOPE.EQ.4)WRITE(6,1031)U7R,FPA,U9R,XM9,UI9R
        IF(AFBURN.NE.0.)WRITE(6,1024)A7AFR
        WRITE(6,1026)
        WRITE(6,1027)
        N=7
        IF(NSCOPE.GT.2)N=9
        DO37I=1,N
        WRITE(6,1028)I,VF(I,1),VF(I,2)
        CONTINUE
        WRITE(6,1026)
```

```
        WRITE(6,*)'New parameters? '
        NNEW=0
        READ(5,1001)IANS
        IF(IANS.EQ.'Y'.OR.IANS.EQ.'y')GOTO14
        WRITE(6,*)'New Gasturbine type? '
        READ(5,1001)IANS
        IF(IANS.EQ.'y'.OR.IANS.EQ.'Y')GOTO10
        STOP
50      WRITE(6,1032)VF(3,1),VI(2)
        GOTO 14
1001    FORMAT(A)
1022    FORMAT(1x,'T7/Tinf=',1pe10.3,'Heatinp=',1pe10.3,'Heatrej='
        1,1pe10.3,'Workjet=',1pe10.3,'F/(mair*ainf)=',1pe10.3,'sfc*='
        2,1pe10.3)

1023    FORMAT(1x,'etather=',1pe10.3,'etaprop=',1pe10.3,'etatot='
        1,1pe10.3,'M7=',1pe10.3)
1024    FORMAT(1X,'A7AB/A7=',1pe10.3)
1026    FORMAT(1H)
1027    FORMAT(3X,'I',3X,'T0*',7X,'P0*')
1028    FORMAT(2X,I2,1P2E10.3)
1030    FORMAT(1x,'Uinf/U7=',1pe10.3,'F/(pinf*A7AB)=',1pe10.3)
1031    FORMAT(1x,'Uinf/U7=',1pe10.3,'F/(pinf*A7AB)=',1pe10.3,'Uinf/',
        1 'U7=',1pe10.3,'M9=',1pe10.3,'U9/U7=',1pe10.3)
1032    FORMAT(1x,'error since compr. stagn. temp=',1pe10.3,'is larger'
        1,'than comb. exit temp.=',1pe10.3)
2001    FORMAT(1x,'1.Appr.Mach=',F6.3,/,1x,'2.Temp.ratio=',F6.3)
2002    FORMAT(1x,'1.Appr.Mach=',F6.3,/,1x,'2.Temp.ratio=',F6.3,/,1x,
        1 '3.Overall compr.ratio=',F6.3,/,1x,'4.Afterburner delta theta'
        2,'=',F6.3)
2003    FORMAT(1x,'1.Appr.Mach=',F6.3,/,1x,'2.Temp.ratio=',F6.3,/,1x,
        1 '3.Overall compr. ratio=',F6.3,/,1x,'4.
        Afterburner delta theta'
        2,'=',F6.3,/,1x,'5.Bypass ratio=',F6.3)
2004    FORMAT(1x,'1.Appr.Mach=',F6.3,/,1x,'2.
        Temp.ratio=',F6.3,/,1x,
        1 '3.Overall compr. ratio=',F6.3,/,1x,'4.
        Afterburner delta theta'
        2,'=',F6.3,/,1x,'5.Bypass ratio=',F6.3,/,1x,'6.
        Fan compr.ratio'
        3,'=',F6.3)
        END
```

2.3   Exercises

1. Explain graphically the difference in the cyclic process in a two-stroke and a four-stroke engine. What are the advantages and disadvantages of each of them? Given no. of cylinders = 6, displacement volume per cylinder = 0.995 l, weight = 148.0 kgf, specific fuel consumption = 72.7 µg/J, rpm = 2,600 min^{-1}. Assume a combustion chamber temperature. Compute the gas state (p, T, V) at the four corners of the ideal thermodynamic cycle and determine per cylinder the work done, heat added, heat rejected, air and fuel mass flow rates, and specific fuel consumption, and compare the results given in the piston engine databank.

2. For the Kawasaki KT5311A Japanese turboprop/turboshaft engine for helicopter applications, the following data are given in the databank for aircraft turboprop engines: mass flow rate = 5.0 kg/s, takeoff thrust = 600 N, shaft power = 819 kW, overall compression ratio in two spools = 6.1, weight = 225 kgf, specific fuel consumption = 116.3 µg/J, no. of axial compressor stages = 5 + 1, rpm = 25,200 and 21,200 min^{-1}, and no. of turbine stages = 1 + 1. Compute the work done, heat added, heat rejected, air and fuel mass flow rate, and the specific fuel consumption, and compare these with the values in turboprop databank.

3. Running the interactive code PAGIC given in Section 2.2.5, investigate the performance parameters of the various types of jet engines.

4. A ramjet is to propel an aircraft at Mach 3 (entry is designed to be shock-free under design condition), where the ambient pressure is 0.085 bar and the ambient temperature is 220 K. The combustion chamber exit temperature is 1,800 K and the usual aviation gasoline (heating value: ΔH_p = 42,700 kJ/kg) is used. Compute the following: (a) the fuel–air ratio f (make sure that $f < f_{stoich.}$), (b) exhaust velocity u_e, (c) specific impulse, (d) thermodynamic efficiency, (e) propulsive efficiency, and (f) overall efficiency.

5. The performance of an ideal ramjet burning aviation gasoline at stoichiometric fuel–air ratio is to be calculated for different supersonic Mach numbers. The engine is to fly at an altitude of 15 km, where the ambient pressure is 0.116 bar and the ambient temperature is 205 K.

6. Let's consider the following specification of the Canadian straight-jet engine from United Aircrafts Canadian Ltd. (UACL), model no. JT115D-4, mass flow rate = 34.1 kg/s, takeoff thrust = 10,600 N, overall compression ratio = 10.0, turbine inlet temperature = 960°C. Consider running the engine near the ground with ambient pressure = 1.0 bar, ambient temperature = 298 K, approaching flow velocity = 250 m/s. Compute the gas state at the characteristic points of the engine for an ideal cycle analysis, the thrust, specific thrust, fuel-to-air mass ratio, work output, heat added and heat released, and thermodynamic, propulsive, and overall efficiency. Examine if the compression ratio is optimum for the given temperature ratio. Draw the cycle in a (T, s) chart.

7. The idling engines of a landing turbojet produce forward thrust when operating in a normal manner, but they can produce reverse thrust if the jet is properly deflected.

Suppose that while the aircraft rolls down the runway at 150 km/h, the idling engine consumes air at 50 kg/s and produces an exhaust velocity of 150 m/s.

(a) What is the forward thrust of this engine?
(b) What are the magnitude and direction (that is, forward or reverse) if the exhaust is deflected 90°?
(c) What are the magnitude and direction of the thrust (forward or reverse) after the plane has come to a stop, with 90° exhaust deflection and an airflow of 40 kg/s? (Problem taken from Hill and Peterson (1992), p. 209, with thanks.)

8. Consider the following specifications of the French SNECMA low-bypass turbojet engine LARZAC04: mass flow rate = 27.6 kg/s, takeoff thrust = 13.2 kN, overall compression ratio = 10.7, turbine inlet temperature = 1,130°C, SFC = 20.10 mg/N·s, bypass ratio = 1.1. For the given bypass ratio, obtain the fan pressure ratio. Compute the gas state at the characteristic points of the engine for an ideal cycle analysis, the thrust, specific thrust, SFC, fuel-to-mass ratio, work output, heat added and heat released, and thermodynamic, propulsive, and overall efficiency. Draw the cycle in a (T, s) chart.

9. Consider the specifications of the Pratt & Whitney P&W JT9D-3A high-bypass turbofan engine: mass flow rate = 684.0 kg/s, takeoff thrust = 169.9 kN, overall compression ratio = 21.5, fan pressure ratio = 1.6, turbine inlet temperature = 1,243°C, SFC = 17.84 mg/N·s, and bypass ratio = 5.2. Compute the gas state at the characteristic points of the engine for an ideal cycle analysis, the thrust, specific thrust, SFC, fuel-to-mass ratio, work output, heat added and heat released, and thermodynamic, propulsive, and overall efficiency. How much air is sent through the core engine and how much through the outer fan? Draw the cycle in a (T, s) chart.

Chapter 3
Friction, Work, and Heat Addition in a One-Dimensional Channel

In this chapter, we will discuss the various components of the losses in a one-dimensional channel and their application in various components of a gas-turbine engine.

As the simplest flow case, we consider a steady $(\partial/\partial t = 0)$ one-dimensional flow in a channel of arbitrary cross-sectional distribution without any change in the mass flow rate. For this purpose, we write first the equation for the conservation of mass for a two-dimensional case as

$$\frac{\partial}{\partial x}(\rho u) + \frac{\partial}{\partial y}(\rho v) = 0$$

We now multiply the equation by b, which is the channel width being constant in the plane perpendicular to the plane of this paper, and integrate over the y-direction from $y = -h/2$ to $y = +h/2$, where h is the (local) height of the channel, to get

$$b \int_{-h/2}^{+h/2} \frac{\partial(\rho u)}{\partial x}\,\mathrm{d}y = b\frac{\partial}{\partial x}\int_{-h/2}^{h/2}(\rho u)\mathrm{d}y = -b\int_{-h/2}^{h/2}\frac{\partial}{\partial y}(\rho v)\mathrm{d}y$$

Introducing an average mass flux,

$$\overline{\rho u} = \frac{b}{A}\int_{-h/2}^{h/2}\rho u\mathrm{d}y \quad A = \text{ cross - sectional area}$$

we can write

$$b\frac{\partial}{\partial x}\int_{-h/2}^{h/2}\rho u\mathrm{d}y = \frac{\mathrm{d}}{\mathrm{d}x}(\overline{\rho u}A) = \frac{\mathrm{d}\dot{m}}{\mathrm{d}x} = 0$$

T. Bose, *Airbreathing Propulsion: An Introduction*,
Springer Aerospace Technology, DOI 10.1007/978-1-4614-3532-7_3,
© Springer Science+Business Media, LLC 2012

Fig. 3.1 Channel flow with
friction, heat, and work

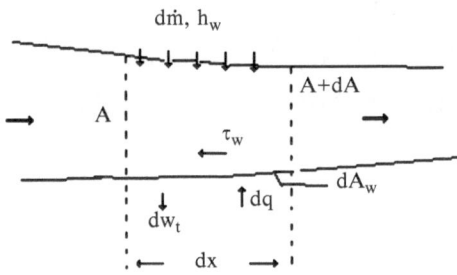

where, in the *one-dimensional sense*, the definition of the *mass flow rate* is

$$\dot{m} = \overline{\rho u} A$$

which, according to the *principle of conservation of mass* (without explicit *mass addition*), does not change in the flow direction.

On the other hand, according to one-dimensional theory, the *mean transverse mass flux* is

$$\overline{\rho v} = \frac{b}{A} \int\limits_{-h/2}^{h/2} \frac{\partial}{\partial y}(\rho v)\mathrm{d}y = 0$$

Having explained the essence of one-dimensional flow, the mean mass flux in the flow, and transverse flow directions, we no longer need to keep the *average* sign to designate them, since there will be no difficulty in understanding them as such.

As the gas flows through a gas turbine engine, one must either bring in work externally for the compressor or extract work in a turbine. In addition, the flow is subjected to friction everywhere in either a stationary channel or a rotating channel, and there is heat and mass addition in the combustion chamber. In this chapter, therefore, the flow of an ideal gas in a one-dimensional channel is studied, which is subjected to friction, work, mass addition, and heat addition; the heat of reaction due to chemical reaction in the combustion chamber is substituted by the heat addition, and the external mass addition is allowed to cause changes in the mass flow rate and the specific enthalpy of the gas flowing through the channel. The flow is shown in Fig. 3.1, where A is the cross-sectional area, τ_w is the shear stress on the wall, $\mathrm{d}A_w$ is the differential wall area, and $\mathrm{d}w_t$ is the (technical) work extracted from the flow. The continuity equation for the one-dimensional channel flow is thus

$$\frac{\mathrm{d}\dot{m}}{\dot{m}} = \frac{\mathrm{d}\rho}{\rho} + \frac{\mathrm{d}u}{u} + \frac{\mathrm{d}A}{A} \qquad (3.1a)$$

where $\dot{m} = \rho u A$, ρ is the density, and u is the gas speed.

Now, let dw_t (in J/kg) be the specific (technical) work in the flow process per unit mass being extracted from the flow to convert the flow energy into the mechanical work (for example, in a turbomachine). Then the equivalent force applied on the surface (in N/m²) is

$$dF = \rho dw_t$$

which has the unit of pressure from the thermodynamic definition, whereas the technical work is $dw_t = \rho^{-1} dp$; by multiplying both sides by ρ, we get the pressure or force equivalent of the technical work. For $dw_t > 0$, the work is considered to be extracted as in a *turbine*, and for $dw_t > 0$, work is considered to be added to the system as in a *compressor*. From thermodynamics, the sign of the extracted work is in the direction opposite the pressure change. Further, since the shear stress τ_w acts on the wall surface dA_w that is in the direction opposite the flow direction (that is, opposite the momentum flux direction), the *momentum balance equation* is

$$A dp + d(\dot{m}u) + \tau_w dA_w + dF = A dp + d(\dot{m}u) + \tau_w dA_w + \rho dA w_t = 0$$

Further, the following definitions are introduced:

Hydraulic diameter: $D = \dfrac{4A}{(dA_w/dx)}$

That is, $dA_w = \frac{4A}{D} dx$ and

Friction coefficient: $c_f = \dfrac{\tau_w}{\frac{1}{2}\rho u^2} > 0$

The above definition of the friction coefficient assumes that there is a uniform shear stress along the circumference c; if it is not so, one can replace it by an average shear stress $\overline{\tau_w}$ defined as follows:

$$\overline{\tau_w} = \frac{1}{c} \int \tau_w dc$$

Therefore, the term $\tau_w dA_w$ in the momentum balance equation can be substituted with a term $\rho u^2 A(2c_f/D)dx$, and then the *momentum balance equation* becomes

$$A dp + \rho u A + u d\dot{m} + \rho u^2 A(2c_f/D)dx + \rho A dw_t = 0$$

Dividing the above equation by ρA, and again by stating that $p/\rho = u^2/(\gamma M^2)$, the above momentum balance equation becomes

$$\frac{dp}{p} + \gamma M h^2 \left[\frac{du}{u} + \frac{d\dot{m}}{\dot{m}} + \left(\frac{2c_f}{D} \right) dx \right] + \frac{dw_t}{(RT)} = 0 \tag{3.1b}$$

Furthermore, from the equation of state, $p = \rho R T$, and from the definition of the Mach number, $M = u/\sqrt{\gamma R T}$, we write the two equations in the differential form as follows:

State:

$$\frac{dp}{p} + \frac{d\rho}{\rho} + \frac{dT}{T} \tag{3.1c}$$

Mach number:

$$\frac{dM}{M} = \frac{du}{u} - \frac{1}{2}\frac{dT}{T} \tag{3.1d}$$

Now we turn our attention to the energy equation. While various term in the energy equation are written in terms of unit mass of the gas (in J/kg), since the external mass is entering the system, we write

$$(\dot{m} + d\dot{m})c_p(T + dT) - \dot{m}c_p T - h_w d\dot{m} = 0$$

which reduces (after neglecting product of differentials) to

$$c_p\, dT + c_p T\left[\frac{1 - h_w}{c_p T}\right]\frac{d\dot{m}}{\dot{m}} = 0$$

In the usual energy equation without mass addition, $d\dot{m} = 0$, and we generally have a term $c_p dT$, which is now replaced by the entire left-hand side of the above equation. For injection into high-temperature gases (for example, in the combustion chamber), one could take the limit of $h_w/(c_p T) \to 0$. On the other hand, if the external flow is kept locally at the same temperature as in the chamber, then we have the limit of $h_w/(c_p T) \to 1$.

The energy equation can be written separately for a stationary observer, who observes the flow passing by and makes an energy balance from energy input and output at two control places, and for a moving observer, who moves with the flow coming probably with a tiny thermometer to measure the temperature and a tiny pressure measuring device to measure the pressure. Therefore, we write the two energy equations, as noted by the two observers, as follows:

Stationary observer (the observer is outside the flow and makes an energy balance outside a control space):

$$dq = c_p dT + u\,du + dw_t + c_p T\left[1 + \frac{\gamma - 1}{2}M^2 - \frac{h_w}{c_p T}\right]\frac{d\dot{m}}{\dot{m}}$$

Moving observer (the observer is moving with the fluid and observes the energy change and reports ourside):

$$dq = c_p dT - \frac{1}{\rho}dp - dw_f + c_p T\left[1 - \frac{h_w}{c_p T}\right]\frac{d\dot{m}}{\dot{m}}$$

where the *friction work* is

$$dw_f = \left(\frac{\tau_w dA}{\dot{m}}\right)u = \left(\frac{4}{D}\right)\frac{u^2}{2}c_f dx = \left(\frac{2}{D}c_f dx\right)u^2$$

in which c_f is the nondimensional *friction coefficient*.

The value of the fluid friction coefficient is different for different situations. For a laminar of a straight turbulent flow, it can be derived easily for *Poiseuille flow* as

$$c_f = \frac{64}{Re}$$

whereas for a straight turbulent smooth pipe,

$$\frac{1}{\sqrt{c_f}} = 2\ln(Re\sqrt{c_f} - 0.8)$$

Similar expressions for a straight rough turbulent pipe flow, a flow in a gap between two disks, one of them being stationary and the other rotating, or between two rotating disks are needed. However, to give a complete list of all these formulas would be outside the scope of this book, and the reader can look for them in the literature. Suffice it to say that a rotating flow due to azimuthal speed of the rotating blades would have a much larger friction coefficient than for the straight flow.

We will now explain this last term due to friction work. The force acting on the surface is $\tau_w dA_w$, which, when divided by the mass flow rate $\dot{m}$, leads to a force per unit mass flow rate. Further multiplication with the flow speed u gives the friction work dw_f. Note further that dw_f may be added to the external heat addition dq, and thus dw_f becomes the equivalent heat due to the irreversible nature of the friction.

The energy equation for the stationary observer therefore becomes

$$\frac{dq}{c_p T} = \frac{dT}{T} + (\gamma - 1)M^2\frac{du}{u} + \left[1 + \frac{\gamma - 1}{2}M^2 - \frac{h_w}{c_p T}\right]\frac{d\dot{m}}{\dot{m}} + \frac{\gamma - 1}{\gamma RT}dw_t$$

from which it follows that

$$\frac{dT}{T} = \frac{dq}{c_p T} - (\gamma - 1)M^2\frac{du}{u} - \left[1 + \frac{\gamma - 1}{2}M^2 - \frac{h_w}{c_p T}\right]\frac{d\dot{m}}{\dot{m}} - \frac{\gamma - 1}{\gamma RT}dw_t \qquad (3.1e)$$

Herein h_w is the specific enthalpy of air (J/kg) that enters the flow channel through side slots.

Further, from the definition of stagnation temperature and pressure,

$$\frac{T^o}{T} = 1 + \frac{\gamma - 1}{2}M^2, \quad \frac{p^o}{p} = \left[1 + \frac{\gamma - 1}{2}M^2\right]^{\frac{\gamma}{\gamma - 1}}$$

we may write the two expressions after taking the logarithm and then differentiating as follows:

$$\frac{\mathrm{d}T^o}{T^o} = \frac{\mathrm{d}T}{T} + \left[\frac{(\gamma-1)M^2}{\{1+\frac{\gamma-1}{2}\}M^2}\right]\frac{\mathrm{d}M}{M} \tag{3.1f}$$

and

$$\frac{\mathrm{d}p^o}{p^o} = \frac{\mathrm{d}p}{p} + \left[\frac{\gamma M^2}{\{1+\frac{\gamma-1}{2}\}M^2}\right]\frac{\mathrm{d}M}{M} \tag{3.1g}$$

Now, combining (3.1a), (3.1c), and (3.1e), we get

$$\frac{\mathrm{d}p}{p} = \frac{\mathrm{d}\rho}{\rho} + \frac{\mathrm{d}T}{T} = \frac{\mathrm{d}\dot{m}}{\dot{m}} - \frac{\mathrm{d}u}{u} - \frac{\mathrm{d}A}{A} + \frac{\mathrm{d}T}{T}$$

$$= -\left[1+(\gamma-1)M^2\right]\frac{\mathrm{d}u}{u} - \frac{\mathrm{d}A}{A} + \frac{\mathrm{d}q}{c_pT} - \frac{(\gamma-1)}{\gamma RT}\mathrm{d}w_t - \left[\frac{(\gamma-1)}{2}M^2 - \frac{h_w}{c_pT}\right]\frac{\mathrm{d}\dot{m}}{\dot{m}}$$

Further, from (3.1b),

$$\frac{\mathrm{d}p}{p} = -\gamma M^2\left[\frac{\mathrm{d}u}{u} + \frac{\mathrm{d}\dot{m}}{\dot{m}} + \left(\frac{2c_f}{D}\right)\mathrm{d}x\right] - \frac{\mathrm{d}w_t}{RT}$$

Combining the above two equations gives

$$\frac{\mathrm{d}u}{u} = \frac{1}{1-M^2}\left[\left(\frac{\gamma+1}{2}M^2 + \frac{h_w}{c_pT}\right)\frac{\mathrm{d}\dot{m}}{\dot{m}} - \frac{\mathrm{d}A}{A} + \frac{\mathrm{d}q}{c_pT} + \gamma M^2\left(\frac{2c_f}{D}\right)\mathrm{d}x\right] \tag{3.2a}$$

Substituting (3.2a) back into the previous equation and rearranging yields

$$\frac{\mathrm{d}p}{p} = \frac{\gamma M^2}{M^2-1}\left[\left\{1+\frac{\gamma-1}{2}M^2 + \frac{h_w}{c_pT}\right\}\frac{\mathrm{d}\dot{m}}{\dot{m}} - \frac{\mathrm{d}A}{A} + \frac{\mathrm{d}q}{c_pT} + \frac{\mathrm{d}w_t}{\gamma M^2 RT}\right.$$

$$\left. + \{1+(\gamma-1)M^2\}\left(\frac{2c_f\mathrm{d}x}{D}\right)\right] \tag{3.2b}$$

Combining (3.2a) with (3.1a), we get further

$$\frac{\mathrm{d}\rho}{\rho} = \frac{1}{M^2-1}\left[\left\{M^2-1+\frac{\gamma+1}{2}M^2 + \frac{h_w}{c_pT}\right\}\frac{\mathrm{d}\dot{m}}{\dot{m}} - M^2\frac{\mathrm{d}A}{A} + \frac{\mathrm{d}q}{c_pT} - \frac{\mathrm{d}w_t}{\gamma RT} + \gamma M^2\left(\frac{2}{D}c_f\mathrm{d}x\right)\right]$$

$$\tag{3.2c}$$

Similarly, further from (3.1c), (3.2b), and (3.2c),

$$
\begin{aligned}
\frac{dT}{T} = \frac{1}{M^2-1} \Bigg[&\left\{ 1 + \frac{\gamma-3}{2}M^2 + \frac{\gamma(\gamma-1)}{2}M^4 + (\lambda M^2 - 1)\frac{h_w}{c_pT} \right\}\frac{d\dot{m}}{\dot{m}} \\
&- (\gamma-1)M^2\frac{dA}{A} + (\gamma M^2 - 1)\frac{dq}{c_pT} + \frac{(\gamma-1)}{\gamma RT}dw_t \\
&+ \gamma(\gamma-1)M^4\left(\frac{2c_f}{D}\right)dx \Bigg]
\end{aligned} \tag{3.2d}
$$

Note that the coefficient of $d\dot{m}/\dot{m}$ is always larger than 1 at all Mach numbers. Further from (3.1d), we have

$$
\begin{aligned}
\frac{dM}{M} = \frac{1}{2(M^2-1)} \Bigg[&-\left\{ 1 + \frac{(\gamma=1)}{2}M^2 + \frac{h_w}{c_pT}(\lambda M^2) \right\}\frac{d\dot{m}}{\dot{m}} \\
&+ (2 + (\gamma-1)M^2)\frac{dA}{A} - (\gamma M^2 + 1)\frac{dq}{c_pT} - \frac{(\gamma+1)}{\gamma RT}dw_t \\
&- \left\{ 2\gamma M^2 + \gamma(\gamma-1)M^4 \right\}\left(\frac{2c_f}{D}dx\right) \Bigg] \tag{3.2e}
\end{aligned}
$$

Finally, by combining (3.2b), (3.2d), and (3.2e) into (3.1f) and (3.2g), we get the two equations for the change of stagnation temperature and stagnation pressure as follows:

$$
\begin{aligned}
\frac{dT^o}{T^o} = \frac{1}{M^2-1}\Bigg[&1 - \frac{\gamma(\gamma-1)}{2}M^4 + \left\{ (\gamma M^2 - 1) - \frac{\gamma(\gamma-1)M^4}{2+(\gamma-1)M^2} \right\}\frac{h_w}{c_pT} \Bigg] \\
&\times \frac{d\dot{m}}{\dot{m}} + \frac{(\gamma-1)}{\gamma RT^o}(dq - dw_t)
\end{aligned} \tag{3.2f}
$$

and

$$
\begin{aligned}
\frac{dp^o}{p^o} = &\frac{\gamma M^2}{2(M^2-1)}\left[1 + (\gamma-1)M^2 + (2-M^2)\frac{h_w}{c_pT^o} \right] - \gamma M^2\left(\frac{2c_f}{D}dx\right) \\
&- \frac{\gamma M^2}{2}\frac{dq}{c_pT^o} - \frac{dw_t}{RT^o}
\end{aligned} \tag{3.2g}
$$

Note that in both (3.2f) and (3.2g), for $(d\dot{m}/\dot{m}) = 0$, the stagnation states do not change whether the Mach number is larger or smaller than 1. The effect of friction is seen only for the stagnation pressure.

Furthermore, the change of entropy is

$$
\frac{ds}{c_p} = \frac{dT^o}{T^o} - \frac{\gamma-1}{\gamma}\frac{dp^o}{p^o} \tag{3.2h}
$$

Equations 3.2a, 3.2b, 3.2c, 3.2d, 3.2e, 3.2f, 3.2g, and 3.2h now all have a change in one of the thermodynamic or flow state variables, such as $u, p, \rho, T, M, T^o, p^o$, and s, as dependent variable on the left-hand side, and there are, on the right-hand side, independent variables due to a change in the mass flow rate $(d\dot{m}/\dot{m})$, change in cross section (dA/A), addition (or removal) of heat $dq/(c_p T)$, extraction (or addition) of work $dw_t/(RT)$, and the friction coefficient c_f—the last coefficient is always positive except for the stagnation pressure, where it is always negative (*pressure loss*). Except for the five independent variables, it is significant that for the equations for stagnation temperature, stagnation pressure, and entropy, there is no effect of the cross-sectional area change. The stagnation temperature is not dependent on friction either, although the friction always causes a loss in the stagnation pressure. The introduction of cooler gas from outside always causes a decrease in the stagnation temperature, which also causes a loss in stagnation pressure due to the introduction of an external mass flow rate that has to be accelerated; the external gas may be hotter or cooler than the gas in the channel. For positive work (such as that in the turbine), $dw_t > 0$, the effect is to reduce both the stagnation temperature and the stagnation pressure, or the effect is opposite for the compressor. Thus, component-wise in an aircraft gas-turbine engine, one can draw the following conclusions just by considering the signs of the terms:

1. Compressor:

$$dw_t<0, \ d\dot{m} \approx 0, \ dq \approx 0 : dT^o>0, \ dp^o>0 \ \text{and} \ ds>0$$

due to friction only.
2. Turbine:

$$dw_t>0, \ d\dot{m} \approx 0, \ dq \approx 0 : dT^o<0, \ dp^o<0 \ \text{and} \ ds>0$$

due to friction only.
3. Combustion chamber (including afterburner):

$$dw_t = 0, \ d\dot{m} \geq 0, \ dq>0 : dT^o>0, \ dp^o<0 \ \text{and} \ ds>0,$$

due to heat addition, $dT^o>0$, $dp^o<0$, and due to mass addition of cooler air entering the combustion chamber through film-cooling slits, $dT^o<0$, $dp^o<0$. Note that in the combustion chamber, the pressure loss is due to both friction and heat addition.

A combustion chamber is built to have stable combustion inside the chamber. It is a two-channel system in which there is an outside channel and there is an inner channel as the flame tube. Initially, only a small quantity (about 15%) of air is introduced around the fuel burner in a vortex motion, so that a hot zone of recirculating flow of very hot combustion gases with stoichiometric proportion is created. The hot gas temperature is then reduced to fit the requirement of the allowable temperature of the turbine blades with the help of cooling air entering from the outer channel into the flame tube through side holes. Some of these holes are also so directed to create a cooling air film to protect the flame tube.

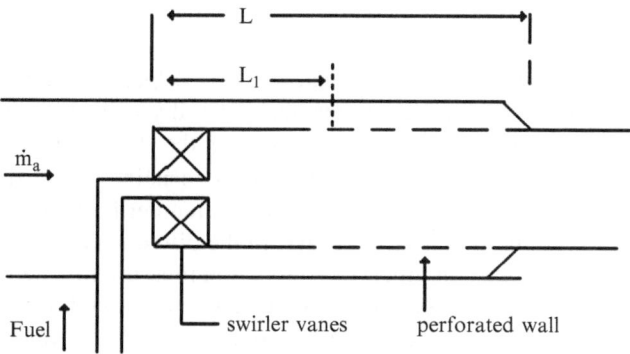

Fig. 3.2 Schematic sketch of a combustion chamber

In older days with a smaller air mass flow rate, a series of combustor cans (*can combustors*) were built with interconnected holes through which all flames in each can could be ignited almost simultaneously with the help of one or two spark-plugs placed strategically in one or two cans. With larger mass flow rates in gas turbines, only the individual inner flame tubes are placed inside outer annular channels (*cannular combustor*), and further both the outer channel and the inner flame channel are of the annular type (*annular combustor*). References to these three types of combustors can be read in the description of engines in the literature.

The change in the stagnation pressure in the combustion chamber is due to mass addition, friction, and heat addition—all three courses act in the same direction. This will now be explained on the basis of a simple calculation of a combustion chamber, a schematic sketch of which is given in Fig. 3.2.

We consider a can-type combustor of the engine type De Havalland Goblin with the following set of data among others: mass flow rate = 28.5 kg/s, overall compression ratio = 3.7, turbine inlet temperature = 800°C (estimated) = 1,073 K, ambient pressure = 1 bar, ambient temperature = 298 K, percentage of air through swirl vanes = 15. From measurements, the estimated design data are flame tube diameter = 8 cm = 0.08 m, length of the combustion zone L_1 = 20 cm = 0.2 m, and length of the flame tube = 0.6 m.

Computed data: before combustor can, pressure = 3.7 bar, isentropic temperature = 814.9 K, assuming compressor efficiency = 0.84, temperature = 913.3 K, density = 1.409 kg/m³.

Temperature at the end of combustion zone = 0.15(1,073 − 913.3) + 913.3 = 937.3 K, temperature at the exit of combustor = 1,073 K. Consequently, the respective densities are 1.373 and 1.2 kg/m³.

Average density in combustion zone = (1.409 + 1.373)/2 = 1.391 kg/m³, average density in dilution zone = (1.373 + 1.2)/2 = 1.282 kg/m³.

Mass flow rate in combustion zone = 28.5 × 0.15/18 = 0.2375 kg/s, average mass flow rate in dilution zone = 28.5 × 0.925/18 = 1.465 kg/s.

Since the flame tube cross-sectional area = 0.005 m², the average flow velocity in the combustion zone = 0.2375/(1.391 × 0.005) = 34.15 m/s, and the same in the dilute zone $= 1.465/(1.282 \times 0.005) = 228.5$ m/s. The corresponding Mach numbers are 0.056 and 0.359, respectively.

In the combustion zone, heat added $= 24.07$ kJ/kg. Assume the friction coefficient $= 0.003$. Now in the combustion zone, $d\dot{m} = 0$, $dw_t = 0$, and in the dilute zone, $dq = 0$, $dw_t = 0$. As seen from (3.2g), the stagnation pressure loss $= \Delta p^o / p^o = -10^{-5}(0.015 + 0.013) = -2.8 \times 10^{-7}$ consisting of loss due to friction and heat addition. However, since the flow Mach number is small, the stagnation pressure loss is very small. On the contrary, in the dilute zone (let $h_w \approx 0$) with a much higher average Mach number, the stagnation pressure loss here due to mass addition and friction $= -0.0879(0.94883 + 0.00054) = -0.0835$ due to mass addition (which has to be accelerated) and friction, the former contributing very significantly. A pressure loss of 8.3% is quite significant in this combustor, but the approximate pressure loss in modern-day combustors is about 5%.

3.1 Discussion on Efficiencies

On the basis of the discussion in previous pages about various losses in different components in a gas-turbine engine, we will now estimate the efficiency of these components.

3.1.1 Diffuser Efficiency

Generally, for inlet diffusers, even if there is a turbulent boundary layer, the thickness of the boundary layer will be much smaller than the radius of the diffusion. For such a case, the boundary-layer thickness and the friction coefficient for the turbulent boundary layer are given by the two relations for a flat plate, which can be taken approximately as

$$\frac{\delta_t}{x} = 0.380 \text{Re}^{-0.2} \tag{3.3a}$$

and

$$c_f = 0.0592 \text{Re}^{-0.2} \tag{3.3b}$$

where the *Reynolds number,* Re, is based on the length along the wall from the leading edge 1999 of the diffuser. Substituting (3.3b) into the friction term in (3.2g) and integrating numerically, one can find the loss in the total pressure fairly accurately for a subsonic diffuser. Now, let the diffuser inlet's static and stagnation

Fig. 3.3 Schematic sketch of the gas states in a diffuser

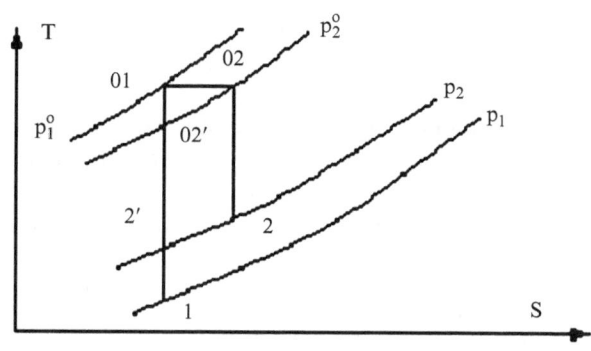

Table 3.1 Subsonic diffuser efficiencies

M_1	0.2	0.4	0.6	0.8
η_d	0.988	0.990	0.992	0.994
M_2	0.096	0.085	0.112	0.123

states be denoted as 1 and 01, and let the corresponding static and stagnation states at the diffuser's exit be 2 and 02 (Fig. 3.3).

The diffuser's efficiency is defined as

$$\eta_d = \frac{(T_{2'}^o - T_1)}{(T_1^o - T_1)} \tag{3.4}$$

Noting that

$$\eta_d = \frac{(p_2^o/p_1)^{\frac{(\gamma-1)}{\gamma}} - 1}{\left(\frac{T_1^o}{T_1} - 1\right)}; \frac{T_1^o}{T_1} = 1 + \frac{\gamma-1}{2}M_1^2$$

we get

$$\eta_d = \frac{(p_2^o/p_1)^{\frac{(\gamma-1)}{\gamma}} - 1}{\left(\frac{\gamma-1}{2}M_1^2\right)} \tag{3.4a}$$

Let's consider a typical axisymmetric subsonic diffuser designed for 20 kg/s at $M_1 = 0.5$, $p_1 = 1$ bar, and $T_1 = 298$ K, whose diameter increases by 50% linearly within the length $L = D_1$. We assume the fully developed turbulent boundary-layer relation $c_f = 0.05/\text{Re}_D^{0.2}$, and we will calculate the diffuser's efficiency, where the inlet Mach number M_1 is changed from 0.2 to 0.8. Let the diffuser's inlet diameter be $D_1 = 0.25$ m, and locally let the diffuser's diameter be given by $D = D_1[1 + x/(2L)]$. These results are given in Table 3.1.

For supersonic entry ($M_1 > 1$), the diffuser's efficiency depends mainly on the shock loss, that is, on the approaching flow Mach number and the type of shock

Table 3.2 Supersonic diffuser with normal shock

M_1	1.0	1.5	2.0	2.5	3.0	3.5	4.0
η_d	1.0	0.934	0.799	0.676	0.576	0.497	0.437

(normal, oblique—number of oblique shocks). The diffuser has to be carefully designed for a supersonic entry to minimize this shock loss. However, for a normal shock entry, it can be estimated easily.

Note that for a *normal shock*, the following relations are valid:

$$\frac{p_2^o}{p_2} = \left(1 + \frac{\gamma - 1}{2} M_2^2\right)^{\frac{\gamma}{\gamma-1}}; \quad \frac{p_2}{p_1} = 1 + \frac{2\gamma}{\gamma+1}(M_1^2 - 1) = 2\gamma M_1^2 - \frac{\gamma - 1}{\gamma + 1}$$

and

$$M_2^2 = \frac{2 + (\gamma - 1)M_1^2}{2\gamma M_1^2 - (\gamma - 1)} \rightarrow M_2 = \sqrt{\frac{2 + (\gamma - 1)M_1^2}{2\gamma M_1^2 - (\gamma - 1)}} \rightarrow \frac{p_2^o}{p_2} = 1 + \frac{1 + \frac{\gamma-1}{2}M_1^2}{\frac{2\gamma}{\gamma-1}M_1^2 - 1}$$

Therefore, for a supersonic entry diffuser with *normal entry shock*,

$$\eta_d = \frac{p_1^o - p_2^o}{p_1^o} = 1 - \frac{p_2^o}{p_2}\frac{p_2}{p_1}\frac{p_1}{p_1^o} = 1 - \frac{\frac{2\gamma M_1^2}{\gamma+1} - \frac{\gamma-1}{\gamma+1}}{\frac{2\gamma M_1^2}{\gamma-1} - 1} \tag{3.5}$$

which has been computed for $\gamma = 1.4$; the results are given in Table 3.2. It shows that for a supersonic entry with normal shock, the entry Mach number cannot be more than 1.5 since the loss will be too high. For higher Mach numbers, the diffuser with one or more oblique shocks has to be designed carefully to enable operation of the diffuser in design and off-design conditions.

3.1.2 Combustion Chamber Efficiencies

Stagnation pressure loss due to friction, heat addition, and mass addition in a combustion chamber has already been discussed earlier in this chapter. An efficiency based on the pressure loss

$$\eta_{comb} = 1 - \frac{\Delta p^o}{p^o} \tag{3.6}$$

is often used, where $\eta_{comb} \approx 0.95$ can be taken as a starting point, but it can be estimated by the method described earlier for losses due to friction, heat addition, and mass addition, at least for the forward combustion chamber and the afterburner.

Fig. 3.4 Definition of nozzle efficiency due to friction in a (T, s) diagram

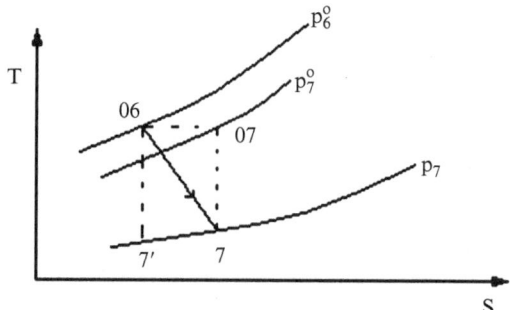

3.1.3 Exhaust Nozzle Efficiencies

For exhaust nozzles, one can again estimate the stagnation pressure loss due to friction; the entire process is shown in Fig. 3.4 in a (T, s) diagram.

The *nozzle efficiency* is then defined as

$$\eta_{\text{Noz}} = \frac{T_7^o - T_7}{T_7^o - T_{7'}} \tag{3.7}$$

where state 6 is at the nozzle entry and state 7 is at the nozzle exit; for optimum nozzle pressure at the nozzle exit, it is equal to the ambient pressure.

As a result of the friction loss in the stagnation pressure, there is a loss in the exhaust jet speed also. While $c_{\text{p}}\left(T_7^o - T_7'\right) = u_7^{*2}/2$ is the jet exhaust kinetic energy under isentropic condition, the actual jet exhaust kinetic energy is, and one can then write the *nozzle efficiency* as

$$\eta_{Noz} = \left(\frac{u_7}{u_7^*}\right)^2 \tag{3.7a}$$

3.2 Compressor and Turbine Stage and Polytropic Efficiencies

While discussing the efficiency of individual components in a gas-turbine or jet engine, we have considered the entire component. However, for both turbines and compressors we seldom have a single stage, and therefore, we have to seek a relationship between the efficiency of the individual stage and the global efficiency.

In both compressors and turbines, the initial and final values will be designated by the subscripts i and f, and the stagnation and static quantities will be designated with the help of the superscript o; the primed subscript f' refers to "find value under isentropic condition." (Fig. 3.5) Therefore, with reference to Fig. 1.21b, we can

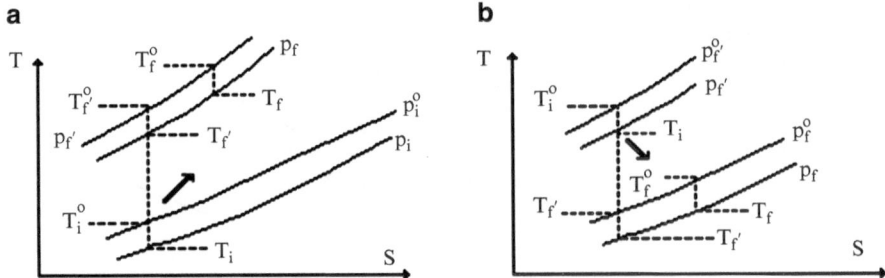

Fig. 3.5 To explain efficiencies in (**a**) a compressor and (**b**) a turbine

define for compressors and turbines the adiabatic efficiency based on stagnation values as follows:

For compressors:

$$\eta^o_{ad,c} = \frac{T^o_{f'} - T^o_i}{T^o_f - T^o_i} = \frac{\left(p^o_f - p^o_i\right)^{\frac{\gamma-1}{\gamma}} - 1}{\left\{\left(T^o_f - T^o_i\right) - 1\right\}} \tag{3.8a}$$

For turbines:

$$\eta^o_{ad,t} = \frac{T^o_i - T^o_f}{T^o_i - T^o_{f'}} = \frac{\left\{1 - \left(T^o_f / T^o_i\right)\right\}}{1 - \left(p^o_f - p^o_i\right)^{\frac{\gamma-1}{\gamma}}} \tag{3.8b}$$

While we have based the adiabatic efficiency on the stagnation state values, we can additionally define these based on static state values as follows:

For compressors:

$$\eta^0_{ad,c} = \frac{T_{f'} - T_i}{T_f - T_i} = \frac{\left(p_f - p_i\right)^{\frac{\gamma-1}{\gamma}} - 1}{\left\{\left(T_f - T_i\right) - 1\right\}} \tag{3.9a}$$

For turbines:

$$\eta_{ad,t} = \frac{T_i - T_f}{T_i - T_{f'}} = \frac{\left\{1 - \left(T_f - T_i\right)\right\}}{1 - \left(p_f - p_i\right)^{\frac{\gamma-1}{\gamma}}} \tag{3.9b}$$

Further, let $\pi^o = p^o_f / p^o_i$ and $\pi = p_f / p_i$ be designated as the *stagnation* and *static pressure ratio*, respectively, and let the corresponding temperature ratio be $\tau^o = T^o_f / T^o_i$ and $\tau = T_f / T_i$. Note that for compressors both ratios are greater than 1, and for turbines both are less than 1. Therefore, we can write the *adiabatic efficiency* based on the stagnation and static variables in terms of temperature and pressure ratios as follows:

For compressors:

$$\eta_{ad,c}^o = \frac{\left(\pi_c^o\right)^{\frac{\gamma-1}{\gamma}} - 1}{\left\{\left(\tau_c^o\right) - 1\right\}} \tag{3.10a}$$

and

$$\eta_{ad,c}^o = \frac{\left(\pi_c\right)^{\frac{\gamma-1}{\gamma}} - 1}{\left\{\left(\tau_c\right) - 1\right\}} \tag{3.10b}$$

For turbines:

$$\eta_{ad,t}^o = \frac{\left\{1 - \left(\tau_t^o\right)\right\}}{1 - \left(\pi_t^o\right)^{\frac{\gamma-1}{\gamma}}} \tag{3.10c}$$

and

$$\eta_{ad,t} = \frac{\left\{1 - \left(\tau_t\right)\right\}}{1 - \left(\pi_t\right)^{\frac{\gamma-1}{\gamma}}} \tag{3.10d}$$

While one can derive expressions linking η_{ad}^o and η_{ad} through the *initial* and *final kinetic energy*, they are quite complex expressions. It is much easier to write an expression for these using π and π^o, and τ and τ^o.

From the relations

$$T_f^o - T_f = \frac{c_c^2}{2c_p} \quad \text{and} \quad T_i^o - T_i = \frac{c_i^2}{2c_p}$$

we may write (where c_i and c_f are initial and final velocity, respectively)

$$\frac{T_f}{T_f^o} = 1 - \frac{c_f^2}{2c_p T_f^o} \quad \text{and} \quad \frac{T_i}{T_i^o} = 1 - \frac{c_i^2}{2c_p}$$

Furthermore,

$$\frac{p_f}{p_f^o} = \left(\frac{T_f}{T_f^o}\right)^{\frac{\gamma}{\gamma-1}} = \left(1 - \frac{c_f^2}{2c_f T_f^o}\right)^{\frac{\gamma}{\gamma-1}}$$

and

$$\frac{p_i}{p_i^o} = \left(\frac{T_i}{T_i^o}\right)^{\frac{\gamma}{\gamma-1}} = \left(1 - \frac{c_i^2}{2c_p T_i^o}\right)^{\frac{\gamma}{\gamma-1}}$$

Now,

$$\tau = \frac{T_f}{T_i} = \left(\frac{T_f^o}{T_i^o}\right) - \left(\frac{T_f}{T_f^o}\right)\left(\frac{T_i^o}{T_i}\right) = \tau^o \left[\frac{1 - \frac{c_f^2}{2c_p T_f^o}}{1 - \frac{c_i^2}{2c_p T_i^o}}\right]^{\frac{\gamma}{\gamma - 1}} \tag{3.11a}$$

and

$$\pi = \frac{\pi_f}{\pi_i} = \left(\frac{p_f^o}{p_i^o}\right) - \left(\frac{p_f}{p_f^o}\right)\left(\frac{p_i^o}{p_i}\right) = \pi^o \left[\frac{1 - \frac{c_f^2}{2c_p T_f^o}}{1 - \frac{c_i^2}{2c_p T_i^o}}\right]^{\frac{\gamma}{\gamma - 1}} \tag{3.11b}$$

We will now discuss another type of efficiency, namely the *polytropic efficiency*, which is in terms of the differential change of state k, which is more relevant for compressors, because the change in the absolute value of the pressure difference across each stage in the compressor is much smaller than in the turbine, and hence we derive it for the compressor first.

For compressors, it is the ratio of the differential change in the stagnation enthalpy under isentropic condition to the actual differential change of the total enthalpy, that is,

$$\eta_{\text{pol,c}} = \frac{\Delta h^{o*}}{\Delta h^o} \approx \frac{\Delta T^{o*}}{\Delta T^o} \tag{3.12}$$

In (3.12), the asterisk refers to isentropic change. From the first and second laws of thermodynamics, we have

$$T^o \, ds - dh^o = -\frac{1}{\rho^o} dp^o$$

which for constant entropy implies $dh^{o*} = (\rho^o)^{-1} dp^o$. Now for isentropic case, we also have $p^o / T^{o\frac{\gamma}{\gamma-1}} = k$, which means

$$dp^o = \frac{\gamma}{\gamma - 1} \frac{p^o}{T^o} dT^{o*}$$

and hence,

$$dh^{o*} = (\rho^o)^{-1} dp^o = \frac{\gamma}{\gamma - 1} \frac{p^o}{\rho^o T^o} dT^{o*} = c_p dT^{o*}$$

or

$$dT^{o*} = \frac{1}{\rho^o c_p} dp^o \quad \text{or} \quad dh^{o*} = \frac{1}{\rho^o} dp^o$$

On the other hand, for the nonisentropic case, $dh^o \approx c_p \, dT^o$. Therefore,

$$\eta_{\text{pol,c}} = \frac{\Delta h^{o*}}{\Delta h^o} \frac{1}{\rho^o c_p} \frac{dp^o}{dT^o} = \frac{RT^o}{p^o c_p} \frac{dp^o}{dT^o} = \frac{\gamma - 1}{\gamma} \frac{T^o}{p^o} \frac{dp^o}{dT^o}$$

Noting further that

$$\Delta h^{o*} \approx \frac{1}{\rho^o} dp^o; \quad \Delta h^o \approx c_p dT^o$$

Rearranging the previous expression gives

$$\frac{dT^o}{T^o} = \frac{\gamma - 1}{\gamma} \eta_{\text{pol,c}} \frac{dp^o}{p^o}$$

which by integration gives

$$\tau_c^o = \left(\frac{p_f^o}{p_c^o} \right)^{\frac{\gamma - 1}{\gamma \eta_{\text{pol,c}}}}$$

and from (3.10a), we can write

$$\eta_{\text{ad,c}}^o = \frac{\left\{ \{\pi_c^o\}^{\frac{\gamma-1}{\gamma}} - 1 \right\}}{\{\tau_c^o - 1\}} = \frac{\left\{ \{\pi_c^o\}^{\frac{\gamma-1}{\gamma}} - 1 \right\}}{\left\{ \{\pi_c^o\}^{\frac{\gamma-1}{\gamma \eta_{\text{pol,c}}}} - 1 \right\}} \tag{3.13a}$$

Thus,

$$\{\pi_c^o\}^{\frac{\gamma-1}{\gamma \eta_{\text{pol,c}}}} = 1 + \frac{\{\pi_c^o\}^{\frac{\gamma-1}{\gamma}} - 1}{\eta_{\text{ad,c}}^o}$$

By taking the logarithm on either side, we have

$$\frac{\gamma - 1}{\gamma \eta_{\text{pol,c}}} \ln(\pi_c^o) = \ln\left[1 + \frac{\{\pi_c^o\}^{\frac{\gamma-1}{\gamma}} - 1}{\eta_{\text{ad,c}}^o} \right]$$

from which it follows that

$$\eta_{\text{pol,c}} = \frac{\frac{\gamma-1}{\gamma} \ln(\pi_c^o)}{\ln\left[1 + \frac{\{\pi_c^o\}^{\frac{\gamma-1}{\gamma}} - 1}{\eta_{\text{ad,c}}^o} \right]} = \frac{\gamma - 1}{\gamma} \frac{\{\ln(\pi_c^o)\}}{\ln\{\tau_c^o\}} \tag{3.13b}$$

By a similar procedure like the one above, we can derive expressions for turbines linking the polytropic efficiency with the turbine efficiency based on change in stagnation state variables, and the equivalent expressions of (3.13a) and (3.13b) are

$$
\eta_{\text{ad},t}^o = \frac{\{1 - \tau_t^o\}}{\left\{1 - \{\pi_t^o\}^{\frac{\gamma-1}{\gamma}}\right\}} = \frac{\left\{1 - \{\pi_t^o\}^{\frac{(\gamma-1)\eta_{\text{pol},t}}{\gamma}}\right\}}{\left\{1 - \{\pi_t^o\}^{\frac{\gamma-1}{\gamma}}\right\}}
\tag{3.14a}
$$

and

$$
\eta_{\text{pol},t} = \frac{\gamma}{\gamma-1} \frac{\ln\left[1 - \eta_{\text{ad},t}^o\left(1 - \{\pi_t^o\}^{\frac{\gamma-1}{\gamma}}\right)\right]}{\ln(\pi_t^o)} = \frac{\gamma}{\gamma-1} \frac{\ln\{\tau_t^o\}}{\{\ln(\pi_t^o)\}}
\tag{3.14b}
$$

We will now explain a practical application of the *polytropic efficiency* for a multistage compressor or turbine. Since the polytropic efficiency refers to a differential change of state, it is expected that this will be very near the stage efficiency. Our discussion will be first for compressors; let there be N stages. Now let the stagnation compressor ratio in each stage, τ_{cs}^o, be the same for all stages, and let the stagnation temperature ratio in each stage, τ_{cs}^o, also be the same for all stages, whereas let π_c^o and π_c^o be the *overall stagnation pressure ratio* and *overall stagnation temperature ratio*, respectively. Then the above assumption implies that

$$
\pi_c = \prod_N \pi_{\text{cs}} \quad \text{and} \quad \tau_c = \prod_N \tau_{\text{cs}}
$$

Now, if we write an expression similar to (3.13a) for each stage, it would be

$$
\eta_{\text{ad},\text{cs}}^o = \frac{\left\{\{\pi_{\text{cs}}^o\}^{\frac{\gamma-1}{\gamma}} - 1\right\}}{\{\tau_{\text{cs}}^o - 1\}} = \frac{\left\{\{\pi_{\text{cs}}^o\}^{\frac{\gamma-1}{\gamma}} - 1\right\}}{\left\{\{\pi_{\text{cs}}^o\}^{\frac{\gamma-1}{\gamma\eta_{\text{pol},c}}} - 1\right\}}
\tag{3.15a}
$$

and we may further write

$$
\tau_c^o = \prod_N \tau_{\text{cs}}^o = \prod_N \left[\left\{1 + \frac{1}{\eta_{\text{ad},\text{cs}}^o}\right\}\left(\{\pi_{\text{cs}}^o\}^{\frac{\gamma-1}{\gamma}} - 1\right)\right] = \left[\left\{1 + \frac{1}{\eta_{\text{ad},\text{cs}}^o}\right\}\left(\{\pi_{\text{cs}}^o\}^{\frac{\gamma-1}{\gamma}} - 1\right)\right]^N
$$

Thus from (3.15a), we write

$$
\eta_{\text{ad},c}^o = \frac{\left\{\{\pi_c^o\}^{\frac{\gamma-1}{\gamma}} - 1\right\}}{\tau_c^o - 1} = \left\{\frac{\{\pi_{\text{cs}}^o\}^{\frac{N(\gamma-1)}{\gamma}} - 1}{\left[1 + \frac{1}{\eta_{\text{ad},\text{cs}}^o}\left(\{\pi_{\text{cs}}^o\}^{\frac{\gamma-1}{\gamma}} - 1\right)\right]^N}\right\}
\tag{3.15b}
$$

On the other hand, the *polytropic efficiency* for the overall compressor and for the stage can be written as

$$\eta_{\text{pol,c}} = \frac{\gamma - 1}{\gamma} \frac{\{\ln \pi_c^o\}}{\left\{ \ln\left(\left\{1 + \{\pi_c^o\}^{\frac{\gamma-1}{\gamma}}\right\} - \frac{1}{\eta_{\text{ad,c}}^o}\right)\right\}} \tag{3.16a}$$

and

$$\eta_{\text{pol,c}} = \frac{\gamma - 1}{\gamma} \frac{\{\ln \pi_c^o\}}{\left\{ \ln\left(\left\{1 + \{\pi_c^o\}^{\frac{\gamma-1}{\gamma}}\right\} - \frac{1}{\eta_{\text{ad,cs}}^o}\right)\right\}} \tag{3.16b}$$

Noting that

$$\tau_c^o = 1 + \frac{\pi_c^{o\frac{\gamma-1}{\gamma}} - 1}{\eta_{\text{ad,c}}^o} = 1 + \frac{\{\pi_{cs}^o\}^{\frac{N(\gamma-1)}{\gamma}} - 1}{\eta_{\text{ad,c}}^o}$$

and

$$\tau_{cs}^o = 1 + \frac{\pi_{cs}^{o\frac{\gamma-1}{\gamma}} - 1}{\eta_{\text{ad,cs}}^o} = \tau_c^{o1/N} = 1 + \frac{\pi_{cs}^{o\frac{N(\gamma-1)}{\gamma}} - 1}{\eta_{\text{ad,c}}^o} = + \frac{\pi_c^{o\frac{\gamma-1}{\gamma}} - 1}{\eta_{\text{ad,c}}^{o(1/N)}}$$

we write

$$\eta_{\text{pol,cs}} = \frac{\gamma-1}{\gamma}\left\{\frac{\ln \pi_{cs}^o}{\left[\left\{1 + \frac{\{\pi_{cs}^o\}^{\frac{\gamma-1}{\gamma}}-1}{\eta_{\text{ad,cs}}^o}\right\}\right]}\right\} = \frac{\gamma-1}{\gamma}\left\{\frac{\ln \pi_{cs}^o}{\frac{1}{N}\ln\left[1 + \left[\frac{\pi_c^{o\frac{\gamma-1}{\gamma}}-1}{\eta_{\text{ad,c}}^o}\right]\right]}\right\}$$

$$= \frac{\gamma-1}{\gamma}\left\{\frac{\ln \pi_{cs}^{oN}}{\left\{\ln\left[1 + \left\{\frac{\pi_c^{o\frac{\gamma-1}{\gamma}}}{\eta_{\text{ad,c}}^o}-1\right\}\right]\right\}}\right\} = \frac{\gamma-1}{\gamma}\left\{\frac{\ln \pi_c^o}{\ln\left[1 + \left\{\frac{\pi_c^{o\frac{\gamma-1}{\gamma}}-1}{\eta_{\text{ad,c}}^o}\right\}\right]}\right\}$$

Therefore, the polytropic efficiency in each stage is the same as the overall polytropic efficiency, but the same is not true for the adiabatic efficiency. Now noting that if $f = g^f$, then $f = f \ln g$, from which it follows that $f = g^f = \exp^{f \ln g} = 1 + f \ln g + O(1/f^2)$.
Thus, if

$$x = \left[\left\{1 + \frac{1}{\eta_{\text{ad,cs}}^o(\pi_{cs}^{o\frac{\gamma-1}{\gamma}} - 1)}\right\}\right]^N$$

and

$$y = \pi_c^{o\,\frac{\gamma-1}{\gamma}} \eta_{ad,cs}^o$$

we write

$$x^{1/N} = 1 + \frac{1}{\eta_{ad,cs}^o \left(\pi_{cs}^{o\,\frac{\gamma-1}{\gamma}} - 1\right)} = 1 + \frac{1}{\eta_{ad,cs}^o \left(\pi_c^{o\,\frac{\gamma-1}{\gamma N}} - 1\right)} = 1 + \frac{1}{\eta_{ad,cs}^o \left(y^{\eta_{ad,cs}^o/N} - 1\right)}$$

$$= 1 + \frac{1}{\eta_{ad,cs}^o}\left[\left\{\left\{1 + \frac{\eta_{ad,cs}^o}{N}\ln y + O\left(\frac{1}{N^2}\right)\right\} - 1\right\}\right] = \exp\{\tfrac{1}{N}\ln x\}$$

$$= 1 + \frac{1}{N}\ln x + O\left(\frac{1}{N^2}\right) = 1 + \frac{1}{N}\ln y + \frac{1}{\eta_{ad,cs}^o} + O\left(\frac{1}{N^2}\right)$$

Thus, as $N \to \infty$

$$x^{1/N} = 1 + \frac{1}{N}\ln x = 1 + \frac{1}{N}\ln y$$

which means that $x \to y$

Now, from (3.15b), we have

$$\eta_{ad,c}^o = \frac{\pi_c^{o\,\frac{\gamma-1}{\gamma}} - 1}{x - 1}$$

and from (3.15a), we have

$$\eta_{ad,c}^o = \frac{\left\{\{\pi_c^o\}^{\frac{\gamma-1}{\gamma}} - 1\right\}}{\left\{\{\pi_c^o\}^{\frac{\gamma-1}{\gamma\eta_{pol,c}}} - 1\right\}}$$

Since as $N \to \infty$, $x \to y$ we have therefore proved that as $N \to \infty$, $\eta_{ad,cs}^o = \eta_{pol,c}$.

Let's now consider a sample 16-stage axial compressor, that is, $N = 16$, which has a measured stage efficiency $\eta_{ad,cs}^o = 0.93$ and overall compression ratio $\pi_c^o = 25$ Therefore, $\pi_{cs}^o = 25^{1/16} = 1.223$ Taking $\gamma = 1.4$ and from (3.13b), we get $\eta_{pol,cs} = \eta_{pol,c} = 0.932$ On the other hand, from (3.15b), we get

$$\eta_{ad,c}^o = \left\{\frac{\{\pi_c^o\}^{\frac{N(\gamma-1)}{\gamma}} - 1}{\left[1 + \frac{1}{\eta_{ad,cs}^o}\left(\{\pi_c^o\}^{\frac{\gamma-1}{N\gamma}} - 1\right)\right]^N} - 1\right\} = 0.896$$

and by using (3.15a), we get $\eta_{ad,c}^o = 0.895$.

Fig. 3.6 A typical variation in adiabatic efficiency as a function of the pressure ratio

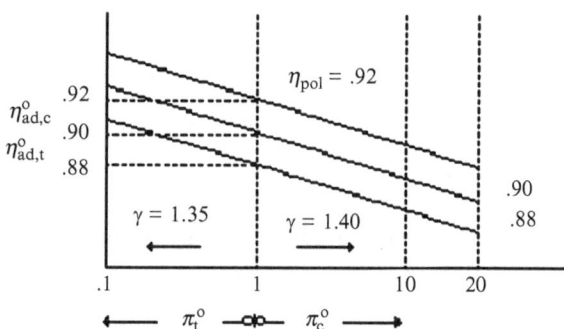

The general conclusion about adiabatic stage efficiency being almost equal to the polytropic efficiency for compressors is also valid for turbines. Equations 3.14a and 3.14b can be used to evaluate the overall adiabatic turbine efficiency and turbine polytropic efficiency. Analogously to (3.14a), the adiabatic stage efficiency can be obtained for turbines from the relation

$$\eta^o_{ad,cs} = \frac{\left\{1 - \tau_t^{o^{1/N}}\right\}}{\left\{1 - \{\pi_t^o\}^{\frac{\gamma-1}{\gamma N}}\right\}} = \frac{\left\{1 - \{\pi_t^o\}^{\frac{(\gamma-1)\eta_{pol,t}}{\gamma N}}\right\}}{\left\{1 - \{\pi_t^o\}^{\frac{\gamma-1}{\gamma N}}\right\}} \tag{3.17}$$

Different values of $\eta^o_{ad,c}$ and η_{pol} have been calculated from (3.13a), (3.13b) and $\eta^o_{ad,t}$ from (3.14a); the results are shown in Fig. 3.6.

3.3   Heating Factor During Compression and Expansion

Because of diverging isobar lines in a temperature-entropy diagram, the total of the isentropic total enthalpy change in each stage may be considerably larger than the overall isentropic total enthalpy change.

For compressors, if T_i^o is the *stagnation temperature* before the ith stage and $\pi_i^o \approx \pi_c^{o^{1/N}}$ is the *compression ratio* in the ith stage, where N is the total number of stages, then the isentropic total enthalpy change in each stage is given by

$$H_i^* = c_p T_i^o \left(\pi_i^{o^{\frac{\gamma-1}{\gamma}}} - 1\right) = c_p T_i^o \left(\pi_c^{o^{\frac{\gamma-1}{\gamma N}}} - 1\right) \tag{3.18a}$$

and the initial temperature before the ith stage is

$$T_i^o = T_{i-1}^o \left[1 + \frac{\pi_c^{o\,\frac{\gamma-1}{\gamma N}}}{\eta_{ad,cs}^o} \right]_{i-1} \tag{3.18b}$$

Therefore, the sum total of the isentropic total enthalpy change at each stage in the compressor is

$$\sum_{i=1}^{N} H_i^* = c_p T_1^o \left(\pi_c^{o\,\frac{\gamma-1}{\gamma N}} - 1 \right) \sum_{i=1}^{N} \left[1 + \frac{\pi_c^{o\,\frac{\gamma-1}{\gamma N}}}{\eta_{ad,cs}^o} \right]_{i-1}$$

$$= c_p T_1^o \eta_{ad,cs}^o \frac{\left[1 + \left\{ \frac{\left\{ \{\pi_c^o\}^{\frac{\gamma-1}{\gamma N}} - 1 \right\}}{\eta_{ad,cs}^o} \right\} \right]^N - 1 \right]}{\left[\frac{\{\pi_c^o\}^{\frac{\gamma-1}{\gamma N}} = 1}{\eta_{ad,cs}^o} \right]}$$

$$\approx c_p T_1^o \eta_{ad,cs}^o \left[1 + \left\{ \frac{\{\pi_c^o\}^{\frac{\gamma-1}{\gamma N}} - 1}{\eta_{ad,cs}^o} \right\}^N - 1 \right] \tag{3.19a}$$

On the other hand, the *overall isentropic total enthalpy change* is

$$H^* = c_p T_1^o \left(\pi_c^{o\,\frac{\gamma-1}{\gamma}} - 1 \right) \tag{3.19b}$$

and the ratio of the two, called the *heating factor for a compressor,* is given by the equation

$$1 + f_c = \frac{\sum H_i^*}{H*} = \frac{\eta_{ad,cs}^o}{\{\pi_c^o\}^{\frac{\gamma-1}{\gamma}} - 1} \left[\left\{ 1 + \frac{\{\pi_c^o\}^{\frac{\gamma-1}{\gamma N}} - 1}{\eta_{ad,cs}^o} \right\}^N - 1 \right] \tag{3.20a}$$

We can show that (3.15a) can lead to

$$1 + f_c = \left(\frac{\eta_{pol,c}}{\eta_{ad,c}^o} - 1 \right) \left(1 - \frac{1}{N} \right) + 1 \tag{3.20b}$$

and for $N \to \infty : 1 + f_c = \eta_{pol,c} / \eta_{ad,c}^o$.

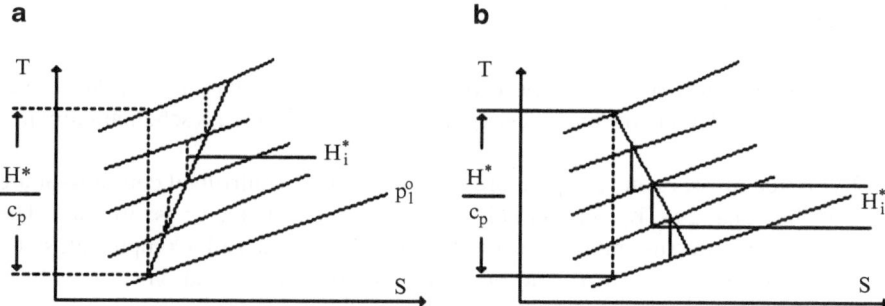

Fig. 3.7 To explain heating factor for (**a**) a compressor and (**b**) a turbine

Similar to the above equation for compressors, one can write equivalent equations for turbines (Fig. 3.7). Equivalent expressions of (3.18a) and (3.18b) are

$$H_i^* = c_p T_1^o \left(1 - \{\pi_{t,i}^o\}^{\frac{\gamma-1}{\gamma}} \right) = c_p T_1^o \left(1 - \{\pi_t^o\}^{\frac{\gamma-1}{\gamma N}} \right) \tag{3.21a}$$

and

$$T_i^o = T_1^o \left[\left\{ 1 - \left(\left\{ 1 - \{\pi_t^o\}^{\frac{\gamma-1}{\gamma N}} \right\} \right) \eta_{ad,ts}^o \right\} \right]_{i-1} \tag{3.21b}$$

Therefore, the sum of the total enthalpy of each stage is

$$\sum H_i^* = c_p T_1^o \left(\left\{ 1 - \{\pi_t^o\}^{\frac{\gamma-1}{\gamma N}} \right\} \right) \sum_{i=1}^{N} \left[\left\{ 1 - \left(1 - \{\pi_t^o\}^{\frac{\gamma-1}{\gamma N}} \right) \eta_{ad,ts}^o \right\} \right]_{i-1}$$

$$= c_p T_1^o \left(\left\{ 1 - \{\pi_t^o\}^{\frac{\gamma-1}{\gamma N}} \right\} \right) \frac{\left[\left\{ 1 - \left(\{\pi_t^o\}^{\frac{\gamma-1}{\gamma N}} \right) \eta_{ad,ts}^o \right\}^N \right\} - 1 \right]}{-\left\{ \left(\left\{ 1 - \{\pi_t^o\}^{\frac{\gamma-1}{\gamma N}} \right\} \right) \eta_{ad,ts}^o \right\}}$$

$$= c_p T_1^o \frac{1}{\eta_{ad,ts}^o} \left[\left\{ 1 - \left\{ 1 - \left(1 - \{\pi_t^o\}^{\frac{\gamma-1}{\gamma N}} \right) \eta_{ad,ts}^o \right\} \right\}^N \right] \tag{3.22a}$$

The overall change in isentropic total enthalpy for the turbine, on the other hand, is

$$H^* c_p T_1^o \left(1 - (\pi_t^o)^{\frac{\gamma-1}{\gamma}} \right) \tag{3.22b}$$

Therefore, the ratio of the two is the *heating factor for turbines* (see Fig. 3.7):

$$1 + f_t = \frac{\sum H_i^*}{H^*} = \frac{\left[1 - \left\{ \left\{ 1 - \left(1 - \pi_t^{o \frac{\gamma-1}{\gamma N}} \right) \right\} \eta_{ad,ts}^o \right\}^N \right]}{\eta_{ad,ts}^o \left(\left\{ 1 - \pi_t^{o \frac{\gamma-1}{\gamma}} \right\} \right)} \tag{3.22c}$$

3.4   Exercises

1. The initial states of a subsonic diffuser are $p_1 = 1.01$ bar, $T_1 = 298$ K, and $M_1 = 0.8$. Calculate the exit states (p, T) and show the process schematically in a (T, s) chart.
2. The initial pressure and temperature of a single-stage centrifugal compressor are 1.01 bar and 298 K, respectively, and the initial velocity is 30 m/s. The compression pressure ratio based on stagnation values is 0.84. Compute all static and stagnation states (p, T) and the adiabatic efficiency based on static values. Also, compute the polytropic efficiency.
3. For a 12-stage axial compressor with $p_1^o = 1.01$ bar, $T_1^o = 298$ K, $\eta_{ad,cs} = 0.88$, compute the heating factor and explain it by making a small sketch in a (T, s) chart.
4. The friction coefficient for a fully developed shear layer inside a tube is given by the equation

$$c_f = 0.3164 \mathrm{Re}^{-0.25}$$

where $\mathrm{Re} = \rho u D / \mu$ is the *Reynolds number*, $D =$ tube diameter.

For a flow condition of air at the entry to a constant-diameter tube (inside diameter $= 4$ cm) at $p_1 = 4.0$ bar, $T_1 = 300$ K, $M_1 = 0.4$, and the *dynamic viscosity coefficient* $\mu = 1.8 \times 10^{-5}$ kg/ms, and assuming a fully developed turbulent shear layer inside the tube, calculate the change in the state variables for air at a distance of 1 m from the entry. (Suggestion: To calculate the Reynolds number, the inlet Re can be taken as a first approximation.) Furthermore, since the tube length required for choking is about 2.52 m, the flow inside the tube can be considered fully subsonic. (Answer: $M = 0.466, T = 297.03$ K, $p = 3.966$ bar.)
5. For a long tube of length 100 m and an internal tube area of 10 cm^2, attached to a reservoir (condition of the reservoir: $p^o = 10$ bar, $T^o = 300$ K), the mass of air flowing through the tube to the free atmosphere is 1.5 m/s. Calculate the Mach number at two ends and the average friction coefficient. (Answer: $M_1 = 0.411$, $M_2 = 1.0, \bar{c}_f = 0.00019$.)

Chapter 4
Flow Through a Turbomachine

A turbomachine consists of rows of blades, some of which rotate (rotor) while others remain stationary (stator). As explained earlier, work can be extracted or added in a rotating row of blades only, where the stagnation pressure and stagnation temperature change, while in the stationary row of blades, the stagnation temperature and stagnation pressure do not change (except for the small effect of friction on the stagnation pressure). Hence, in a compressor, the air is accelerated in the rotor (the stagnation pressure is increased) and is decelerated in the stator, where the kinetic energy of air is converted into the pressure (same total pressure; the static pressure is increased). On the other hand, air is accelerated in a nozzle or stator blades' row at constant stagnation temperature and static pressure with increasing air (gas) speed, and subsequently the kinetic energy is converted into mechanical energy in a rotating row of blades, where the stagnation pressure and temperature decrease. For a single stage of turbomachinery blades, we thus have a row of rotating blades and a row of stationary blades; for compressors, the rotating blades come first, but for turbines, the stationary blades come first. We therefore have two rows of blades, and depending on the direction of motion of air (gas), they act as a compressor or turbine stage, as shown in Fig. 4.1. They are numbered 1–3 from the rotor side. If it is a compressor, air is scooped at blade edge 1 in the rotor and is pushed toward edge 2 of the rotor after accelerating it; it is subsequently decelerated in the stator row from 2 to 3. On the other hand, as turbine air (or gas) enters the stator at 3, it is accelerated up to 2 while the static pressure is lowered, but the stagnation temperature and stagnation pressure remain the same (except for a small loss in stagnation pressure due to friction), and then while passing through the rotor and exhausting at 1, it converts the kinetic energy into mechanical energy.

While the arrangement of blades in a turbomachinery stage with one stationary row of blades and a rotating row of blades is generally valid, there can be some special exceptions. One exception is the installation of guiding vanes or a prewhirler, which are stationary, before the first compressor stage to enable the flow to be directed in a certain direction, so that the rotor blades accept it optimally (minimum loss). The other exception is a propeller without any following stator blades.

T. Bose, *Airbreathing Propulsion: An Introduction*,
Springer Aerospace Technology, DOI 10.1007/978-1-4614-3532-7_4,
© Springer Science+Business Media, LLC 2012

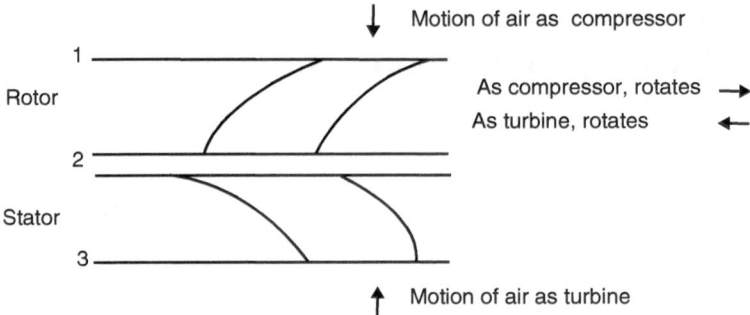

Fig. 4.1   A pair of blade rows in a turbomachinery stage

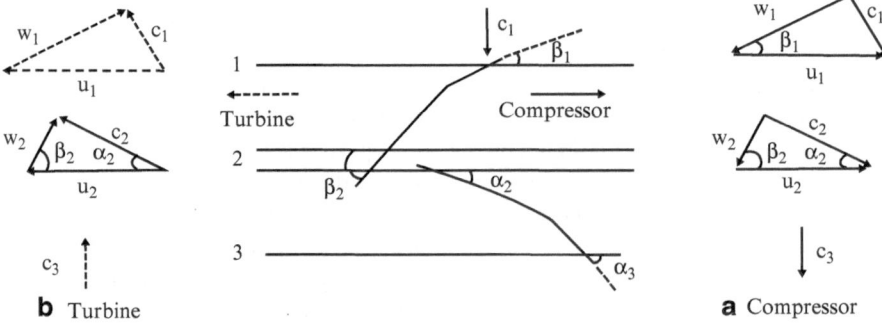

Fig. 4.2   Velocities in a pair of blade rows in a turbomachine

For rotating blades, it is useful to differentiate between the *absolute velocity*, **c**, as seen by an external observer, and the *relative velocity*, **w**, with respect to the rotating blades with *azimuthal velocity*, **u**. Therefore, we write formally

$$\mathbf{c} = \mathbf{w} + \mathbf{u} \qquad\qquad (4.1)$$

These velocities are shown in a *velocity triangle* in Fig. 4.2a, b, depending on whether the stage is acting as a compressor or a turbine.

While acting as a compressor, the leading edge 1 of the rotor blade scoops air, which at least ideally must be in the tangential direction of the blade (leading-edge blade angle) although if c_1 is in the meridian direction (perpendicular to the direction of azimuthal motion), then perhaps the relative velocity angle may be smaller than the *blade angle* β_1. For the present, however, we consider β_1 to be the angle of the flow direction, and it is also the leading blade angle, and the relative speed at the leading edge is w_1. For this purpose, the blades have to be sufficiently near each other, which can, of course, lead to high friction losses. The rotating blade then pushes the gas (or air) at its trailing-edge *blade angle* β_2 with relative velocity w_2, but seen from the outside, it would have absolute flow velocity c_2 and *flow angle* α_2.

For a smooth operation of the stator and for a sufficient number of blades, α_2 also has to be the stator blade's leading-edge angle. Finally, the gas emanates at the edge 3 tangential to the stator blades with flow angle α_3. Thus, for the following stage, c_1 will the same as c_3 of the previous stage.

If the stage acts as a turbine, then the flow direction is just the opposite. Air or gas enters the stator at 3 with the direction of the blade angle at α_3. If there are a sufficient number of blades, it expands to get an absolute speed $(c_2 > c_3)$, emanates tangentially to the stator blade trailing-edge angle α_2 at point 2, and enters the rotor blade entry angle β_1 with relative speed w_2. By passing through the rotor blade and changing direction, it imparts a rotary motion to the rotor, which may pick up sufficient speed to have relative velocity emanating at the other end with angle β_1 $(w_1 < w_1)$, and the flow is decelerated. We are therefore discussing here the two cases of a compressor and a turbine, where a steady situation has arisen, and where the flow everywhere is tangential to the local blade angle.

4.1 Elementary Theory of a Stage

From our earlier discussion, it is evident that in a stator row, the total enthalpy $h^o = h + c^2/2$ must remain unchanged; therefore, in the stator,

$$h_2^o = h_2 + \frac{1}{2}c_2^2 = h_3 + \frac{1}{2}c_3^2 = h_3^o \qquad (4.2)$$

As we discussed earlier, even if there is friction, (4.2) remains valid, although the gas speed at the trailing-edge side (c_3 for compressor and c_2 for turbine) will be smaller than the corresponding value under isentropic change of stage. For a rotating channel (Fig. 4.3), we can write for the *centrifugal* and *coriolis* *accelerations* of a fluid element

$$\text{Centrifugal acceleration}: b_{\text{cent}} = r \times \omega^2 \qquad (4.3a)$$

$$\text{Coriolis acceleration}: b_{\text{coriolis}} = 2[w \times \omega] \qquad (4.3b)$$

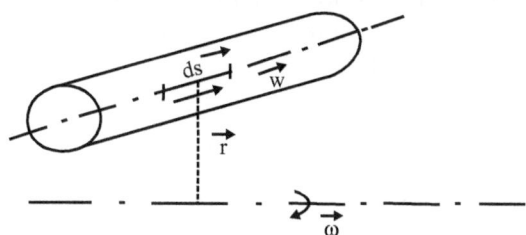

Fig. 4.3 To explain the flow in a rotating channel

The specific work due to the *centrifugal force* (= force × distance) along a line = acceleration × distance element ds to move in a radial direction by dr is

$$dW_{cent} = \mathbf{b}_{cent}\mathbf{dr} = \omega^2 r dr$$

By integration, the specific work (note: here we are designating specific work with a capital W instead of a lowercase w, as in an earlier chapter, in order to avoid confusion with the relative speed) due to the centrifugal force is

$$W_{cent} = \omega^2 \int_{r_1}^{r_2} r dr = \omega^2 \left(r_2^2 - r_1^2\right)/2 = \left(u_2^2 - u_1^2\right)/2 \qquad (4.4)$$

where u_1 and u_2 are *azimuthal speeds*.

On the other hand, the specific work of the *Coriolis force* along the line element ds is

$$dW_{coriolis} = \mathbf{b}_{coriolis} \bullet \mathbf{ds} = 2[(w \times \omega) \bullet \mathbf{ds}] = 2[(w \times \omega) \bullet w]dt = 0$$

We can consider the above results with an alternate formulation also. From (4.3a), we can write for the centrifugal force $\mathbf{F}_{cent}$ acting on a mass element $dm = \rho dA dr$, where dA is the elemental area $\mathbf{F}_{cent} = \omega^2 r dm = \rho \omega^2 r dA dr$.

This centrifugal force gives rise to a pressure differential dp on dA normal to dr. Thus, under steady and equilibrium conditions,

$$\mathbf{F}_{cent} = dp \bullet dA = \rho \omega^2 r dA dr$$

and, therefore, $dp = \rho \omega^2 r dr$.

Now, *specific work* from thermodynamics is

$$|W_{cent}| = \omega^2 \int_1^2 r dr = \frac{1}{2}\omega^2 \left(r_2^2 - r_1^2\right) = \frac{1}{2}\left(u_2^2 - u_1^2\right)$$

which is again (4.4).

Therefore, in a rotating channel, the stagnation enthalpy based on static enthalpy and a relative velocity must remain constant, except by taking into account the above specific work due to centrifugal force. Thus, the energy equation for a rotating system is

$$h_2 - h_1 = \frac{1}{2}\left[\left(w_1^2 - w_2^2\right) + \left(u_2^2 - u_1^2\right)\right] \qquad (4.2b)$$

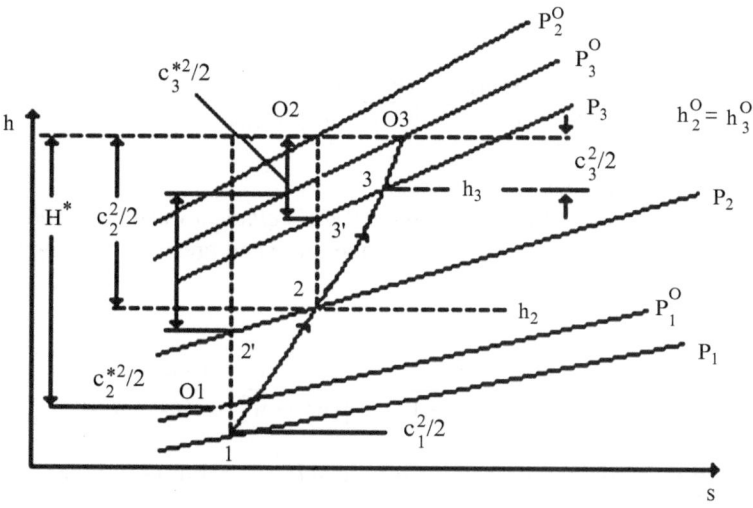

Fig. 4.4 Flow process in a single-stage compressor shown in an (h, s) chart

Once again, here we are not talking about a friction effect due to which the trailing-edge velocity (w_2 for a compressor and w_1 for a turbine) has to be different from that due to an isentropic change of state.

We first consider the situation for a compressor in a temperature-entropy diagram, in which w_2 with associated absolute velocities c_2 and c_3 are the respective trailing-edge velocities in the rotor and the stator, whereas w_2^* with associated absolute velocities c_2^* and c_3^* are the respective trailing-edge velocities under an isentropic change-of-state condition (Fig. 4.4).

Now for a compressor, from (4.2b), we have

$$h_2 + \frac{1}{2}w_2^2 - h_1 = \frac{1}{2}\left[(w_1^2) + (u_2^2 - u_1^2)\right]$$

Generally, $h_2 > h_1$ (even if $u_1^2 \approx u_2^2$ for an axial compressor) and $w_1 > w_2$. Thus, the enthalpy change in the compressor rotor is

$$\Delta h_{\text{rot,c}} = h_2 - h_1 = \frac{1}{2}\left[(w_1^2 - w_2^2) + (u_2^2 - u_1^2)\right] \tag{4.5a}$$

In case the change of state is isentropic, then

$$\Delta h_{\text{rot,c}}^* = h_2' - h_1 = \frac{1}{2}\left[(w_1^2 - w_2^{*2}) + (u_2^2 - u_1^2)\right] \tag{4.5b}$$

Hence, we can define a *rotor efficiency of the compressor*:

$$\eta_{\text{rot,c}} = \frac{\Delta h^*_{\text{rot,c}}}{\Delta h_{\text{rot,c}}} = \frac{\left[\left(w_1^2 - w_2^{*2}\right) + \left(u_2^2 - u_1^2\right)\right]}{\left[\left(w_1^2 - w_2^2\right) + \left(u_2^2 - u_1^2\right)\right]} \tag{4.6a}$$

Since $\eta_{\text{rot,c}} < 1$, obviously $w_2^2 < w_2^{*2}$.

The above definition of the *compressor rotor's efficiency* can be linked easily to the thermodynamic state variables like temperature and pressure by using the well-known definition

$$\eta_{\text{rot,c}} = \frac{T_2' - T_1}{T_2 - T_1} = \frac{\left(\frac{p_2}{p_1}\right)^{\frac{\gamma-1}{\gamma}}}{\frac{T_2}{T_1} - 1} = \frac{\left[\left(w_1^2 - w_2^{*2}\right) + \left(u_2^2 - u_1^2\right)\right]}{\left[\left(w_1^2 - w_2^2\right) + \left(u_2^2 - u_1^2\right)\right]} \tag{4.6b}$$

One could also define a *kinetic energy efficiency*. From (4.5a) and (4.5b), we write

$$\frac{1}{2} w_2^2 = \frac{1}{2}\left[w_1^2 + \left(u_2^2 - u_1^2\right)\right] - \Delta h_{\text{rot,c}} \tag{4.7a}$$

and

$$\frac{1}{2} w_2^{*2} = \frac{1}{2}\left[w_1^2 + \left(u_2^2 - u_1^2\right)\right] - \Delta h^*_{\text{rot,c}} \tag{4.7b}$$

Since $\Delta h^*_{\text{rot,c}} < \Delta h_{\text{rot,c}}$, obviously, once again, $w_2^2 < w_2^{*2}$.

Now we'll define the *kinetic energy efficiency* for a compressor rotor as

$$\begin{aligned}
\eta^k_{\text{rot,c}} &= \frac{w_2^2}{w_1^{*2}} = \frac{w_1^2 + \left(u_2^2 - u_1^2\right) - 2\Delta h_{\text{rot,c}}}{w_1^2 + \left(u_2^2 - u_1^2\right) - 2\Delta h^*_{\text{rot,c}}} \\
&= \frac{\left[w_1^2 + \left(u_2^2 - u_1^2\right)\right]/\left(2\Delta h_{\text{rot,c}}\right) - 1/\eta_{\text{rot,c}}}{\left[w_1^2 + \left(u_2^2 - u_1^2\right)\right]/\left(2\Delta h^*_{\text{rot,c}}\right) - 1}
\end{aligned} \tag{4.7c}$$

which can be inverted to become

$$\eta_{\text{rot,c}} = \frac{1}{\left[\left(w_1^2 + u_2^2 - u_1^2\right)/\left(2\Delta h^*_{\text{rot,c}}\right)\right] - \left(1 - \eta^k_{\text{rot,c}}\right)} + \eta^k_{\text{rot,c}} \tag{4.7d}$$

One can, of course, obtain the term in brackets above from the gas state. Since

$$\frac{1}{2}\left[\left(w_1^2 - w_2^{*2}\right) + \left(u_2^2 - u_1^2\right)\right] = c_{\text{p}}\left(T_\gamma - T_1\right)\Delta h^* = c_{\text{p}} T_1 \left[\left(\frac{p_2}{p_1}\right)^{\frac{\gamma-1}{\gamma}} - 1\right]$$

we therefore write

$$\left(w_1^2 - u_2^2 - u_1^2\right) = w_2^{*2} 2\Delta h^* = 2c_p T_1 \left[\left(\frac{p_2}{p_1}\right)^{\frac{\gamma-1}{\gamma}} - 1\right]$$

and thus, further,

$$\frac{\left(w_1^2 + u_2^2 - u_1^2\right)}{2\Delta h_{\text{rot,c}}^*} = 1 + \frac{w_2^{*2}}{2\Delta h_{\text{rot,c}}^*} \tag{4.7e}$$

The reason for introducing both the *compressor rotor efficiency* and the *compressor kinetic energy efficiency* is that while the former gives the change of state across the rotor, the latter is obtained by considering losses in the rotor affecting the gas velocity due to friction.

We now continue the analysis for the *compressor stator*. From (4.2), we write for the compressor stator and for the frictionless case

$$h_3 - h_2 = \Delta h_{\text{stat,c}} = \frac{1}{2}\left(c_2^2 - c_3^2\right) \tag{4.8a}$$

Further, for an isentropic change of state in a stator,

$$h_3^* - h_2 = \Delta h_{\text{stat,c}}^* = \frac{1}{2}\left(c_2^2 - c_3^{*2}\right) \tag{4.8b}$$

Therefore, the *compressor stator's efficiency* (including the relation with state variables) is

$$\eta_{\text{stat,c}} = \frac{\Delta h_{\text{stat,c}}^*}{\Delta h_{\text{stat,c}}} = \frac{\left(c_2^2 - c_3^{*2}\right)}{\left(c_2^2 - c_3^2\right)} = \frac{T_{3'} - T_2}{T_3 - T_2} = \frac{(p_3/p_2)^{\frac{\gamma-1}{\gamma}} - 1}{(T_3/T_2) - 1} \tag{4.9a}$$

Since $c_3^2 < c_3^{*2}$, we conclude that $\eta_{\text{stat,c}} < 1$ and $\Delta h_{\text{stat,c}}^* < \Delta h_{\text{stat,c}}$. Now, from the relations

$$c_3^2 = c_2^2 - 2\Delta h_{\text{stat,c}} \quad and \quad c_3^{*2} = c_2^2 - 2\Delta h_{\text{stat,c}}^*$$

we note first that

$$\frac{c_2^2}{2\Delta h_{\text{stat,c}}^*} = 1 + \frac{c_3^{*2}}{2\Delta h_{\text{stat,c}}^*} > 1$$

Further, by defining a *compressor stator's kinetic energy efficiency* as

$$
\eta^k_{\text{stat,c}} = \frac{c_3^2}{c_3^{*2}} = \frac{c_2^2 - 2\Delta h_{\text{stat,c}}}{c_2^2 - 2\Delta h^*_{\text{stat,c}}} = \frac{c_2^2/(2\Delta h_{\text{stat,c}}) - (1/\eta_{\text{stat,c}})}{c_2^2/\left(2\Delta h^*_{\text{stat,c}}\right) - 1} \tag{4.9b}
$$

from which, by inversion, we get

$$
\eta_{\text{stat,c}} = \frac{1}{\frac{c_2^2}{\left[(2\Delta h^*_{\text{stat,c}})\right]}} \left(1 - \eta^k_{\text{stat,c}}\right) + \eta^k_{\text{stat,c}} \tag{4.9c}
$$

As before for the rotor, we also write for the stator

$$
c_2^2 - c_3^{*2} = 2\Delta h^*_{\text{stat,c}} = 2c_p\left[(p_3/p_2)^{\frac{\gamma-1}{\gamma}} - 1\right]
$$

from which we write further

$$
\frac{c_2^2}{\left(2\Delta h^*_{\text{stat,c}}\right)} = 1 + \frac{c_3^{*2}}{2\Delta h^*_{\text{stat,c}}} \tag{4.9d}
$$

For the turbine, we now repeat the analysis from the stator side in the same way we did for the compressor. From (4.2a), we have

$$
h_3^o = h_3 + \frac{1}{2}c_3^2 = h_2 + \frac{1}{2}c_2^2 = h_2^o
$$

and thus,

$$
\Delta h_{\text{stat,t}} = h_3 - h_2 = \frac{1}{2}\left(c_2^2 - c_3^2\right) \tag{4.10a}
$$

For the friction case, on the other hand, it is

$$
\Delta h^*_{\text{stat,t}} = h_3 - h_{2'} = \frac{1}{2}\left(c_2^{*2} - c_3^2\right) \tag{4.10b}
$$

In the above, $\Delta h^*_{\text{stat,t}} > \Delta h_{\text{stat,t}}$ and hence $c_2^{*2} > c_2^2$.
We now define again the two efficiencies for a turbine stator as follows:

Turbine stator's efficiency:

$$
\eta_{\text{stat,t}} = \frac{\Delta h_{\text{stat,t}}}{\Delta h^*_{\text{stat,t}}} = \frac{c_2^2 - c_3^2}{c_2^{*2} - c_3^2} = \frac{T_3 - T_2}{T_3 - T_{2'}} = \frac{1 - (T_2/T_3)}{(T_{2'}/T_3)} = \frac{1 - (T_2/T_3)}{1 - (p_2/p_3)^{\frac{\gamma-1}{\gamma}}} \tag{4.11a}
$$

and

Turbine stator's kinetic energy efficiency:

$$\eta_{\text{stat,t}}^k = \frac{c_2^2}{c_2^{*2}} = \frac{2\Delta h_{\text{stat,t}} + c_3^2}{2\Delta h_{\text{stat,t}}^* + c_3^2} = \frac{\eta_{\text{stat,t}} + c_3^2/\left(2\Delta h_{\text{stat,t}}^*\right)}{1 + c_3^2/\left(2\Delta h_{\text{stat,t}}^*\right)} \tag{4.11b}$$

from which, by inversion, we get

$$\eta_{\text{stat,t}} = \eta_{\text{stat,t}}^k - \left(1 - \eta_{\text{stat,t}}^k\right)\frac{c_3^2}{2\Delta h_{\text{stat,t}}^*} \tag{4.11c}$$

Obviously, $\eta_{\text{stat,t}} < \eta_{\text{stat,t}}^k$. Further, since $0 < \eta_{\text{stat,t}} < 1$, we get

$$1 > \eta_{\text{stat,t}}^k > \frac{c_3^2/\left(2\Delta h_{\text{stat,t}}^*\right)}{1 + \left\{c_3^2/\left(2\Delta h_{\text{stat,t}}^*\right)\right\}}$$

For a turbine rotor, we write, analogously to (4.2),

$$\Delta h_{\text{rot,t}} = h_2 - h_1 = \frac{1}{2}\left[\left(w_1^2 - w_2^2\right) + \left(u_2^2 - u_1^2\right)\right] \tag{4.12a}$$

and from an *isentropic change of state,* we have

$$\Delta h_{\text{rot,t}}^* = h_2 - h_1^* = \frac{1}{2}\left[\left(w_1^{*2} - w_2^2\right) + \left(u_2^2 - u_1^2\right)\right] \tag{4.12b}$$

We define again the two efficiencies for turbine rotor as follows:

Turbine rotor's efficiency:

$$\eta_{\text{rot,t}} = \frac{\Delta h_{\text{rot,t}}}{\Delta h_{\text{rot,t}}^*} = \frac{\left(w_1^2 - w_2^2\right) + \left(u_2^2 - u_1^2\right)}{\left(w_1^{*2} - w_2^2\right) + \left(u_2^2 - u_1^2\right)} = \frac{T_2 - T_1}{T_2 - T_{1'}} = \frac{1 - (T_1/T_2)}{1 - (p_1/p_2)^{\frac{\gamma-1}{\gamma}}} \tag{4.13a}$$

Turbine rotor's kinetic energy efficiency:

$$\eta_{\text{rot,t}}^k = \frac{w_1^2}{w_1^{*2}} = \frac{\left(u_1^2 - u_2^2 + w_2^2\right) + 2\Delta h_{\text{rot,t}}}{\left(u_1^2 - u_2^2 + w_2^2\right) + 2\Delta h_{\text{rot,t}}^*}$$

$$= \frac{\eta_{\text{rot,t}} + \left(u_1^2 - u_2^2 + w_2^2\right)/\left(2\Delta h_{\text{rot,t}}^*\right)}{1 + \left(u_1^2 - u_2^2 + w_2^2\right)/\left(2\Delta h_{\text{rot,t}}^*\right)} \tag{4.13b}$$

from which, by inversion, we get

$$\eta_{\text{rot,t}} = \eta_{\text{rot,t}}^k - \left(1 - \eta_{\text{rot,t}}^k\right)\left(w_2^2 + u_1^2 - u_2^2\right)/\left(2\Delta h_{\text{rot,t}}^*\right) \qquad (4.13c)$$

Let's now calculate the combined work done in the stator and rotor in a stage. For both the compressor and the turbine, we write from (4.5a) and (4.8a), or from (4.10a) and (4.12a), the *combined enthalpy change in a stage*:

$$\Delta h_{\text{stage}} = \Delta h_{\text{rot}} + \Delta h_{\text{stat}} = \frac{1}{2}\left[\left(w_1^2 - w_2^2\right) + \left(u_2^2 - u_1^2\right) + \left(c_2^2 - c_3^2\right)\right] \qquad (4.14a)$$

On the other hand, by taking the complete stage globally, we get

$$h_3 + \frac{1}{2}c_3^2 = h_3^o = h_1 + \frac{1}{2}c_1^2 + H = h_1^o + H \qquad (4.14b)$$

where H is the *specific work in the stage*. Thus,

$$H = (h_3 - h_1) + \frac{1}{2}\left(c_3^2 - c_1^2\right) = \Delta h_{\text{stage}} + \frac{1}{2}\left(c_3^2 - c_1^2\right)$$

$$= \frac{1}{2}\left[\left(w_1^2 - w_2^2\right) + \left(u_2^2 - u_1^2\right) + \left(c_2^2 - c_1^2\right)\right] \qquad (4.14c)$$

which is also called the *principal equation of turbomachinery*. What is remarkable about this equation is that all blade losses are included in the respective kinetic energy, each of them is additive, and the stage-specific work depends only on the velocity of the rotor, and not the stator. In practice, however, the losses that take place in the compressor stator go into the performance of the next stage, and if it is already the last stage, there is static pressure recovery in the diffuser. Similarly, losses in the stator blades (nozzle block) for the turbine ultimately contribute to the irreversibility in the process (higher reject heat and higher static temperature of the exhaust gas).

Defining again the *stage efficiency* (bases on stagnation enthalpy), we write

For compressor:

$$\eta_{\text{s,c}}^o = \frac{H_{\text{s,c}}^*}{H_{\text{s,c}}} \qquad (4.15a)$$

and *for turbine*:

$$\eta_{\text{s,t}}^o = \frac{H_{\text{s,c}}^*}{H_{\text{s,t}}} \qquad (4.15b)$$

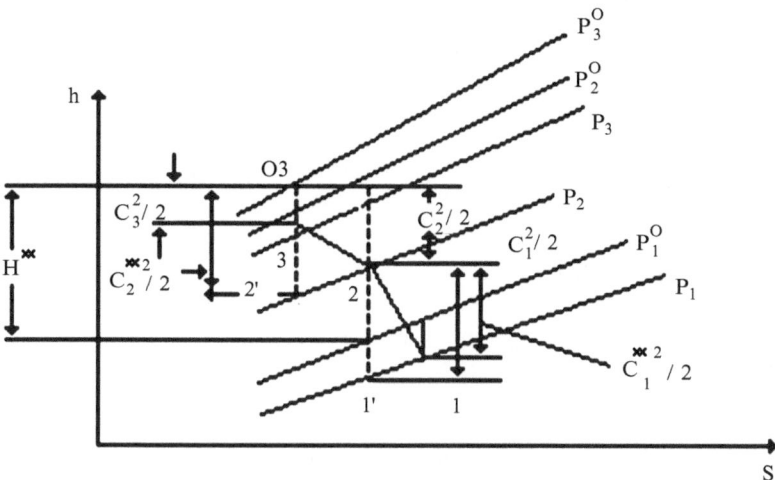

Fig. 4.5 Flow process in an (h, s) chart of a single-stage turbine

where the *isentropic specific work* for an ideal gas can be written from a change of state of variables in a stage as follows:

For compressor:

$$H_c^* = \frac{\gamma R T_1^o}{\gamma - 1} \left[\left(\frac{p_3^o}{p_1^o} \right)^{\frac{\gamma-1}{\gamma}} - 1 \right] \tag{4.16a}$$

and *for turbine*:

$$H_t^* = \frac{\gamma R T_1^o}{\gamma - 1} \left[1 - \left(\frac{p_1^o}{p_3^o} \right)^{\frac{\gamma-1}{\gamma}} \right] \tag{4.16b}$$

Note that the stage efficiency, defined in (4.15a) or (4.15b), is the same as the adiabatic efficiency based on stagnation values as given in (3.9a) as applied to a single-stage compressor, or as given in (3.12a) as applied to a single-stage turbine (Fig. 4.5).

In (4.14c), we have an expression for the specific work in a stage, which is dependent on the three velocities of the velocity triangle at the two ends of the rotor. We can considerably simplify the equation if we take into account the various relations between the velocities in a velocity triangle (Fig. 4.6). In Fig. 4.6, w_u and c_u are components of w and c toward u, and c_m is the meridian velocity perpendicular to the azimuthal velocity u. In addition, β is the rotor blade angle.

Thus, the following relations are obvious:

$$c_m^2 = w^2 - w_u^2 = w^2 - (u - c_u)^2 = w^2 - u^2 - c^2 + 2uc_u$$

Fig. 4.6 Explaining relations
in a velocity triangle

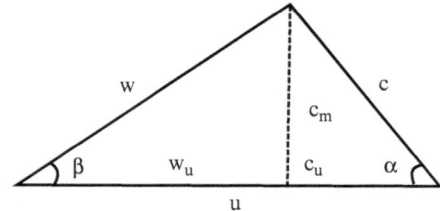

from which it follows that

$$c_m^2 + c^2 = w^2 - u^2 + 2uc_u$$

and

$$c^2 - w^2 + u^2 = 2uc_u$$

Writing the above equation for the two ends of the rotor and substituting it in (4.14c), we get

$$H = \frac{1}{2}\left[(w_1^2 - w_2^2) + (u_2^2 - u_1^2) + (c_2^2 - c_1^2)\right] = \frac{1}{2}(2u_2c_{u2} - 2u_1c_{u1})$$
$$= (u_2c_{u2} - u_1c_{u1}) \tag{4.17a}$$

Since $u = r\omega$, where r is the radius and ω is the *azimuthal radian speed*, (4.17a) can also be written as

$$H = \omega(r_2c_{u2} - r_1c_{u1}) \tag{4.17b}$$

Now, for a *single-stage compressor*,

$$H_s = \frac{H_s^*}{\eta_s^o} = c_p(T_3^o - T_1^o) = \frac{\gamma}{\gamma - 1}\frac{RT_1^o}{\eta_s^o}\left[(\pi_{cs}^o)^{\frac{\gamma-1}{\gamma}} - 1\right] = \omega(r_2c_{u2} - r_1c_{u1})$$

where

$$\pi_{cs}^o = \frac{p_3^o}{p_1^o} = \text{stagnation pressure ratio across a stage}$$

Therefore, for the *single-stage compressor*,

$$\frac{T_3^o}{T_1^o} = \frac{1}{\eta_{cs}^o}\left[(\pi_s^o)^{\frac{\gamma-1}{\gamma}} - 1\right] = 1 + \frac{\omega}{c_pT_1^o}(r_2c_{u2} - r_1c_{u1})$$
$$= 1 + \frac{\omega r_1/(\gamma - 1)}{\gamma RT_1}\frac{T_1}{T_1^o}(r_2c_{u2} - r_1c_{u1})$$

Noting for an ideal gas that

$$\frac{T_1^o}{T_1} = 1 + \frac{\gamma - 1}{2} M_1^2$$

and defining an *azimuthal Mach number*

$$M_{u1} = \frac{\omega r_1}{\sqrt{\gamma R T_1}} = \frac{u_1}{\sqrt{\gamma R T_1}}$$

we can write the previous equation as

$$\frac{T_3^o}{T_1^o} = 1 + \frac{1}{\eta_s^o}\left[\{\pi_{cs}^o\}^{\frac{\gamma-1}{\gamma}} - 1\right] = 1 + \frac{(\gamma - 1)M_{u1}^2}{\omega r_1\left[1 + \frac{\gamma-1}{2}M_1^2\right]}\left[\left(\frac{r_2}{r_1}\right)c_{u2} - c_{u1}\right] \quad (4.18a)$$

Thus, the stagnation temperature ratio can be increased by increasing either $M_{u1}^2/(\omega r_1) \approx r_1/(\gamma R T_1)$ or $(r_2/r_1)c_{u2} - c_{u1}$. By setting $c_{u1} < 0$, one can, of course, reduce the requirement for a higher azimuthal Mach number M_{u1} for the same stagnation temperature ratio. We can inquire into the problem in another manner. Let's divide H by the *characteristic azimuthal kinetic energy* $u_2^2/2$ and call this a *nondimensional specific work number* Ψ. Therefore,

$$H = \frac{\Psi u_2^2}{2} = \frac{\Psi u_1^2}{2}\left(\frac{r_2}{r_1}\right)^2 = \frac{\gamma}{\gamma - 1}\frac{R T_1^o}{\eta_s^o}\left(\{\pi_{cs}^o\}^{\frac{\gamma-1}{\gamma}} - 1\right)$$

Multiplying the above equation by

$$\frac{T_1}{T_1^o}\left[1 + \frac{\gamma - 1}{2}M_1^2\right] = 1$$

we get

$$\frac{\Psi}{2}\left(\frac{r_2}{r_1}\right)^2 = \frac{\gamma}{\gamma - 1}\frac{R T_1^o}{\eta_s^o u_1^2}\frac{T_1}{T_1^o}\left[1 + \frac{\gamma - 1}{2}M_1^2\right]\left(\{\pi_{cs}^o\}^{\frac{\gamma-1}{\gamma}} - 1\right)$$

Therefore,

$$\Psi = \frac{2}{\gamma - 1}\left(\frac{r_1}{r_2}\right)^2\frac{1}{\eta_s^o M_{s1}^2}\left[1 + \frac{\gamma - 1}{2}M_1^2\right]\left(\{\pi_{cs}^o\}^{\frac{\gamma-1}{\gamma}} - 1\right) \quad (4.18b)$$

Let's examine (4.18b) for an axial compressor with $r_1 \approx r_2$. Assuming typical values for a compressor with titanium alloy blades, $u = u_2 \approx u_1 = 550$ m/s, $M_1 \approx 0$ .7, $M_{u1} \approx 1.7$, $\pi_{st}^o \approx 1.6$, and $\eta_s^o = 0.86$, we get for a single-stage axial compressor

$$\Psi = 5\left[1 + 0.2 \times 0.7^2\right] \frac{\left(1.6^{0.286} - 1\right)}{0.86 \times 1.7^2} = 0.3176$$

As will be shown later in Table 6.1, the optimum work values for an axial compressor are 0.26–0.32. Further, it is interesting to note that Ψ is inversely proportional to the square of the azimuthal Mach number!

4.2 Reaction and Shape of Blades

We have so far discussed the state across a stage without really specifying the state between the rotor and the stator, especially p_2. In order to define p_2, we introduce the *concept of degree of reaction*, or simply *the reaction*, as defined in the following:

$$\hat{r} = \frac{\Delta h_{\text{rot}}}{\Delta h_{\text{rot}} + \Delta h_{\text{stat}}} = \frac{h_2 - h_1}{h_3 - h_1} \approx \frac{T_2 - T_1}{T_3 - T_1} = \frac{(T_2 - T_1) - 1}{(T_3 - T_1) - 1} \tag{4.19a}$$

Assuming that the change of state is polytropic with exponent n, then

$$\hat{r} = \frac{\left(\frac{p_1 + \Delta p_{\text{rotor}}}{p_1}\right)^{\frac{n-1}{n}} - 1}{\left(\frac{p_1 + \Delta p_{\text{rotor}} + \Delta p_{\text{stat}}}{p_1}\right)^{\frac{n-1}{n}} - 1} = \frac{\left(\frac{p_1 + \frac{n-1}{n}\Delta p_{\text{rot}} + \cdots}{p_1} - 1\right)}{\left(\frac{p_1 + \frac{n-1}{n}(\Delta p_{\text{rot}} + \Delta p_{\text{stat}}) + \cdots}{p_1} - 1\right)} \approx \frac{\Delta p_{\text{rot}}}{\Delta p_{\text{rot}} + \Delta p_{\text{stat}}} \tag{4.19b}$$

Taking the definition of Δh_{rot} and Δh_{stat} into account, which are

$$\Delta h_{\text{rot}} = \frac{1}{2}\left[\left(w_1^2 - w_2^2\right) + \left(u_2^2 - u_1^2\right)\right]$$

and

$$\Delta h_{\text{stat}} = \frac{1}{2}\left(c_2^2 - c_3^2\right)$$

gives the difference in the kinetic energies. It becomes particularly simpler if we set $c_3 = c_1$, the latter strictly for the following stage, and take into account (4.17a):

$$\hat{r} = \frac{\left(w_1^2 - w_2^2\right) + \left(u_2^2 - u_1^2\right)}{2H + \left(c_1^2 - c_3^2\right)} \approx \frac{\left(w_1^2 - w_2^2\right) + \left(u_2^2 - u_1^2\right)}{2H} \tag{4.19c}$$

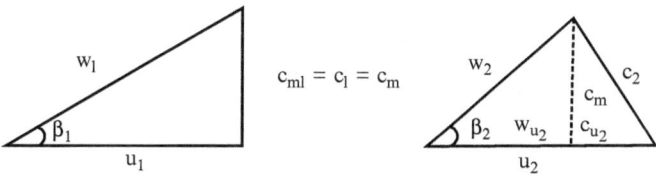

Fig. 4.7 Velocity triangles for $c_{u1} = 0$ and $c_{m1} = c_{m2}$

Now since $H = \Delta h_{rot} + \Delta h_{stat}$ (for the special case of $c_3 = c_1$), we can write

$$\Delta h_{rot} = h_2 - h_1 = \hat{r}H \qquad (4.20a)$$

and

$$\Delta h_{stat} = h_3 - h_2 = (1 - \hat{r})H \qquad (4.20b)$$

It is obvious that the reaction $\hat{r}$ is relevant if it is between 0 and 1. For $\hat{r} = 0$ and thus $\Delta h_{rot} = 0$, there is no change in the static state values in the rotor, whereas there is no change in the static state values in the stator if $\hat{r}$ and thus $\Delta h_{stat} = 0$.

Let's now examine the two special cases.

(a) Firstly, we set $c_{u1} = 0$ (no azimuthal component for entry into the compressor or no exit azimuthal component for the turbine) and $c_{m1} = c_{m2}$ (meridian velocity does not change). The relevant velocity triangle is explained in Fig. 4.7.

Now, from simple geometrical considerations, we write

$$
\begin{aligned}
c_1^2 = c_m^2 &= w_1^2 - u_1^2, w_2^2 - c_m^2 \\
&= (u_2 - c_{u2}), (w_1^2 - w_2^2) + (u_2^2 - u_1^2) - (w^2 - u_2^2) \\
&= c_m^2 - [c_m^2 - 2u_2c_{u2} + c_{u2}^2] \\
&= 2u_2c_{u2} - c_{u2}^2 = c_{u2}(2u_2 - c_{u2})
\end{aligned}
$$

and

$$\tan \beta_2 = \frac{c_{m2}}{2u_2} = \frac{c_{m2}}{u_2 - c_{u2}}$$

Now from (4.17a) for $c_{u1} = 0$, we get $H = u_2c_{u2}$ and from (4.19c) we have

$$
\begin{aligned}
\hat{r} &= \frac{c_{u2}(2u_2 - c_{u2})}{2u_2c_{u2}} = 1 - \frac{c_{u2}}{2u_2} = 1 - \frac{u_2 - w_{u2}}{2u_2} \\
&= \frac{1}{2}\left(1 + \frac{w_{u2}}{u_2}\right) = \frac{1}{2}\left(1 + \frac{c_{m2}}{u_2 \tan \beta_2}\right) \qquad (4.19d)
\end{aligned}
$$

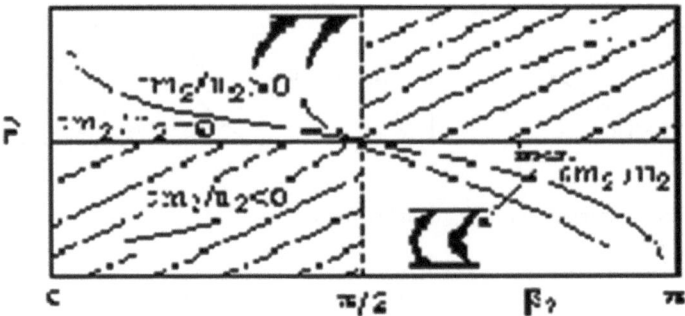

Fig. 4.8 Reaction for $c_{u1} = 0$ and various blade angles β_2

This implies that if

$$\hat{r} = 1 : \frac{c_{m2}}{u_2} = \tan \beta_2$$

and, on the other hand, if

$$\hat{r} = 0 : \frac{c_{m2}}{u_2} = -\tan \beta_2$$

Therefore, in addition to the condition $c_{u1} = 0$, one can have different values of reaction as follows, depending on various situations:

1. $\beta_2 = 90°$ (normal rotor blade exit angle): $\hat{r} = 1/2$;
2. $w_{u2} = u_2$, that is, $c_{u2} = 0$ and $H = 0$, but $\hat{r} = 1$;
3. $\beta_2 < 90°$, that is, $0 < w_{u2}/u_2 < 1 : 0 \le \hat{r} \le 1$; and
4. $\beta_2 > 90°$, that is, $w_{u2}/u_2 < 0 : 0 \le \hat{r} \le 1/2$.

The above results are shown in Fig. 4.8, in which the shaded area for $c_{m2}/u_2 < 0$ is not realistic. Note that for $0 \le \hat{r} \le 1/2$., we must have $\beta_2 = \pi/2$ to π, to π, and we would have forward-moving rotor blades for the compressor. On the other hand, for $1/2 \le \hat{r} \le 1$, we have $\beta_2 = 0$ to π. These are also shown in Fig. 4.8. Further, as we will discuss in a later chapter, the centrifugal compressor blades, because of structural considerations, in general have $\beta_2 = 90°$; this, however, does not mean that $\hat{r} = 1/2$, since $c_{u1} \ne 0$

(b) Secondly, we consider *axial machines only*, where we write approximately $u_1 \approx u_2 = u$. Further, we'll also let $c_{m1} \approx c_{m2} = c_m$, but $c_u \ne 0$. Hence, the reaction is

$$\hat{r}_{axial} = \frac{\left(2_1^2 - 2_2^2\right)}{\left(w_1^2 - 2_2^2\right) + \left(c_2^2 - c_1^2\right)}$$

We define now an azimuthal velocity component $w_{u\infty}, c_{u\infty}$ as the average of the azimuthal components of these velocity components with the help of the relation

$$w_{u\infty} = \frac{1}{2}(w_{u1} + w_{u2}) = \frac{1}{2}(2u - c_{u1} - c_{u2}) = u - \frac{1}{2}(c_{u1} + c_{u2}) = u - c_{u\infty}$$

$$(4.21a)$$

where

$$c_{u\infty} = \frac{1}{2}(c_{u1} + c_{u2}) = u - w_{u\infty} \qquad (4.21b)$$

Further, for *axial machines*,

$$u = w_{u1} + c_{u1} = w_{u2} + c_{u2}$$

from which we write

$$w_{u1} - w_{u2} = c_{u2} - c_{u1}$$

Further,

$$w^2 - c^2 = (c_m^2 + w_u^2) - (c_m^2 + c_u^2) = w_u^2 - c_u^2$$

Thus,

$$w_1^2 - w_2^2 = (c_m^2 + w_{u1}^2) - (c_m^2 + w_{u2}^2) \approx w_{u1}^2 - w_{u2}^2$$

and

$$(w_1^2 - w_2^2) + (c_2^2 - c_1^2) = (w_{u1}^2 - c_{u1}^2) - (w_{u2}^2 - c_{u2}^2) = (c_{u2}^2 - c_{u1}^2) - (w_{u2}^2 - w_{u1}^2)$$
$$= (c_{u2} + c_{u1})(c_{u2} - c_{u1}) - (w_{u2} + w_{u1})(w_{u2} - w_{u1})$$
$$= (w_{u1} - w_{u2})(c_{u2} + c_{u1} + w_{u1} + w_{u2}) = 2u(w_{u1} - w_{u2})$$

Therefore,

$$\hat{r}_{axial} = \frac{(w_1^2 - w_2^2)}{(w_1^2 - w_2^2) + (c_2^2 - c_1^2)} = \frac{(w_{u2} + w_{u1})(w_{u2} - w_{u1})}{2u(w_{u1} - w_{u2})} = \frac{w_{u\infty}}{u} \qquad (4.21c)$$

Let's now examine what happens for different values of $\hat{r}$, first for *axial machines only.*

$$H_{axial} \approx u(c_{u2} - c_{u1}) \qquad (4.21d)$$

From (4.21a) and (4.21b), we get

$$c_{u1} = u(1 - |\hat{r}_{axial}|) - \frac{H_{axial}}{2u} \tag{4.22a}$$

and

$$c_{u2} = u(1 - \hat{r}_{axial}) + \frac{H_{axial}}{2u} \tag{4.22b}$$

Since $w_u = u - c_u$, we additionally find

$$w_{u1} = u - c_{u1} = u\hat{r}_{axial} + \frac{H_{axial}}{2u} \tag{4.22c}$$

and

$$w_{u2} = u - c_{u2} = u\hat{r}_{axial} - \frac{H_{axial}}{2u} \tag{4.22d}$$

We now consider different values of reaction, $\hat{r}_{axial}$ for *axial machines* only:
(1) $\hat{r}_{axial} = 0$; (2) $\hat{r}_{axial} = 0.5$; and (3) $\hat{r}_{axial} = 1$.

1. $\hat{r}_{axial} = 0$:
   From (4.21a), $w_{\infty u} = w_{u1} + w_{u2} = 0$, which means $w_{u1} = -w_{u2}$.
   Therefore, the rotor blade angle

$$\tan \beta_1 = \frac{c_m}{2_{u1}} = -\frac{c_m}{w_{u2}} = -\tan \beta_2$$

which means that $\beta_1 = -\beta_2$. However,

$$\tan \beta_1 = -\tan \beta_2 = \frac{c_m}{w_{u1}} = \frac{2uc_m}{H_{axial}} = \frac{4c_m}{u} \frac{u^2}{2H_{axial}}$$

As we defined earlier, $\beta_1 = -\beta_2 2H_{axial}/u^2 = \varphi_{axial}$ as a specific work number for axial machines, which have *optimum values* for the compressor $\varphi_{axial,comp} \approx$ 0.25–0.9, depending on the reaction, and for the turbine $\varphi_{axial,comp} \approx$ 1.1–6, again depending on the reaction. Therefore, in the above relation,

$$\tan \beta_1 = \frac{4c_m}{u\varphi_{axial}}$$

it is shown that the value of $(c_m/u, \beta_1)$ for compressors is much larger than β for turbines. Typical blades for an axial machine at $(\hat{r} = 0)$ are shown in Fig. 4.9, with the corresponding velocity triangle.

But since $\hat{r} = 0$ means that there is no change in the static pressure in the rotor, it is hardly possible to maintain flow in an axial compressor. However,

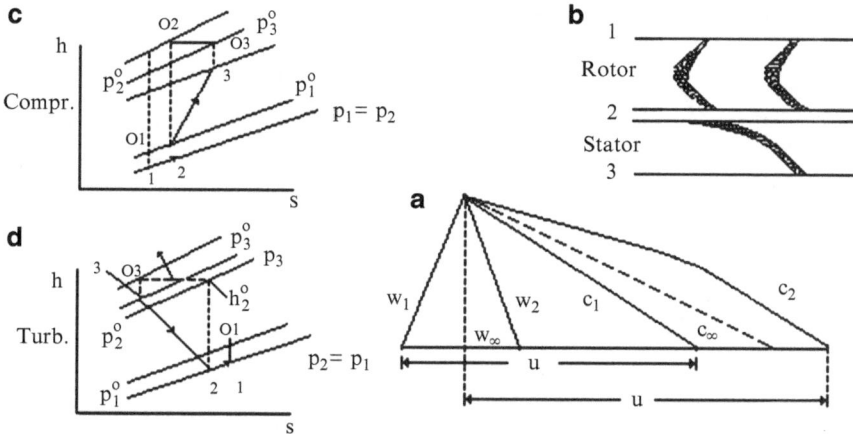

Fig. 4.9 For axial machines with reaction $= 0$: (**a**) velocity triangle; (**b**) typical blades; (**c**) process for compressor; and (**d**) for turbine in an (h, s) chart

since for axial turbines, $\hat{r} = 0$ or a small value means that work is extracted mainly due to the transfer of momentum from the gas to gas to the blades, such machines are called *impulse turbines* in contrast to reaction turbines with $\hat{r}$ larger than zero, or at least larger than 0.25.

2. $(\hat{r}_{\text{axial}}) = 0.5$:

   From (4.22a) to (4.22d), we can show that $w_{u1} = c_{u2}$ and $w_{u2} = c_{u1}$. We therefore have a complete reversal of the role between a rotor and a stator in a multistage machine. Further, $\hat{r}_{\text{axial}} = 0.5$ means $\Delta h_{\text{rot}} = \Delta h_{\text{stat}}$ and $\Delta h_{\text{rot}} \approx \Delta h_{\text{stat}}$. The velocity triangle, the typical blades, and the process for a compressor and a turbine in schematic enthalpy–entropy charts are now shown in this case in Fig. 4.10.

3. $\hat{r}_{\text{axial}} = 1.0$:

   This means $c_{u\infty} = 0$ and $c_{u1} = -c_{u2}$. It means also that $\Delta p_{\text{stat}} \approx 0$. The velocity triangle, typical blades, and typical thermodynamic process for a compressor and a turbine in schematic enthalpy–entropy charts are shown in Fig. 4.11.

Finally, note that the *reaction* or *degree of reaction*, as it is often called, is an important design factor. For aircraft axial compressors, $\hat{r}_{\text{axial}} = 0.5$–0.7, and it may vary from root to tip. For axial turbines for aircraft, $\hat{r}_{\text{axial}} = 0.5$.

Problem: For a single-stage axial compressor for air, let $\pi_c^o = 1.6, \hat{r} = 0.5, \eta_c^o = 0$
.84, $T_{\text{inlet}}^o = 300$ K, $c_{m1} = c_{m2} = 100$ m/s, and $u = u_1 = u_2 = 500$ m/s. Compute the blade angles and all static pressures. (Assume $c_3 = c_1$ and $p_i^o = 1.013$ bar).

Solution:

$$\frac{T_{f'}^o}{T_i^o} = \left(\frac{p_f^o}{p_i^o}\right)^{\frac{\gamma-1}{\gamma}} = \{\pi_c^o\}^{\frac{\gamma-1}{\gamma}} = 1.6^{0.286} = 1.143874$$

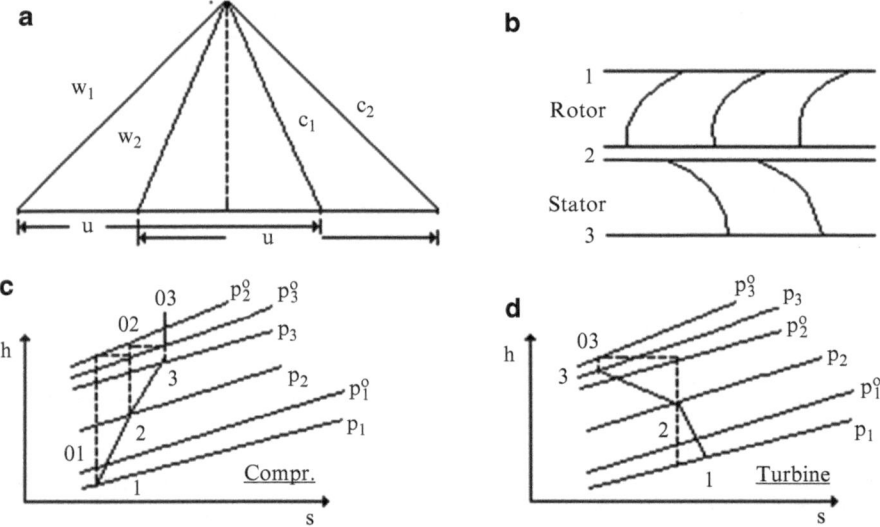

Fig. 4.10 Axial machine blades with 50% reaction: (**a**) velocity triangle; (**b**) typical blades; (**c**) process for a compressor; and (**d**) for a turbine in (h, s) chart

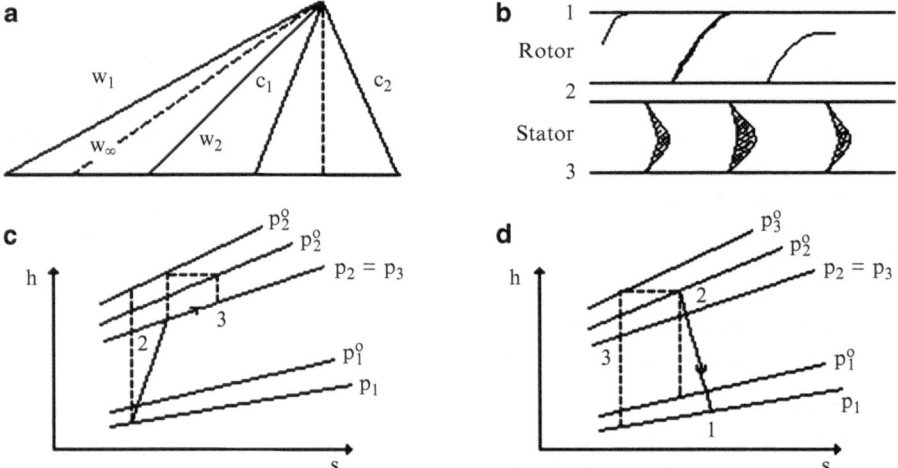

Fig. 4.11 Axial machine blades with 100% reaction: (**a**) velocity triangle; (**b**) typical blades; (**c**) process for a compressor; and (**d**) for a turbine in (h, s) chart

Thus, $T_f^o = 300 \times 1.143874 = 343.16$ K .
Now

$$\eta_c^o = \frac{T_f^{o'} - T_i^o}{T_f^o - T_i^o} = \frac{\left(T_f^{o'}/T_i^o\right) - 1}{\left(T_f^o/T_i^o\right) - 1} = 0.84$$

Thus,

$$\frac{T_{f'}^o}{T_i^o} = 1 + \frac{0.143874}{0.84} = 1.17128 \rightarrow T_f^o = 300 \times 1.17128 = 351.384 \text{ K}$$

$$H = c_p\left(T_f^o - T_i^o\right) = 1{,}005 \ \frac{\text{kJ}}{\text{kgK}} (351.384 - 300.) = 51{,}640.61 \ \frac{\text{kJ}}{\text{kg}}$$

$$c_{u1} = w_{u2} = \frac{1}{2}\left(u - \frac{H}{u}\right) = 198.36 \text{ m/s}$$

$$c_{u2} = w_{u1} = \frac{1}{2}\left(u - \frac{H}{u}\right) = 301.64 \text{ m/s}$$

$$\tan \beta_1 = \tan \alpha_2 = \frac{c_m}{w_{u1}} = \frac{100}{301.64} 0.33152 \rightarrow \beta_1 = 18.34°$$

$$\tan \beta_2 = \tan \alpha_3 = \frac{c_m}{w_{u2}} = \frac{100}{198.36} 0.504 \rightarrow \beta_2 = 26.75°$$

$$c_3 = c_1 = \sqrt{c_{u1}^2 + c_m^2} = \sqrt{198.36^2 + 100^2} = 222.14 \text{ m/s} \rightarrow T_1$$

$$= 300. - \frac{222.14^2}{2{,}010} = 275.45 \text{ K}$$

Now,

$$p_1 = 1.013 \left(\frac{275.45}{300}\right)^{3.5} = 0.751 \text{ bar} \rightarrow p_f^o = p_2^o \approx p_3^o = 1.6 \times 1.013 = 1.620 \text{ bar}$$

$$c_2 = \sqrt{301.64^2 + 100^2} = 317.78 \text{ m/s} \rightarrow T_2 = 51.384 - \frac{317.78^2}{2{,}010} = 301.14 \text{ K}$$

$$p_2 = 1.6208 \left(\frac{301.54}{351.384}\right)^{3.5} = 0.944 \text{ bar}$$

$$T_3 = 351.384 - \frac{222.142^2}{2010} = 326.83K \quad and$$

$$p_3 = 1.6208 \left(\frac{326.83}{351.384}\right)^{3.2} = 1.258 \text{ bar}$$

4.3 Exercises

1. Consider the flow process in a single-stage axial compressor. Let air at pressure p_1, temperature T_1, and relative velocity $w_1 = 130$ m/s enter the rotor with entry angle $\beta_1 = 30°$. Let azimuthal speed $u = u_1 = u_2 = 250$ m/s. Also, let $c_{1m} = c_{2m}$. In addition, the optimum value of H for an axial compressor (explained in a later chapter) is $u^2/4$. Hence, obtain the values of c_1, $c_2, w_1, w_2, \alpha_3 = \alpha_1$ and the degree of reaction, and draw the blade schematically. Compute Δh_{rotor}, Δh_{stator}. Let the efficiency $\eta = H^*/H = 0.86$ and, similarly, $\Delta h_{\text{rotor}}, \Delta h_{\text{stator}}$. With $p_1 = 1$ bar, $T_1 = 288$ K , compute the pressure and temperature at the end of the rotor and the stator, and show the process in a (T, s) chart.
2. Explain why the work number increases inversely proportionally to the square of the azimuthal Mach number.

Chapter 5
Estimating Losses

5.1 Estimating Losses for Axial Turbomachinery Blades

Earlier we defined the blade efficiency in terms of both the static enthalpy change
and the kinetic energy. For axial turbomachinery with or without friction losses, the
total enthalpy should remain constant, and thus the energy equation is written as

$$h^o = h_i + \frac{c_i^2}{2} = h_e + \frac{c_e^2}{2} \qquad (5.1a)$$

where c_i is the entry (inlet) speed, which for the rotor is the relative entry speed and
for the stator is the absolute entry speed. Therefore, for the compressor rotor
$c_i = w_1$, and for the compressor stator $c_i = c_2$, whereas for the turbine stator
$c_i = c_3$, and for the turbine rotor $c_i = w_2$. Similarly, c_e is the exit (final) speed in
the same blade row, and h_i and h_e are the respective specific enthalpies at the entry
and exit. While (5.1a) includes losses, in the case of a loss-free (isentropic) process
(designated with *), we may write

$$h^o = h_i + \frac{c_i^2}{2} = h_e^* + \frac{c_e^{*2}}{2} \qquad (5.1b)$$

Therefore, we may define the change in enthalpy as

$$\Delta h = h_e - h_i = \frac{1}{2}(c_i^2 - c_e^2) \quad \text{and} \quad \Delta h^* = h_e^* - h_i = \frac{1}{2}(c_i^2 - c_e^{*2})$$

and the loss due to friction (shown in the pressure loss) is

$$\Delta h_{loss} = \Delta h - \Delta h^* = \frac{1}{2}(c_e^{*2} - C_e^2)$$

T. Bose, *Airbreathing Propulsion: An Introduction*,
Springer Aerospace Technology, DOI 10.1007/978-1-4614-3532-7_5,
© Springer Science+Business Media, LLC 2012

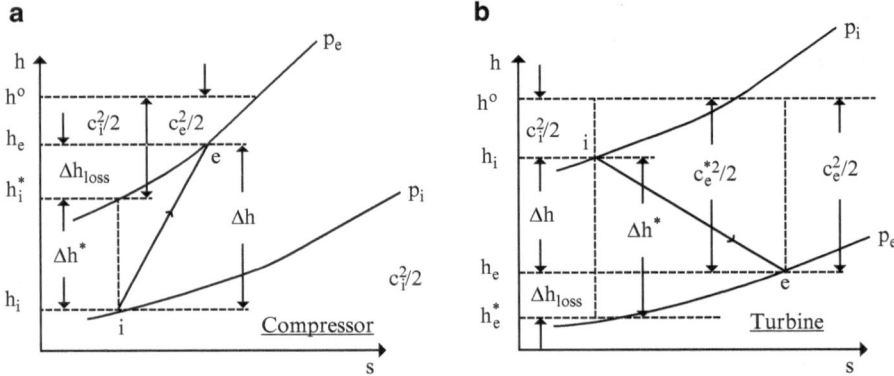

Fig. 5.1 To explain blade losses for (**a**) a compressor and (**b**) a turbine

Note that $\Delta h^* > \Delta h$, corresponding to $c_e^{*2} < c_e^2$, is for the compressor, and $\Delta h^* > \Delta h$, corresponding to $c_e^{*2} < c_e^2$, is for the turbine. Now, since the blade loss is basically due to the *wake defect* (friction effect on the blade) of the blade profile, and also due to the *Mach number effect* (*wave drag*) in the supersonic flow, Traupel (1966) defined the *profile loss coefficient* ζ as follows:

$$\zeta = 1 - \frac{c_e^2}{c_e^{*2}} = \frac{\Delta h - \Delta h^*}{(1/2)c_e^{*2}} = \frac{\Delta h_{\text{loss}}}{1/2)c_e^{*2}} \tag{5.2}$$

Comparing ζ in (5.2) with the definition of the *kinetic energy efficiency*, η^k, discussed earlier in Sect. 4.1, it can be seen that $\zeta = 1 - \eta^k$. For this purpose, let's consider the process for an axial compressor and an axial turbine separately in an (h, s) diagram (Fig. 5.1a, b).

(a) Compressor: From Fig. 5.1a, it is evident that $c_e^2/2 = c_i^2/2 - \Delta h$ and $c_e^{*2}/2 = c_i^2/2 - \Delta h^*$. Therefore,

$$\Delta h - \Delta h^* = \left(c_e^{*2} - c_e^2\right)/2$$

Substituting for $\Delta h_{\text{loss}} + c_e^2/2$ in the right-hand side of (5.2), we find $c_e^2/2 = (1 - \zeta)c_e^{*2}/2$. By comparing the definition of the kinetic energy efficiency, η^k, in Sect. 4.1, we find that $\eta^k = (1 - \zeta)$.

(b) Turbine: From Fig. 5.1b, the following relations are obvious:

$$\Delta h = h_i - h_e = \frac{1}{2}\left(c_e^2 - c_i^2\right), \Delta h^* = h_i - h_e^* = \frac{1}{2}\left(c_e^{*2} - c_i^2\right)$$

Fig. 5.2 To explain
axial blade losses due
to friction

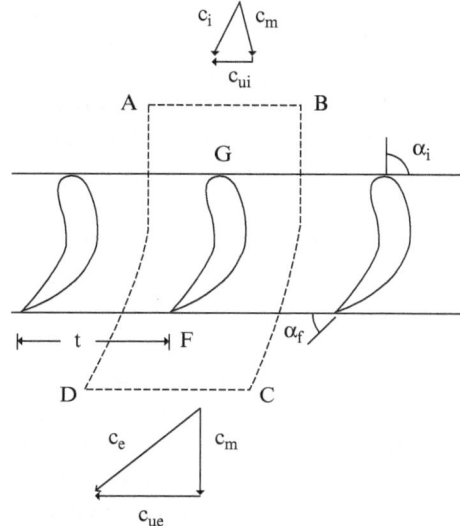

Thus,

$$\Delta h_{\text{loss}} = \Delta h^* - \Delta h = \frac{1}{2}\left(c_e^{*2} - c_e^2\right)$$

Substituting once again for $\Delta h_{\text{loss}} + c_e^2/2$ in the right-hand side of (5.2) and comparing it with the definition of the kinetic energy efficiency, η^k, in Sect. 4.1, we find again that $\eta^k = (1 - \zeta)$.

Having demonstrated the relationship between ζ and η^k, the former consists of losses due to various causes, the foremost being due to friction, and it is designated as ζ_{po}, that is, the basic loss due to friction. This will be discussed separately for an axial turbine and an axial compressor.

5.1.1 Losses in an Axial Turbine

Let's now consider the axial blades given in Fig. 5.2; the analysis is valid for both the stator and the rotor. Now in Fig. 5.2, let the control volume be *ABCD*. Along the boundary lines of the control volume, one can write

$$\Gamma = t(c_{\text{ue}} - c_{\text{ui}}) = t\Delta c_{\text{u}} \tag{5.3a}$$

where Γ is the *circulation of the flow around the blade due to viscous action*, t is the *gap between the blades*, and $\Delta c_{\text{u}}$is the velocity difference in the wake of the blade (*wake defect*).

According to *Stokes' law*, the circulation must be the integral of vorticity within a control volume. It may further be assumed that the flow, in general, has a characteristic of a potential flow (without separation and shocks in subsonic flow). Thus, the *vorticity* arises only inside the boundary layer of the blade profile, where along a small distance ds on the wall, the circulation is d$\Gamma = c\,\mathrm{d}s$ and as such the integrated circulation is

$$\Gamma = \int c\,\mathrm{d}s = \int_F^G c\,\mathrm{d}s_{\text{suction}} - \int_G^F c\,\mathrm{d}s_{\text{compression}} \tag{5.3b}$$

One may additionally note that the *boundary layer* or the *vorticity surface* is not only on the blade surface, but also on the side walls.

Now, let $c = c(s) = c_0(s) \pm \Delta c(s)$, where $c_0(s)$ is the mean velocity between the suction surface and the compression surface. The positive sign is for the suction surface and the negative is for the compression surface. In addition, we'll define the average mean surface velocity as

$$\bar{c}_0 = \frac{1}{s} \int_F^G c_0(s)\,\mathrm{d}s$$

and the average deviation on either surface as

$$\Delta\bar{c} = \frac{1}{2} \int_F^G \Delta c_0\,\mathrm{d}s$$

Therefore,

$$\Gamma = \int_F^G (c_0 + \Delta c)\,\mathrm{d}s_{\text{suction}} - \int_F^G (c_0 - \Delta c)\,\mathrm{d}s_{\text{compression}} = s[(\bar{c}_0 + \Delta\bar{c}) - (\bar{c}_0 - \Delta\bar{c})]$$

$$= \bar{c}_0 s\left[\left(1 + \frac{\Delta\bar{c}}{\bar{c}_0}\right) - \left(1 - \frac{\Delta\bar{c}}{\bar{c}_0}\right)\right] = 2\bar{c}_0 s \frac{\Delta\bar{c}}{\bar{c}_0} \tag{5.3c}$$

Comparing (5.3c) and (5.3a), we get

$$\frac{\Delta\bar{c}_0}{\bar{c}_0} = \frac{t\Delta\bar{c}_u}{2s\bar{c}_0} \tag{5.4}$$

Obviously, when (t/s) is small, which is the case for large *chord length* or small *gap* (*pitch*) between the blades, ($\Delta\bar{c}/c_0$) is small, and when (t/s) is large, ($\Delta\bar{c}/c_0$)

becomes large. This has further repercussions in the following. For this we introduce the concept of the *dissipation power* due to friction:

$$P_D = \frac{1}{2}\bar{c}_f s \rho (\bar{c}_o)^3 \left[\left\{1 + \frac{\Delta\bar{c}}{\bar{c}_o}\right\}^3 + \left\{1 + \frac{\Delta\bar{c}}{\bar{c}_o}\right\}^3 \right] = \bar{c}_f s (\bar{c}_o)^3 \left[1 + 3\left(\frac{\Delta\bar{c}}{\bar{c}_o}\right)^2\right] \quad (5.5)$$

where $\bar{c}_f$ is the *average friction coefficient*. Further, since c_i and c_f are the inlet and exit speeds, and α_i and α_f are the corresponding angles, the mass flow rate is

$$\dot{m} = \rho c_e t \sin \alpha_e$$

where the density ρ is assumed to be approximately constant, which may be considered somewhat valid for an axial compressor only, but not for an axial turbine. Thus, the enthalpy loss is

$$\Delta h_{\text{loss}} = \frac{P_D}{\dot{m}} = \bar{c}_f \rho s \frac{(\bar{c}_o)^3}{\rho c_e t \sin \alpha_e}\left[1 + 3\left(\frac{\Delta\bar{c}}{\bar{c}_o}\right)^2\right] \quad (5.6)$$

Now, defining a basic profile loss coefficient due to friction

$$\zeta_{po} = \frac{\Delta h_{\text{loss}}}{\Delta h_{\text{loss}} + (c_e^2/2)} \quad (5.7)$$

we get

$$\frac{\zeta_{po}}{1 - \zeta_{po}} = \frac{\Delta h_{\text{loss}}}{(c_e^2/2)} = 2\bar{c}_f \left(\frac{s}{t}\right)\left(\frac{\bar{c}_o}{c_e}\right)\frac{1}{\sin \alpha_e}\left[1 + 3\left(\frac{\Delta\bar{c}}{\bar{c}_o}\right)^2\right] \quad (5.8a)$$

Substituting (5.4) into it, we further get

$$\frac{\zeta_{po}}{1 - \zeta_{po}} = 2\bar{c}_f \left(\frac{s}{t}\right)\left(\frac{\bar{c}_o}{c_e}\right)^3 \frac{1}{\sin \alpha_e}\left[1 + \frac{3}{4}\left(\frac{t}{s}\right)\left(\frac{\Delta\bar{c}_u}{\bar{c}_o}\right)^2\right] \quad (5.8b)$$

The above expression depends on s/t, which is actually the ratio of the *blade arc length* to the *blade distance*, but can be approximated by the ratio of the blade chord length to the blade distance. An optimum value of the loss coefficient is obtained by differentiating $\zeta_{po}/(1 - \zeta_{po})$ with respect to (s/t) and setting it equal to zero. This gives

$$\left(\frac{t}{s}\right)_{\text{opt}} = \left(\frac{\bar{c}_o}{\Delta c_u}\right)\sqrt{\frac{4}{3}} = 1.1547\left(\frac{\bar{c}_o}{\Delta c_u}\right) \quad (5.9)$$

and the corresponding optimal loss relation is

$$\left(\frac{\zeta_{po}}{1-\zeta_{po}}\right)_{opt} = \sqrt{\frac{3/4}{3.404}}\,\bar{c}_f \left(\frac{\bar{c}_o}{c_e}\right)^3 \left(\frac{\Delta\bar{c}_u}{\bar{c}_o}\right)\frac{1}{\sin\alpha_e} \tag{5.8c}$$

A meaningful value of $\bar{c}_o$ is $\bar{c}_o = (c_i + c_e)/2$. On the other hand, for the blade profiles with a large mechanical strength, it is shaped so that right after entry, the gas is accelerated to c_e by keeping the channel cross-section constant $(\bar{c}_o = c_e)$. Thus, it is quite in order, if an average of these two values is taken; that is,

$$\bar{c}_o = \frac{1}{4}(c_i + 3c_e) = \frac{1}{4}c_u \left(\frac{1}{\sin\alpha_e + \frac{3}{\sin\alpha_i}}\right)$$

Also,

$$\Delta\bar{c}_u = c_m \left(\frac{1}{\tan\alpha_e} - \frac{1}{\tan\alpha_i}\right)$$

Therefore, the following approximate expressions are used by Traupel (1966):

$$\frac{\Delta\bar{c}_u}{\bar{c}_o} = 4\left(\frac{1}{\tan\alpha_e} - \frac{1}{\tan\alpha_i}\right)\frac{\sin\alpha_e \sin\alpha_i}{\sin\alpha_e + 3\sin\alpha_i} \tag{5.9a}$$

$$\frac{\bar{c}_o}{c_e} = \frac{(\sin\alpha_e + 3\sin\alpha_i)}{4\sin\alpha_i} \tag{5.9b}$$

and

$$\left(\frac{s}{t}\right)_{opt} = 3.464\,\frac{\sin\alpha_e}{\sin\alpha_e + 3\sin\alpha_i} \tag{5.9c}$$

Equation 5.9c has been evaluated and shown in Fig. 5.3 for $\alpha_i = 20$–$160°$ and as parameter $\alpha_e = 20$–$45°$.

The total loss coefficient for an axial turbine consists of several factors:

$$\zeta = \lfloor \zeta_{po}\chi_R\chi_M\chi_\delta + \zeta_\delta + \zeta_f \rfloor + \zeta_a + \zeta_{rest} \tag{5.10}$$

where the first term within brackets is the actual profile loss coefficient consisting of ζ_{po} due to the *basic loss*, $\chi_R = $ *loss correction due to Reynolds number*, $\chi_M = $ *correction due to Mach number*, $(\chi_\delta, \zeta_\delta) = $ *losses due to trailing-edge thickness*, and $\zeta_f = $ *secondary flow losses*. Further terms are $\zeta_a = $ additional losses due to overlapping of rotor and stator blade heights, axial gap between the blades, and due to installation of *blade damping wires*, if any, and ζ_{rest} takes care of losses in the blade root and tip region.

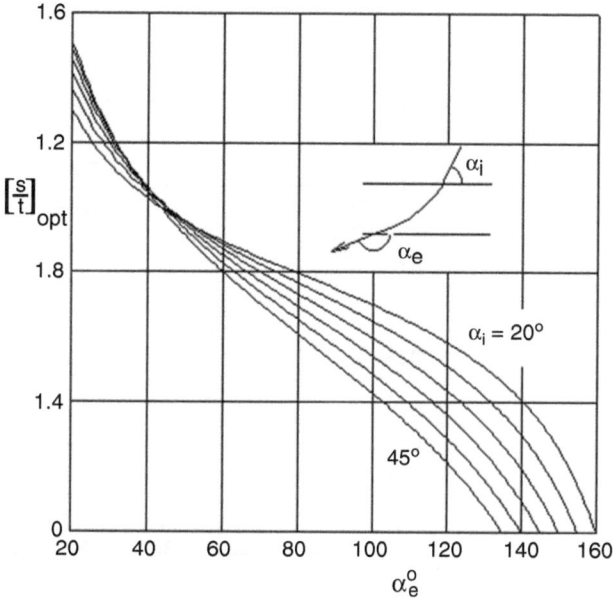

Fig. 5.3 $(s/t)_{opt}$ vs. different exit angles with inlet angles as parameter (According to Traupel (1966), p. 364, Fig. 5.6.1)

The basic loss, which depends approximately on the blade inlet and exit angles, according to Traupel, is reproduced in Fig. 5.4. Traupel obtained these results for an *average friction coefficient* of $c_f = 0.003$.

The *Reynolds number correction factor*, χ_R, considers the influence of the Reynolds number and depends considerably, in the turbulent region, on the ratio of the surface roughness to camber length (k_s/s). A typical dependency has been redrawn from the figure and is shown in Fig. 5.5.

At smaller Reynolds numbers, there is a strong increase in the correction factor, which is due to separation on the blade profile. These values may be shifted somewhat due to strong preturbulence.

Similarly, the Mach number correction factor has been shown in Fig. 5.6. This factor starts growing in the transonic region when the local flow over the blade surface may become supersonic.

Both χ_δ and ζ_δ depend on the trailing-edge thickness-to-pitch ratio (τ/t), shown in Fig. 5.7.

Further, the *secondary loss* ζ_f depends on the ratio of the blade height to mean diameter if the blade is not twisted. Thus, for *untwisted blades*, under the assumption of parabolic distribution of the efficiency over the blade azimuthal speed (blade height b), one gets approximately

$$\zeta_f = \frac{1}{2}\left(\frac{b}{D}\right)^2 \tag{5.11a}$$

Fig. 5.4 Basic loss coefficient for the blade profiles (After Traupel (1966), p. 364, Fig. 8.4.2)

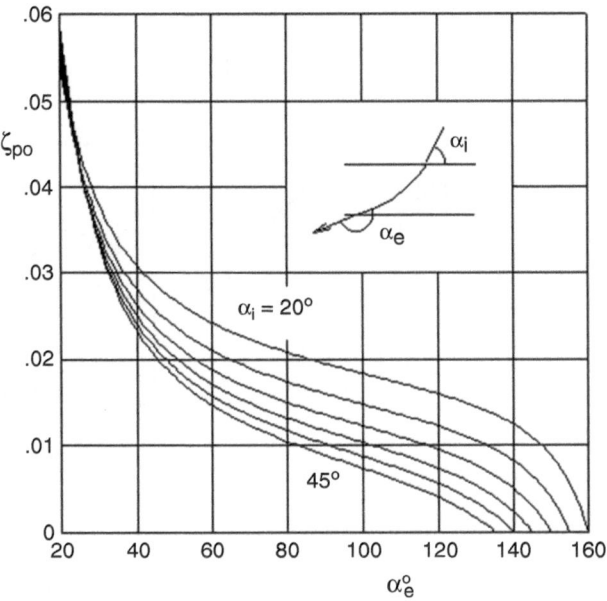

Fig. 5.5 Correction factor on loss coefficient due to Reynolds number (After Traupel (1966), p. 365, Fig. 8.4.3)

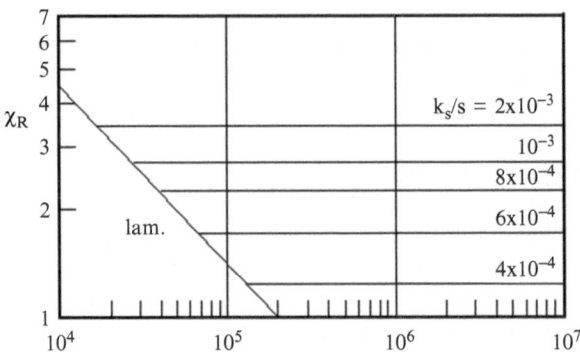

Fig. 5.6 Correction factor on loss coefficient due to Mach number (After Traupel (1966), p. 364, Fig. 8.4.2)

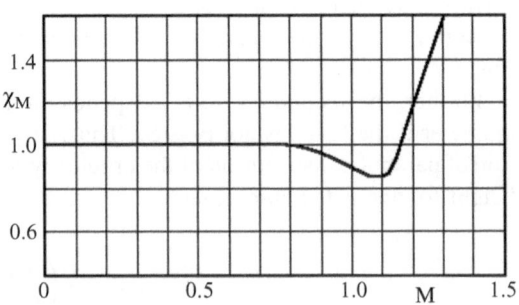

Fig. 5.7 Effect of trailing-edge thickness on the profile loss (After Traupel (1966), p. 364, Fig. 8.4.2)

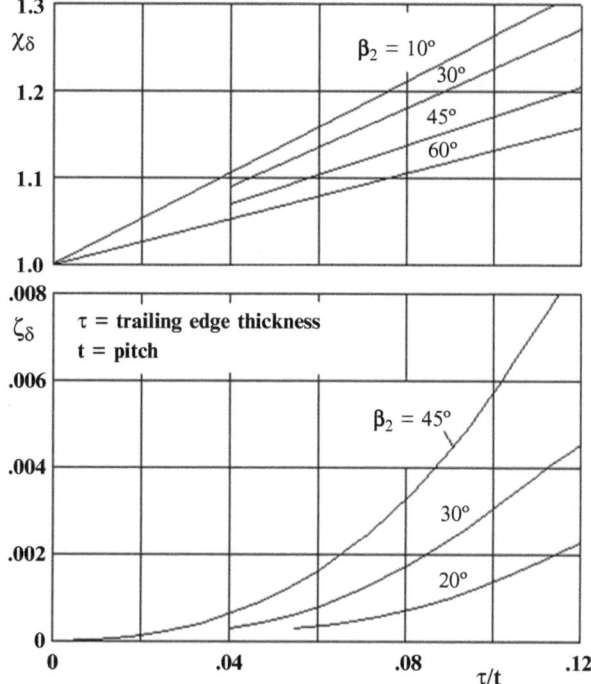

whereas for an NACA-8410 profile by integration over the blade profile height, the expression given is

$$\zeta_f = 8\left(\frac{b}{D_m}\right)^4 \tag{5.11b}$$

For $b/D_m = 0.2$, (5.11b) gives a smaller value than (5.11a). It is, however, recommended that $b/D_m = 0.15$–0.2 for *untwisted blades*.

For the value of ζ_{rest}, we have, according to Traupel (1966), the relation

$$\zeta_{rest} = \chi F(t/b). \tag{5.12}$$

Normally, $\chi \approx 1$, whereas the factor F depends on the velocity ratio (c_{inlet}/c_{exit}) and the turning angle of flow in the blade cascade. Values of F are plotted in Fig. 5.8.

Splitting the profile loss into the profile loss ζ_p and additional loss ζ_{rest} is, however, meaningful if the blades are not too short, so that boundary-layer areas merge together, which is the case if

$$(b/t) < (6 \text{ to } 10)\sqrt{\zeta_p} = (b/t)_{critical}$$

Fig. 5.8 Function F for computation of rest losses (After Traupel (1966), p. 368, Fig. 8.4.6)

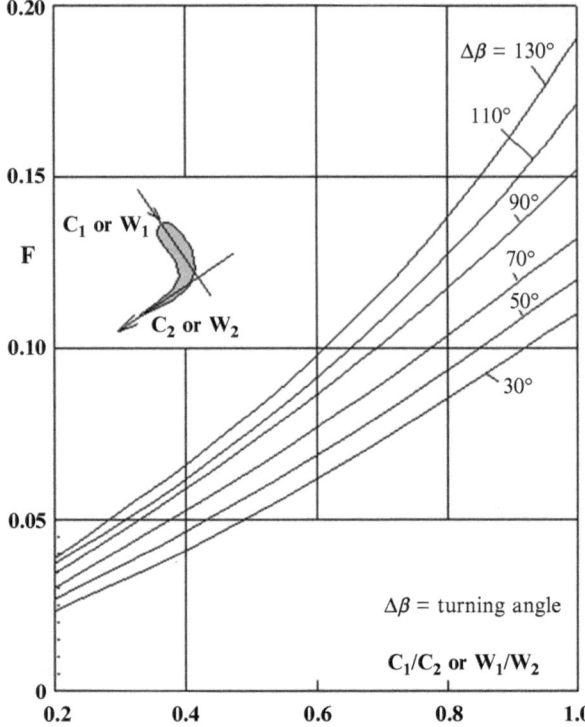

The lower value is for cascades with large acceleration. For (b/t) larger than the above value, the expression can be

$$\zeta_p + \zeta_{rest} = \zeta_p + \frac{\chi F}{\{(b/t)\}_{exit}} + 0.016\left[\frac{l}{b} - \frac{(l/t)}{(b/t)_{exit}}\right]$$

where l is the profile length (*chord length*).

Additional losses ζ_a are due to overlapping of the rotor and stator *blade heights*, *axial gap,* the presence of *damping wire*, and so forth (Fig. 5.9). The last type occurs since in certain long blades, one has to put wire azimuthally through blades to damp the blade oscillations.

Correction due to axial gap:

$$\zeta_{a1} = \frac{0.05 \text{ to } 0.06}{\sin \alpha_{e,previous}}\left(\frac{\Delta_a}{b_{previous}}\right)\left(\frac{c_{exit,previous}}{c_{inlet}}\right)^2 \qquad (5.13a)$$

Correction due to *overlapping* (suction if $\Delta_a/\Delta_b < 1.5$):

$$\zeta_{a2} = 1.1\left(\frac{\Delta_b}{b_{previous}}\right)\left(\frac{c_{inlet}}{c_{exit}}\right)^2 \qquad (5.13b)$$

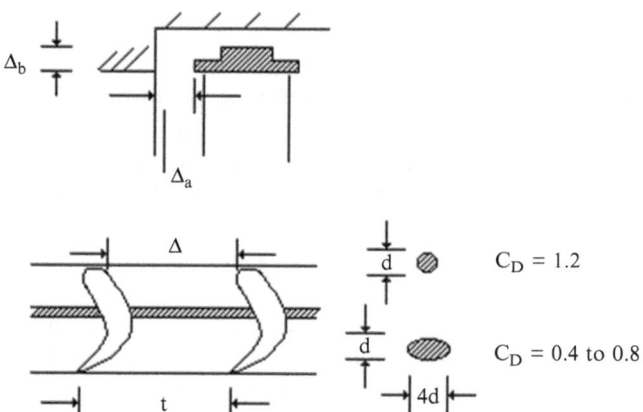

Fig. 5.9 To explain additional blade losses

Correction due to damping wire:
First, we can calculate the pressure loss due to the damping wire (Fig. 5.8) as

$$\Delta p_{\text{loss}} = \text{Drag over}\{\text{cross} - \text{sectional area}\} = \frac{3}{2} c_{\text{D}} \rho c^2 \frac{\pi D_{\text{d}} d}{\frac{\pi}{4}\left(D_{\text{tip}}^2 - D_{\text{root}}^2\right)}$$

where D_{d} is the diameter at which the damping wire is placed and d is the characteristic dimension normal to the flow direction.
Since

$$c^2 = \left(\frac{t}{b}\right)^2 c_i^2 \sin^2 \alpha_i$$

we get

$$\zeta_{a3} = \frac{\Delta p_{\text{loss}}}{\frac{1}{2}\rho c_i^2 + \Delta p_{\text{loss}}} \approx \frac{2\Delta p_{\text{loss}}}{\rho c_i^2} = 4c_{\text{D}}\sin^2\alpha_i \left(\frac{t}{b}\right)^2 \frac{D_{\text{d}} d}{\left(D_{\text{tip}}^2 - D_{\text{root}}^2\right)}$$

As Traupel reported (1966), Lieblein replaced the factor 4 by 8 in the right-hand side of the above equation and got

$$\zeta_{a3} = 8c_{\text{D}}\sin^2\alpha_i \left(\frac{t}{b}\right)^2 \frac{D_{\text{d}} d}{\left(D_{\text{tip}}^2 - D_{\text{root}}^2\right)} \tag{5.13c}$$

Thus, the total *additional blade loss* is

$$\zeta_a = \zeta_{a1} + \zeta_{a2} + \zeta_{a3} \tag{5.13}$$

5.1.2 Losses in an Axial Compressor

The losses in an axial compressor closely follow the losses in an axial turbine, except that the profile losses in an axial compressor with an adverse pressure gradient are quite different and are best taken from experimental results. For an axial compressor, the basic loss coefficient is given by the relation

$$\zeta = \zeta_p + \zeta_{trans} = \zeta_{po}\chi_R\chi_M + \zeta_{trans} \tag{5.14}$$

in which ζ_p is the profile loss consisting of ζ_{po} being the basic profile loss, χ_R being the correction due to Reynolds number, χ_M is the correction due to the Mach number, and ζ_{trans} is the loss in the transonic compressor due to shock. While χ_M and χ_R follow the rule due to the axial turbine, ζ_{po}, according to Lieblein's very systematic experiments (as given by Traupel), is given by the relation

$$\zeta_{po} = \frac{2\theta}{\frac{t}{l}\sin\alpha_i}\left(\frac{\sin\alpha_i}{\sin\alpha_e}\right)^2\left(1 - \theta\frac{1.08}{(t/l)\sin\alpha_e}\right)^{-3} \tag{5.15a}$$

where t is the pitch, l is the chord length, α_i and α_e are the inlet and exit blade angles, and θ is the *momentum thickness* given by

$$\theta = 8 \times 10^{-4}\exp^{1.62G} \tag{5.15b}$$

where G is of the order of 1–2.2 and is defined as

$$G = (\text{max.speed on blade surface})/(\text{exit speed})$$

The value of G can be calculated approximately by the relation

$$G = \frac{c_i}{c_e}\left[1.12 + 0.61\sin^2\alpha_i\left(\frac{t}{l}\right)(\cot\alpha_i - \cot\alpha_e)\right] \tag{5.15c}$$

Further, ζ_{shock} is the additional contribution in the transonic region when the leading edge of the blade is very sharp, due to shock, as given in Fig. 5.10.

5.2 Estimating Losses for Centrifugal Compressors

Although with water as the fluid, there are *radial turbines,* such as Francis turbines, and with steam as the fluid, there are radial turbines, such as Ljungstrom turbines, for aircraft applications, there are not too many applications of radial turbines, except probably the classical jet engine developed by Whittle. We are therefore going

Fig. 5.10 Transonic shock losses for entering a transonic blade row (After Traupel (1966), p. 381, Fig. 8.5.9)

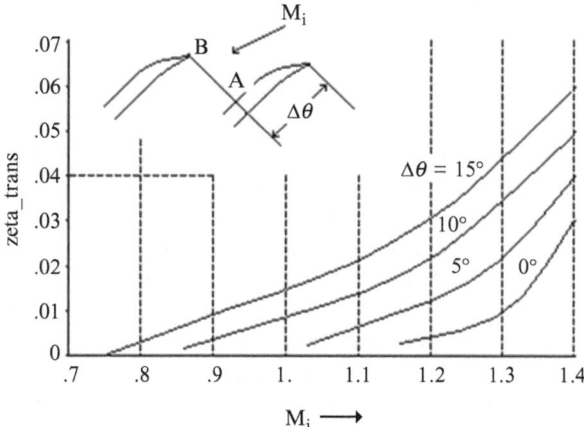

to discuss only the losses in radial compressors, better known as centrifugal compressors (Eckert 1953).

In the rotor entry (with or without blades), there is loss because of the change in the flow direction, and the loss coefficient is given by the relation

$$\zeta_1 = \frac{\Delta h_{loss1}}{\frac{1}{2}c_1^2} = 0.1 \text{ to } 0.15 \tag{5.16a}$$

Further, in the rotor itself, the loss coefficient is given by the relation

$$\zeta_2 = \frac{\Delta h_{loss2}}{\frac{1}{2}w_1^2} = 0.2 \text{ to } 0.25 \tag{5.16b}$$

Further, the friction loss coefficient in the diffuser (with blades as in a stator, or as a stator and spiral chamber) for entry speed c_3 and exit speed c_4, assuming that no flow separation takes place in the diffuser, is given by

$$\zeta_3 = \frac{\Delta h_{loss3}}{\frac{1}{2}\left(c_3^2 - c_4^2\right)} = 0.25 \tag{5.16c}$$

In case the diffuser, in the form of a stator, is put directly behind the rotor, then $c_3^2 \approx c_2^2 = c_{m2}^2 + c_{u2}^2$. Now, for $c_{m1} \approx c_{m2} = c_m$, which is found to be quite satisfactory for a good efficiency, we have

$$\zeta_3 = \frac{\Delta h_{loss3}}{\frac{1}{2}\left[\left(c_m^2 + c_{u2}^2\right) - c_m^2\right]} = \frac{\Delta h_{loss3}}{\frac{1}{2}c_{u2}^2} \tag{5.16d}$$

5.3   Other Losses

In addition to the blade losses, there are losses in labyrinth seals (placed as a seal between rotating and stationary components of a turbomachine) and due to friction between the rotating disk and the stationary casing wall. For the *labyrinth seal losses*, one has to consider first the mass flow rate of the leakage, which not only reduces the actual quantity of gas passing thought the blades, but also, while passing through the seals acting as throttle, causes an irreversible heating, with the result that behind the labyrinth seal, the hot leaked gas mixes with the gas going through the blades. This is particularly true for axial turbines, in which the mass flow rate through the blades and the leakage mass flow rate are in the same direction. Examples of some labyrinth seals are given in Fig. 5.11, in which the left figure is a typical example of the seal placed at the entry of the rotor for a centrifugal compressor, and the figure on the right is an example of the seal brought on the tip band of an axial turbomachine rotor. Other labyrinth seals can be placed on the backside of the centrifugal compressor rotor back-plate, or for axial machines fixed on the stationary casing against the moving shaft. For turbine blades, let $\dot{m}$ be the total mass flow rate of the gas and $\dot{m}_L$ be the mass flow rate through the *labyrinth* ($\dot{m}_L \ll \dot{m}$).

As the gas at entry pressure p_i and entry enthalpy h_i passes through the blade to exit pressure p_e and exit enthalpy under (near) isentropic change of state h_e^*, the gas passing through the labyrinth at the other end is $h_e = h_i$; the specific mixing enthalpy, when both flows meet and mix, is obtained from the energy balance

$$\dot{m}_L h_i + (\dot{m} - \dot{m}_L)h_e^* = \dot{m}h_{\text{mix}}$$

where

$$\frac{h_i}{h_e^*} = \left(\frac{p_i}{p_e}\right)^{\frac{\gamma-1}{\gamma}}$$

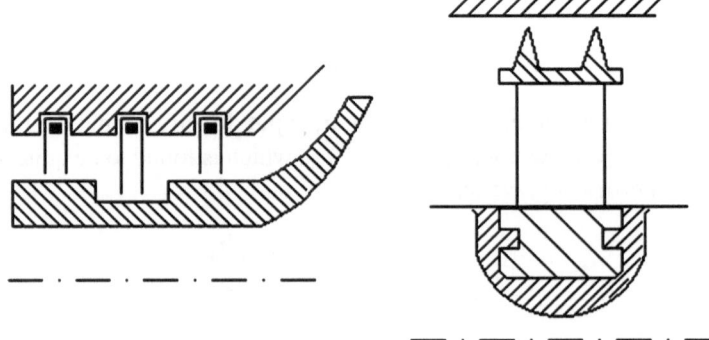

Fig. 5.11   Schematic sketch of a labyrinth seal

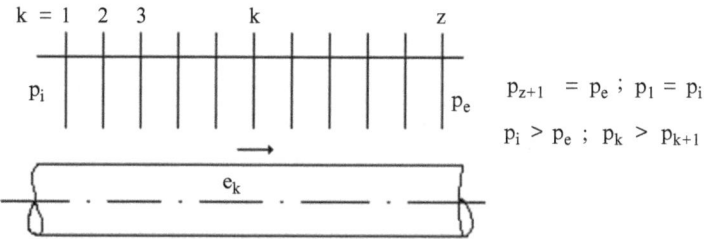

Fig. 5.12 A general analysis of flow in z labyrinths

By rearranging the above equation, we get a measure of the *labyrinth loss*:

$$\frac{\dot{m}_L}{\dot{m}} = \frac{h_{mix} - h_e}{h_i - h_e^*} = \frac{\Delta h_{lab.loss}}{h_e^* \left[\left(\frac{p_i}{p_e} \right)^{\frac{\gamma-1}{\gamma}} - 1 \right]}$$ (5.17)

due to heating for leaks in the labyrinth. Therefore, there is not only a loss in power due to a loss in the mass flow rate through the blades, but also there is loss due to an irreversible throttle effect in the seals.

We can now estimate the mass flow rate $\dot{m}_L$ through the labyrinth seals when there are a number of seals (Fig. 5.12). For this purpose, let there be z labyrinths, in which the jth labyrinth has diameter D_j and a gap s_j.

Further, let $p_{j'}$ and $p_{j''}$ be the pressures across the jth labyrinth, and let p_i and p_e be the two pressures before and after the entire labyrinth block; ρ_j is the average density across the jth labyrinth. Therefore, $p_i = p_{1'}$ and $p_e = p_{z''}$. Now if we consider the jth labyrinth across which there is subsonic flow, then the mass flow rate, as it is known from subsonic orifice relation, is

$$\dot{m}_L = (\pi D_j s_j) \sqrt{\frac{(p_{j''} - p_{j'})}{\rho_j}}$$

or

$$p_{j''} - p_{j'} = \frac{\dot{m}_L^2}{(\pi D_j s_j)^2 / \rho_j}$$

Now, for the *subsonic flow* in the labyrinth throughout, by adding the above equation over all labyrinths, we have

$$\sum_{j=1}^{z} (p_{j''} - p_{j'}) = p_i - p_e = \left(\frac{\dot{m}_L}{\pi} \sum_{j=1}^{z} \frac{1}{(D_j s_j)^2 / \rho_j} \right)$$

and hence,

$$\dot{m}_L = \pi \sqrt{\frac{(p_i - p_e)}{\sum \frac{1}{(D_j s_j)^2 / \rho_j}}} \approx \pi D_m s_m \sqrt{(p_i - p_e)\rho_m}, \quad \left[\frac{\text{kg}}{\text{s}}\right] \tag{5.17a}$$

where the index m denotes some average value.

On the other hand, for a *choked flow* (sonic speed in the gap), we need to have a somewhat more elaborate analysis. If we consider the situation between the jth and the $(j + 1)$st labyrinths, the kinetic energy between the two for the isentropic change of state is

$$\frac{c_j^2}{2} = \Delta h_j^* = \frac{\gamma R T_j}{\gamma - 1} \left[1 - \left(\frac{p_{j+1}}{p_j}\right)^{\frac{\gamma-1}{\gamma}} \right]$$

In addition,

$$\dot{m}_L = \rho_{j+1} c_j A_j$$

$$\rho_{j+1} = \rho_j \left(\frac{p_{j+1}}{p_j}\right)^{\frac{1}{\gamma}}$$

and hence,

$$\dot{m}_L = \rho_j c_j A_j \left(\frac{p_{j+1}}{p_j}\right)^{\frac{1}{\gamma}} = A_j \left[\frac{2\gamma}{\gamma - 1} p_j \rho_j \left\{ \left(\frac{p_{j+1}}{p_j}\right)^{\frac{2}{\gamma}} - \left(\frac{p_{j+1}}{p_j}\right)^{\frac{\gamma-1}{\gamma}} \right\} \right]^{1/2} \tag{5.17b}$$

With $\Delta p_j = p_{j+1} - p_j$, and expanding the above expression, we get

$$\dot{m}_L = A_j \left[\frac{2p_j}{RT_j} \Delta p_j \right]^{1/2} \tag{5.17c}$$

and then the mass flow rate across a single jth labyrinth is

$$\dot{m}_L = K \pi D_j s_j \sqrt{p_j \rho_j} \tag{5.17d}$$

where K is a constant giving the distribution of pressure and density through the labyrinth as a function of the *specific heat ratio* γ.

Kruschik (1960) gives the following formulas:
For

$$p_e < p_i \frac{0.87}{\sqrt{z + 0.68}}$$

and *sonic choking*:

$$\dot{m}_{\mathrm{L}} = \pi D_m s_m \sqrt{\frac{p_i \rho_i}{z + 0.68}} \tag{5.18a}$$

For

$$p_e > p_i \frac{0.87}{\sqrt{z + 0.68}}$$

and *subsonic*:

$$\dot{m}_{\mathrm{L}} = \pi D_m s_m \sqrt{\frac{\rho_m}{p_i}(p_i^2 - p_e^2)} \tag{5.18b}$$

For the normal gap, in which the shaft rpm is subcritical (less than resonance rpm), the minimum gap, as per the industry standard, is

$$s_{\min}[\mathrm{mm}] = 0.6 \frac{D_m[\mathrm{mm}]}{1,000} + 0.1 \tag{5.19}$$

and for an elastic rotating shaft, it should have double the value.

In addition to the labyrinth losses, other losses include disk friction loss. If the rotor consists of only disks, then on either side of the disk wall, the *torque moment* on the rotating disk is

$$M = 2C_{\mathrm{M}} \rho u_{\max}^2 D_{\max}^2 \tag{5.20}$$

where C_{M} is the *moment coefficient* given by the approximate relation

$$C_{\mathrm{M}} = 0.0030 \mathrm{Re}^{-1/5} \tag{5.21}$$

where $\mathrm{Re} = u_{\max} D_{\max}/\nu$, $\nu = $ kinematic viscosity of the gas.

For a typical value of $C_{\mathrm{M}} \approx 0.0002$, equivalent to $\mathrm{Re} \approx 10^6$, the *equivalent power loss due to friction* is

$$P_{\mathrm{loss}} = \frac{M\pi n}{30} = 0.0002 \frac{2n\pi}{30} \rho \left(\frac{\pi D_{\max} n}{60}\right)^2 D_{\max}^2$$

$$\approx 0.115 \left(\frac{n}{1,000}\right)^3 D_{\max}^5, \ [\mathrm{kW}] \tag{5.22a}$$

In case only a part of the disk is subjected to friction, one can calculate the loss in power first for the full disk, from which that due to the smaller disk size is to be subtracted.

For axial machines, usually the diameter up to the middle of the blade D_m is considered. Since this is larger than the diameter of a disk without blades, the values are therefore to be reduced accordingly, and the power loss due to friction is

$$P_{\text{loss}} \approx 0.04 \left(\frac{n}{1,000} \right)^3 D_{\text{max}}^5, [\text{kW}] \qquad (5.22\text{b})$$

5.4   Slip Factor, Diffusion Factor, and Blockage Factor

Because of various losses and imperfect assumptions in design, there can be deviations from the design values. The first of these, the *slip factor*, is actually about the effect of a finite blade number. As shown in Fig. 5.13, while the entry velocity triangle for an axial compressor blade remains unchanged for a finite number of blades, the velocity triangle at the rotor trailing edge deviates, giving a steeper trailing-edge angle.

Let's first assume $c_{u1} = 0$. Therefore, the specific work for the actual number of blades is $H = uc_{u2}$, and for an infinite number of blades it is $H_\infty = uc_{u2'}$. Now, since

$$\tan \beta = \frac{c_m}{u - c_u} \rightarrow c_u = u - \frac{c_m}{\tan \beta}$$

we hence write

$$\frac{H}{H_\infty} = \frac{u_2 - c_{m2}/\tan \beta_2}{u_2 - c_{m2}/\tan \beta_{2'}} \qquad (5.23\text{a})$$

which is less than 1 and which means that $\beta_{2'} > \beta_2$.

The second term, the *diffusion factor* (D.F.), is introduced mainly for axial compressors and not for axial turbines, since in the retarding flow in an axial compressor the flow may easily separate. Therefore, in the American literature it

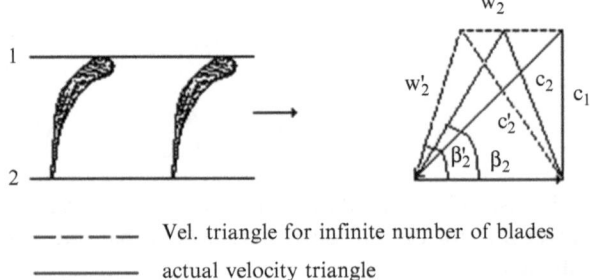

Fig. 5.13 To explain slip factor

‑ ‑ ‑ ‑ ‑   Vel. triangle for infinite number of blades

——————   actual velocity triangle

is suggested that for axial compressors, the ratio (w_2/w_1) for rotors and (c_2/c_1) for stators should not be less than 0.7. American literature states that the criterion for diffusion factor is

$$\text{D.F} = \left(1 - \frac{c_e}{c_i}\right) + \frac{\Delta c_u}{2(l/t)c_e} < 0.5 \tag{5.23b}$$

in which c_i and c_e are the inlet (entry) speed and exit speed, respectively, Δc_u is the change in the azimuthal component of c_i and c_e, l is the chord length, and t is the gap between the blades. For small directional changes $\Delta c_u \approx 0$ and taking the European value of $c_e/c_t > 0.7$, we get D.F. < 0.3, which is somewhat conservative in comparison to American practices for blades with small curvatures.

Furthermore, in axial compressors it is absolutely necessary to introduce a *blockage factor* k, which takes into account the *displacement thickness* δ^* at the hub (subscript "H") and the blade tip band (subscript "S") in the continuity equation. Now if D_H and D_S are the respective diameters, then the ring area without boundary-layer correction is $\pi(D_H^2 - D_S^2)/4$. On the other hand, with displacement thickness (boundary layer) correction, the ring area available is

$$\pi(D_S^2 - D_H^2)/4 - \pi(D_S\delta_S^* + D_H\delta_H^*)/2$$

Therefore, the fractional area available (the so-called blockage factor) is

$$k = 1 - 2\frac{(D_S\delta_S^* + D_H\delta_H^*)}{(D_S^2 - D_H^2)} \tag{5.23c}$$

which is critical generally at the trailing edge of the compressor.

5.5 Exercises

1. Explain *slip factor, diffusion factor,* and *blockage factor.*
2. Calculate the labyrinth mass flow rate where the pressure across a labyrinth seal, at the end of a compressor, of 10 labyrinths is 6 bar and 1.01 bar, respectively, the average diameter is 0.3 m, and the labyrinth gap is 0.12 mm. The required mass density of the air through the labyrinth may be estimated from the adiabatic compression across the compressor.
3. For an axial compressor for which flow data are given in Exercise 1 of Sect. 4.3, and $t/l = 1.0$, compute the profile loss.

Chapter 6
Similarity Rules (on Design Condition)

For the design of turbomachines as a whole, as well as for individual stages, and to understand the performance of stages, certain similarity rules are used. For this purpose, the following notations are used:

p_i^o = inlet stagnation pressure
p_e^o = exit stagnation pressure
T_i^o = inlet stagnation temperature
η = thermodynamic efficiency
D_2 = a characteristic dimension (diameter) at the exit of the rotor (for axial machines, this can be either at the mean rotor exit diameter D_{2m}, or at the rotor blade tip exit diameter D_{20}
H = change in total specific enthalpy
u_2 = a characteristic azimuthal speed at the exit of the rotor
= $(\pi D_2 n/60)$ (for axial machines, corresponding to the characteristic diameter D_{2m} or D_{20}, this can be either u_{2m} or u_{20}, respectively)
n = rpm
R = gas constant
μ = dynamic viscosity coefficient of gas
γ = c_p/c_v
a = speed of sound at a characteristic point

We have already defined the *degree of reaction* in terms of the change in static enthalpy or static pressure as a nondimensional number defined as

$$\hat{r} = \Delta h_{rot}(\Delta h_{rot} + \Delta h_{stat}) \approx \Delta p_{rot}/(\Delta p_{rot} + \Delta p_{stat})$$

T. Bose, *Airbreathing Propulsion: An Introduction*,
Springer Aerospace Technology, DOI 10.1007/978-1-4614-3532-7_6,
© Springer Science+Business Media, LLC 2012

In addition, for axial machines, the rotor blade's exit's root diameter-to-tip diameter ratio

$$v = D_{2,\text{root}}/D_{20}$$

is an important design parameter. In this connection we should discuss for multistage axial compressors the advantages and disadvantages of changing either the root diameter or the tip diameter or both, while the density increases from stage to stage, necessitating a change in the blade height. Among the three main possibilities, the first one, that is, the constant rotor tip diameter, gives a simple shape outwardly. However, in this case, the root diameter has to be made smaller at each succeeding stage; the last stages may have fairly small height blades with not very good stage efficiency. The second choice of changing the outer diameter by keeping the root diameter constant has a better efficiency over the first choice, but u_2 here becomes progressively smaller, and the total enthalpy change per stage, H_s, becomes progressively smaller also, with the result that more stages are required. The third choice of increasing the outer diameter somewhat and the inner diameter strongly has the advantage of keeping the tip Mach number at every stage constant.

Now we derive some of the other important nondimensionless relations from the dimensional relations already derived. For example, from the *principle of the conservation of mass,*

$$\dot{m} = \rho \dot{V} = \rho c_m \Omega \tag{6.1}$$

where $c_m = w \sin \beta$, w is the rotor's relative velocity, β is the blade angle, and Ω is the cross-sectional area. For axial machines, we write $\Omega = \pi D b$, where b is the height of the blade; for radial machines with the same relation, b is the width of the channel. In order to use the same variable b, for axial machines,

$$\Omega = \frac{\pi}{4}\left(D_2^2 - D_{2,\text{root}}^2\right) = \pi \frac{1}{2}\left(D_{2,\text{tip}} + D_{2,\text{root}}\right)\frac{1}{2}\left(D_{2,\text{tip}} - D_{2,\text{root}}\right)$$
$$= \pi D_{2m} b_2 = \frac{\pi}{4} D_{2,\text{tip}}^2 (1 - v^2) \tag{6.2}$$

where $v = D_{2,\text{root}}/D_{2,\text{tip}} < 1$.

Hence, for axial machines, the *discharge coefficient* is given by the relation

$$\varphi_s = \frac{\dot{V}_2}{\pi D_2 b_2 u_2} = \frac{c_{m2}}{u_2} = \frac{\dot{V}_2}{\frac{\pi}{4} D_{2,\text{tip}}^2 (1 - v^2) u_2} = \bar{\varphi}_s = \frac{\hat{\varphi}_s}{(1 - v^2)}$$

For either case, we call b the *transverse dimension.* Therefore, we have

$$\dot{m} = \rho \dot{V} = \pi \rho w D b \sin \beta = \pi \rho c_m D b$$

The above equation needs to be evaluated at a specified point, but conventionally it is done at point 2 (in between rotor and stator) for both the compressor and the turbine. Writing the above equation at point 2, we get

$$\dot{m} = \rho_2 \dot{V}_2 = \pi \rho_2 w_2 D_2 b_2 \sin \beta_2 = \pi \rho_2 c_{m2} D_2 b_2$$

and thus, the volume flow rate is

$$\dot{V}_2 = u_2 \left(\frac{w_2}{u_2}\right) \pi D_2 b_2 \sin \beta_2 = 4 u_2 \left(\frac{\pi D_2^2}{4}\right) \left(\frac{w_2}{u_2}\right) \left(\frac{b_2}{D_2}\right) \sin \beta_2$$

From this a nondimensional *discharge coefficient* can be defined as

$$\varphi_s = \frac{\dot{V}_2}{\pi D_2 b_2 u_2} = \frac{c_{m2}}{u_2} = \left(\frac{w_2}{u_2}\right) \sin \beta_2 \tag{6.3a}$$

An alternate definition of the discharge coefficient is

$$\hat{\varphi}_s = \frac{\dot{V}_2}{\frac{\pi}{4} D_2^2 u_2} = 4 \left(\frac{w_2}{u_2}\right) \left(\frac{b_2}{D_2}\right) \sin \beta_2 = 4 \varphi_s \left(\frac{b_2}{D_2}\right) \tag{6.3b}$$

For axial machines, an alternate definition of the discharge coefficient is

$$\bar{\varphi}_s = \frac{\dot{V}_2}{u_2} \frac{1}{4 \pi D_{2,\text{tip}}^2 (1 - v^2)} \tag{6.3c}$$

where $v = D_{\text{root}} / D_{\text{tip}}$.

Now, the *specific work done per stage* is

$$H_s = \frac{1}{2} \left[(w_1^2 - w_2^2) + (u_2^2 - u_1^2) + (c_2^2 - c_1^2) \right] = (u_2 c_{u2} - u_1 c_{u1})$$

We can now obtain a relation for a nondimensional *work coefficient* as

$$\Psi_s = \frac{H_s}{(u_2^2/2)} = \frac{1}{(u_2^2/2)} \left[(w_1^2 - w_2^2) + (u_2^2 - u_1^2) + (c_2^2 - c_1^2) \right]$$

$$= \frac{2}{u_2^2} \left[u_2 c_{u2} - u_1 c_{u1} \right] \tag{6.4a}$$

In a similar manner, from the *isentropic specific work per stage*, H_s^*, we can define an *isentropic work coefficient*:

$$\Psi_s^* \frac{H_s^*}{(u_2^2/2)} \tag{6.4b}$$

Noting that the stage efficiency from (4.15a) and (4.15b) is $\eta_s^o = H_s/H_s^*$ for the compressor and $\eta_s^o = H_s^*/H_s$ for the turbine, we may write a general relation between two work coefficients as

$$\Psi_s = \eta o_s^{o\pm 1}\Psi_s^* \tag{6.4c}$$

where the upper sign is for the compressor and the lower sign is for the turbine. Since $\eta_s^o \approx 0.8 - 0.9$ for both the turbine and the compressor, it makes little difference which work coefficient is used.

Now, we know that the efficiency is a function of H_s or H_s^*, Mach number M, Reynolds number Re, and mass or volume flow rate $\dot{m}$ or $\dot{V}$, and we can formally write the relation as

$$\eta_s^o = \eta_s^o(H_s \text{ or } H_s^*, M, \text{Re}, \dot{m} \text{ or } \dot{V}) \tag{6.5a}$$

Introducing the nondimensional variables, one may also write

$$\eta_s^o = \eta_s^o(\Psi_s \text{ or } \Psi_s^*, M, \text{Re}, \varphi_s \text{ or } \hat{\varphi}_s) \tag{6.5b}$$

Since the effect of Re and M are not large, as long M is less than 1, we may write

$$\eta_s^o = \eta_s^o(\Psi_s \text{ or } \Psi_s^*, \varphi_s \text{ or } \hat{\varphi}_s) \tag{6.5c}$$

Now, we'll further discuss the rpm n. Since

$$\Psi = \frac{2H}{u_{2m}^2} \rightarrow u_{2m} = \sqrt{\frac{2H}{\Psi}}$$

the rpm is

$$n = \frac{60u_2}{(\pi D_2)}$$

In addition, since

$$\varphi = \frac{\dot{V}_2}{\pi D_{2m}bu_{2m}}$$

we get for the average diameter the relation

$$D_{2m} = \frac{\dot{V}_2}{\pi bu_{2m}\varphi}$$

Alternatively, from another definition of the *discharge coefficient*,

$$\hat{\varphi} = \frac{4\dot{V}_2}{\pi D_{2m}^2 u_{2m}}$$

we can also write another relation for the mean diameter in the rotor's exit plane as

$$D_{2m} = \sqrt{\frac{4\dot{V}_2}{\pi u_{2m}\hat{\varphi}}}$$

We can now write for the rpm

$$n = \frac{60u_2}{\pi D_2} = 120\frac{\varphi_s u_2^2 b_2}{\dot{V}_2} = 120\frac{\varphi_s}{\Psi_s}\frac{H_s b_2}{\dot{V}_2}$$

or alternatively, also

$$n = \frac{60u_2}{\pi D_2} = \frac{60u_2}{\pi}\frac{\sqrt{\pi u^2 \hat{\varphi}_s}}{4\dot{V}_2} = \frac{30}{\sqrt{\pi}}\left(\frac{H_2}{\Psi_2}\right)^{3/4}\frac{\sqrt{\hat{\varphi}_s}}{4\dot{V}_2}$$

From either equation we get the two relations for dimensionless rpm per stage as

$$\hat{K}_{ns} = \frac{30n\dot{V}_2}{b_2 H_s} = 120\frac{\varphi_2}{\Psi_s} \tag{6.6a}$$

and

$$\hat{K}_{ns} = \hat{\varphi}_s^{1/2}\Psi_s^{-3/4} = \left(\frac{\sqrt{\pi}}{30 \times 2^{3/4}}\right)\frac{n\dot{V}_2^{1/2}}{H_s^{3/4}} = 0.03513\frac{n\dot{V}_2^{1/2}}{H_s^{3/4}}$$

$$= \left[\frac{4\varphi_s b_2}{D_2}\right]^{1/2}\Psi_s^{-3/4} = 2\left(\frac{b_2}{D_2}\right)\varphi_s^{1/2}\Psi_s^{-3/4} \tag{6.6b}$$

From the above two relations for dimensionless rpm, the first one requires the value of the characteristic transverse dimension b_2 and may be less suitable to evaluate than the second relation. In addition, we can slightly modify the relation for $\hat{K}_{ns}$ to get

$$\hat{K}_{ns} = \hat{\varphi}_s^{1/2}\Psi_s^{*-3/4} = 0.03513\frac{n\dot{V}_2^{1/2}}{H_s^{*3/4}} \tag{6.6c}$$

Table 6.1 Optimum characteristic values for turbomachines

Type	Reaction $\hat{r}$	φ_{opt}	ψ_{opt}
Impulse turbine	0.05	0.4–0.45	1.9–3.5
Reaction turbine	0.50	0.35–0.45	1.1–2.2
Centripetal turbine	0.4–0.5	0.3–0.65	0.6–1.2
Axial compressor	0.5	0.5	0.26–0.32
		0.7	0.4–0.5
		0.9	0.56–0.6
		1.1	0.72–0.9
Centrifugal compressor:			
Half-open	0.5–0.65	0.25–0.5	0.7–0.8
Closed rotor	0.65–0.9	0.2–0.35	0.6–0.8

Source: From Traupel (1966)

From design considerations, it is well known that an axial compressor stage has a lower H_s^* and a higher $\dot{V}_2$ than a centrifugal compressor. Hence, $\hat{K}_{ns}$ for a centrifugal compressor is much smaller than the one for an axial compressor. On the other hand, between an axial compressor and an axial turbine, the latter has a larger volume flow rate because of a higher temperature level and larger specific work per stage. Hence, $\hat{K}_{ns}$ for both an axial compressor and an axial turbine may be the same order of magnitude. Therefore, each type of machine has a range of $\hat{K}_{ns}$ associated with it.

There are certain optimum values of φ_s or $\hat{\varphi}_s$ and Ψ_s or Ψ_s^* under design condition available, for which the efficiency is at a maximum. A larger value of φ_s means larger exhaust losses, whereas a small value of φ_s means longer blades with higher friction losses. Similarly, for Ψ_s, larger means more exhaust losses, but smaller Ψ_s means more friction losses. Various authors have evaluated the design performance of many turbomachines and have given optimum values. Eckert (1953) gave the following range of $\hat{K}_{ns}$ as optimum:

Centrifugal compressor: $\hat{K}_{ns}^* = 0.1\text{--}0.45$
Axial compressor: $\hat{K}_{ns}^* = 0.45\text{--}2.0$
Axial blower: $\hat{K}_{ns}^* > 2$ (for large blade gap = pitch)

Traupel (1966) gave optimum values mainly for steam turbines and stationary gas turbines; these appear in Table 6.1. For these values by Traupel (1966), values for axial turbines have been computed based on D_{2m}, but for axial compressors, u_2 is based on $D_{2,tip}$.

While the previous data are given mainly for stationary turbomachines, we evaluated nondimensional numbers for aircraft gas-turbine components by using some of the published data on temperature and pressure before and after the compressor and turbine, and estimating the mean diameter and transverse dimension from cutout figures of the engine along with the published maximum diameter of the engine. These are given in Tables 6.2 and 6.3 for various single-spool aircraft jet engines with axial compressors and axial turbines, whereby to keep compatibility with Traupel's result, the characteristic diameter taken is at the root for the compressor and at the midpoint for the turbine.

Table 6.2 Characteristic parameters for axial compressor

No. of stages	π_c^o	π_{cs}^o	φ_s	ψ_s	$\hat{K}_{ns}$	ν
7–8	3.8–4.8	1.21–1.25	0.19–0.33	0.25–0.61	0.63–0.74	0.74–0.89
9–12	5.5–7.5	1.15–1.25	0.17–0.37	0.24–1.19	0.61–0.85	0.62–0.85
15–17	7.0–9.3	1.13–1.15	0.21–0.45	0.36–0.74	0.73–0.80	0.73–0.80

Table 6.3 Characteristic parameters for axial turbine

No. of stages	$1/\pi_t^o$	$1/\pi_{ts}^o$	φ_s	ψ_s	$\hat{K}_{ns}$	ν
1	1.73–2.17	1.73–2.17	0.40–0.52	1.39–1.82	0.21–0.27	0.74–0.88
1–2	2.11–3.14	1.45–2.19	0.38–0.89	0.9–4.0	0.15–0.37	0.71–0.86
2–3	2.89–3.91	1.43–1.84	0.48–0.75	0.48–0.75	0.19–0.31	0.79–0.84

The above nondimensional relations per stage can also be used for multistage engines with some modifications. If z is the number of stage in *series*, then

$$\dot{V}_{2,s} = \dot{V}_2 \rightarrow \hat{\varphi} = \hat{\varphi}_s$$

$$H_s^* = \frac{\Psi_s^*}{\frac{1}{2}u_2^2} = \frac{H^*}{\frac{1}{2}\sum u_2^2} = \frac{H^*}{\frac{zu_2^2}{2}} \rightarrow \Psi_s^* = \frac{\Psi^*}{z}$$

$$\hat{K}_{ns} = \hat{\varphi}_s^{1/2}\Psi_s^{*-3/4} = (\hat{\varphi})^{1/2}\left(\frac{\Psi^*}{z}\right)^{-3/4} = \frac{\hat{K}_n}{z^{3/4}}$$

If z stages are *parallel*, then the relations are

$$\dot{V}_{2,s} = \dot{V}_2/z \rightarrow \hat{\varphi}/z = \hat{\varphi}_s$$

$$H_s^* = H^* = \frac{\Psi_s^*}{\frac{1}{2}u_2^2} = \frac{H^*}{\frac{1}{2}\sum u_2^2} = \frac{H^*}{\frac{zu_2^2}{2}} \rightarrow \Psi_s^* = \frac{\Psi^*}{z}$$

$$\hat{K}_{ns} = \hat{\varphi}_s^{1/2}\Psi_s^{*-3/4} = \left(\frac{\hat{\varphi}}{z}\right)^{1/2}(\Psi^*)^{-3/4} = \frac{\hat{K}_n}{z^{1/2}}$$

Finally, we should mention certain *pseudo non-dimensional parameters* used in Anglo-Saxon literature, where for the characteristic speed here, actually n (*rpm*) is used. To explain the methodology, we'll first list the following dimensional quantities:

H = specific work [J/kg or m^2/s^2] (per stage or for all stages)

$\dot{V}$ = volume flow rate [m^3/s] (mean or at specific point)

n = characteristic speed [s^{-1}] (but actually min^{-1})

ρ = mass density [kg/m^3] (mean or at specific point); P = power [W or kgm^2/s^3]

μ = dynamic viscosity coefficient [kg/ms]

D = a characteristic dimension [m]

Out of the above seven dimensional variables, we have two primary scale variables (length L and mass M); according to the *Buckingham pi theorem*, the following five $(7 - 2 = 5)$ nondimensional variables or combinations of them can be made:

$$\dot{V}/(nD^3), H/(n^2D^2), P/(\rho n^3 D^5), \mu/(\rho n D^2) \text{ and } n\dot{V}^{1/2}/H^{3/4}$$

Out of the five nondimensional variables above, the fifth one, as may be noted, is similar to the already introduced $\hat{K}_n$.

There is another method of "nondimensionalization," especially in the English literature, where the performance of the compressor or turbine is specified by curves of delivery pressure and temperature plotted against mass flow for various fixed values of rotational speed, which are very useful for practical engineers. These are, however, dependent on other variables, such as the entry pressure and temperature conditions. When considering the dimensions of temperature, it is convenient always to associate with it the gas constant R, so that the combined variable is RT, which is the same as the velocity[2].

We may now formally introduce a relation between various quantities that influence the behavior of the turbomachinery as follows:

$$F(D, n, \dot{m}, p_i^o, p_e^o, RT_i^o, RT_e^o) = 0$$

By the principle of dimensional analysis, often referred to as *Buckingham's pi theorem*, it is known that the function of seven variables given above can be reduced to a different function of $7 - 3 = 4$ and time (t) in the dimensions of the original variables, as

$$F'\left(\frac{p_e^o}{p_i^o}, \frac{T_e^o}{T_i^o}, \frac{\dot{m}\sqrt{RT_i^o}}{p_i^o D^2}, \frac{nD}{\sqrt{RT_i^o}}\right) = 0$$

For a particular aircraft engine flying in atmospheric air, R and D are constant, and hence need not always be considered. Hence, when we are further concerned about the performance of a given machine under various off-design conditions, we may drop R and D and write

$$F''\left(\frac{p_e^o}{p_i^o}, \frac{T_e^o}{T_i^o}, \frac{\dot{m}\sqrt{T_i^o}}{p_i^o}, \frac{n}{\sqrt{T_i^o}}\right)$$

The quantities $(\dot{m}\sqrt{T_i^o}/p_i^o)$ and $(n/\sqrt{T_i^o})$ are usually called "nondimensional" mass flow and rotational speed, respectively, in English-language publications although they are not truly dimensionless.

Finally, it is worth noting one particular physical interpretation of the nondimensional mass flow and rotational speed parameters. These can be written as

$$\frac{\dot{m}\sqrt{RT_i^o}}{p_i^o D^2} = \frac{\rho A c_m \sqrt{RT_i^o}}{p_i^o} \approx \frac{p}{RT_i}\frac{A c_m \sqrt{RT_i}}{p_i D^2} \approx \frac{c_m}{\sqrt{RT_i}} \propto M$$

and

$$\frac{nD}{p_i^o D^2} \approx \frac{\mu}{\sqrt{RT_i}} \propto M_u$$

(M, M_u = flight Mach number, azimuthal flight Mach number).

All operating conditions covered by a pair of values of $(\dot{m}\sqrt{T_i}/p_i)$ and $(n/\sqrt{T_i})$ should therefore give rise to similar velocity triangles and have the same performance in terms of the pressure ratio, temperature ratio, and isentropic efficiency. Further, since sonic speed $a_s \sim \sqrt{T}$, at constant Mach number, all velocities $\sim \sqrt{T}$. Therefore, the *thrust at constant Mach number* is

$$F \sim p_\infty / \sqrt{T_\infty}.$$

6.1 Exercise

The following data are available for a fairly recent fanjet engine, P&W JT9D-1:

Mass flow rate = 698 kg/s
Takeoff thrust = 206 kN
Overall compression ratio = 21.5
Turbine inlet temperature = 1,243°C
Diameter = 2.43 m
Weight = 3,902 kg$_f$
SFC = 18.01
No. of spools = 2
rpm = 3,800 and 8,000 for low- and high-pressure compressor, respectively
Fan: compression ratio = 1.2 (1 stage); low-pressure compression ratio = 1.5 (7 stages); high-pressure compression ratio = 11.73 (7 stages); high-pressure turbine inverse pressure ratio = 4.40 (2 stages); low-pressure turbine inverse pressure ratio = 3.82 (4 stages)
Assume efficiencies as follows: inlet 0.98, compressor 0.86, combustion chamber 0.95, turbine 0.90, and exhaust nozzle 0.98

Compute the characteristic state points (temperature, pressure) of the entire thermodynamic cycle and determine the work done by each of the turbomachinery components, heat added in the combustion chamber and heat rejected in the cycle, thrust separately for core and fan part, and how the total thrust agrees with the above given (takeoff) thrust.

Estimate the characteristic numbers for each of the turbomachinery components, and examine how well they agree with the optimum values in the tables in this chapter (assume for each component that the degree of reaction $= 0.5$).

Chapter 7
Axial Compressors and Turbines

With the definition of the work coefficient in (6.4a) and (6.4b), we can now write from (4.22a) to (4.22d) the nondimensional azimuthal velocity component expressions as follows:

$$\frac{c_{u1}}{u} = (1 - \hat{r}) - \frac{\Psi}{4} \tag{7.1a}$$

$$\frac{c_{u2}}{u} = (1 - \hat{r}) + \frac{\Psi}{4} \tag{7.1b}$$

$$\frac{w_{u1}}{u} = \hat{r} + \frac{\Psi}{4} \tag{7.1c}$$

and

$$\frac{w_{u2}}{u} = \hat{r} - \frac{\Psi}{4} \tag{7.1d}$$

Therefore, the ratio of the tangent of the rotor blade angles is

$$\frac{\tan \beta_2}{\tan \beta_1} = \frac{w_{u1}}{w_{u2}} = \frac{\hat{r} + (\Psi/4)}{\hat{r} - (\Psi/4)} \tag{7.2}$$

7.1 Equation of Radial Equilibrium

Let's consider an axial machine stage in which no work is done by the gas in the gap (*pitch*) between the blades and there is no heat addition or loss. Further, an inviscid flow is assumed. Therefore, the energy equation can be written as

T. Bose, *Airbreathing Propulsion: An Introduction*,
Springer Aerospace Technology, DOI 10.1007/978-1-4614-3532-7_7,
© Springer Science+Business Media, LLC 2012

$$dh^{\circ} = dh + d\left(\frac{c^2}{2}\right) = dh + cdc = 0$$

Further, from the first law of thermodynamics, we have

$$dq = 0 = dh - \frac{1}{\rho}\,dp$$

Therefore,

$$\frac{1}{\rho}\,dp + cdc = 0 \qquad\qquad (7.3a)$$

Equation 7.3a should be valid in either direction in the gap. Keeping $\rho = $ constant in the gap, especially along the radius r, we can write

$$\frac{dp^{\circ}}{dr} = \frac{d}{dr}\left(p + \rho\frac{c^2}{2}\right) \qquad\qquad (7.3b)$$

In addition, the force balance along a column of constant cross section dA in the radial direction in the gap is

$$\rho\frac{c_u^2}{r}\,dA\,dr = dp.dA \;\rightarrow\; \rho\frac{c_u^2}{r}\,dr = dp \qquad\qquad (7.3c)$$

Thus, subtracting (7.3c) from (7.3a), we eliminate the pressure term and we additionally derive

$$cdc + \frac{c_u^2}{r}\,dr = 0 \qquad\qquad (7.3d)$$

Now, in a compressor or a turbine stage, the specific work per stage is

$$H_s = u_2 c_{u2} - u_1 c_{u1} = \omega(r_2 c_{u2} - r_1 c_{u1})$$

where ω is the *radian speed of the rotor*.

Further, from the power requirement, the power is

$$P = M_{\tau}\omega = \dot{m}H_s = \dot{m}\omega(r_2 c_{u2} - r c_{u1})$$

where M_{τ} is the *torque moment*. From the previous relation, the torque moment is

$$M_{\tau} = \dot{m}(r_2 c_{u2} - r c_{u1})$$

If there is no torque moment and no power is developed in the free-space gap between the blades, then $M_\tau = 0$, and thus, $r_2 c_{u2} = r_1 c_{u1} = r c_u$ and $r c_u = $ constant along r.

Therefore, in the gap, $d(r c_u) = 0$, from which it follows that

$$r\, dc_u + c_u\, dr = 0$$

Multiplying the above equation by (c_u/r), we get

$$c_u\, dc_u + \frac{c_u^2}{r}\, dr = 0 \tag{7.3e}$$

Combining (7.3d) and (7.3e), we get, therefore,

$$c\, dc - c_u\, dc_u = \frac{1}{2}\left[d(c^2 - c_u^2)\right] = \frac{1}{2} d(c_m^2) = 0$$

and $c_m(r) = c_m$ is therefore constant along r . It has thus been shown that as a consequence of $dh^o(r) = 0$ and $d(r c_u) = 0$ along r in the gap, we find that in the same direction (radial direction) in the gap, $dc_m = 0$.

It is therefore imperative that if we want uniform total enthalpy distribution and uniform c_m distribution in the gap of an axial machine, we must prescribe that $(r c_u) = $ constant must be maintained in the rotor. The above result may also be formulated in another manner. The energy equation for a stationary observer is

$$\frac{1}{\rho} dp + c\, dc = 0$$

$$\frac{dp^o}{dr} = \frac{d}{dr}\left(p + \frac{1}{2}\rho c^2\right) = \frac{d}{dr}\left[p + \frac{1}{2}\rho(c_m^2 + c_u^2)\right] = 0$$

$$\approx \frac{dp}{dr} + \rho\left(c_m \frac{dc_m}{dr} + c_u \frac{dc_u}{dr}\right) \tag{7.3f}$$

Now, for *radial force equilibrium*, (7.3c), $dp/dr = \rho c_u^2/r$, we can write

$$\frac{dp^o}{dr} = \rho\left[\frac{c_u^2}{r} + c_m \frac{dc_m}{dr} + c_u \frac{dc_u}{dr}\right] = 0 \tag{7.3g}$$

and if $dc_m/dr = 0$, it follows that

$$\frac{c_u}{r} + \frac{dc_u}{dr} = \frac{1}{r}\frac{d}{dr}(r c_u) = 0 \tag{7.3h}$$

By either method, we find that rc_u must remain constant along r. Now let $rc_u = k_1$ and $c_m = k_2$. Since the azimuthal speed $u = \omega r$, where ω is the radian speed, we write $w_u = u - c_u = \omega r - (k_1/r)$. Thus,

$$\tan\beta = \frac{c_m}{w_u} = \frac{k_2}{\omega r - (k_1/r)} = \frac{k_2 r}{\omega r^2 - k_1}$$

Let the blade root be denoted by the subscript i; then we can write

$$\frac{\tan\beta_i}{\tan\beta} = \frac{r_i}{r}\left(\frac{\omega r^2 - k_1}{\omega r_i^2 - k_1}\right)$$

Normally, $\omega r^2 \gg k_1$ which means that $\tan\beta_i/\tan\beta = (r/r_i) > 1$; that is, $\beta < \beta_i$, The blade angle is thus largest at the root and flatter toward the tip, which gives a structurally favorable design.

Now let's consider the special case of $c_{u1} = 0$. Hence, $H = uc_{u2}$. If in the gap between the rotor and the stator H is constant, then $rc_{u2} = uc_{u2} = c_{u2}/\omega = k_1$ is constant. Let $\omega k_1 = k'$. Thus, for

$$c_{u1} = 0 : w_{u1} = u, w_{u2} = u - c_{u2} = u - k'/u.$$

For axial machines, the *reaction*

$$\hat{r} = \frac{w_{\infty u}}{u} = \frac{(w_{u1} + w_{u2})}{2u} = \frac{1}{2u}\left[u + \frac{u - k'}{u}\right] = 1 - \frac{k'}{2u^2}$$

Further, since $u_{tip} > u_{root}$, for the special case of $c_{u1} = 0$, the reaction at the tip is therefore smaller than the reaction at the root ($\hat{r}_{tip} < \hat{r}_{root}$). On the other hand, for an axial machine,

$$\Delta c_u = c_{u2} - c_{u1} = (u - w_{u2}) - (u - w_{u1}) = w_{u1} - w_{u2} = -\Delta w_u$$

and from the definition of reaction, we have

$$w_{u\infty} = \frac{1}{2}(w_{u1} + w_{u2}) = \hat{r}u$$

and

$$c_{u\infty} = (1 - \hat{r})u.$$

Thus,

$$c_{u2} = c_{u1} + \Delta c_u = c_{u1} + c_{u2} - c_{u2} + \Delta c_u = 2c_{u\infty} - c_{u2} + \Delta c_u$$

from which it follows that

$$c_{u2} = c_{u\infty} + \frac{1}{2}\Delta c_u = c_{u\infty} - \frac{1}{2}\Delta w_u$$

Similarly,

$$c_{u1} = c_{u\infty} - \frac{1}{2}\Delta c_u = c_{u\infty} + \frac{1}{2}\Delta w_u = u - w_{u\infty} + \frac{1}{2}\Delta w_u$$

By comparing this with (4.22a), (4.22b), (4.22c), and (4.22d), we can therefore write

$$c_{u2} = u(1 - \hat{r}) + \frac{H}{2u} = u - w_{u\infty} + \frac{1}{2}\Delta c_u$$

and

$$c_{u1} = u(1 - \hat{r}) - \frac{H}{2u} = u - w_{u\infty} - \frac{1}{2}\Delta c_u$$

Let the work coefficient at the blade tip be

$$\Psi_{tip} = \frac{2H}{u_{tip}^2} = \frac{2H}{u}\frac{u}{u_{tip}^2} = \frac{2u\Delta c_u}{u_{tip}^2} = -2\left(\frac{u}{u_{tip}}\right)^2 \frac{\Delta w_u}{u} = -2\left(\frac{r}{r_{tip}}\right)^2 \frac{\Delta w_u}{u}$$

and we get

$$c_{u2} = \omega\left[r(1 - \hat{r}) + \frac{\Psi_{tip}}{4}\left(\frac{r_{tip}^2}{r}\right)\right] = \omega r\left[(1 - \hat{r}) + \frac{\Psi_{tip}}{4}\left(\frac{r_{tip}}{r}\right)^2\right]$$

and

$$c_{u1} = \omega\left[r(1 - \hat{r}) - \frac{\Psi_{tip}}{4}\left(\frac{r_{tip}^2}{r}\right)\right] = \omega r\left[(1 - \hat{r}) - \frac{\Psi_{tip}}{4}\left(\frac{r_{tip}}{r}\right)^2\right]$$

For the special case of $c_{u1} = 0$, we get from the above equation

$$\hat{r} = 1 - \frac{\Psi_{tip}}{4}\left(\frac{r_{tip}}{r}\right)^2$$

and since the tip-to-root radius is larger than 1, $\hat{r}_{tip} > \hat{r}_{root}$ which goes to prove again that the reaction at the tip is larger than the reaction at the root. Now for $(dp^o/dr = 0)$, we write from (7.3e) that

$$c_m \frac{dc}{dr} + c_u \left(\frac{c_u}{r} + \frac{dc_u}{dr} \right) = c_m \frac{dc_m}{dr} + \omega^2 r \left[(1 - \hat{r}) \pm \frac{\Psi_{tip}}{4} \left(\frac{r_{tip}}{r} \right)^2 \right]^2 \qquad (7.4)$$

Now, if $dc_m/dr = 0$, we get from (7.4) the relation for distribution of the reaction along r.

7.2   Selection of Airfoil Profile for Axial Machines

While the blades' entry and exit angle, as per the principal equation of turboma-chinery, is enough to determine the specific work in a stage, it must be compatible with the torque moment developed in that stage. For this purpose, let's examine the forces that act on blades (Fig. 7.1) in a stage.

For rotors, one can apply the same analysis if the relative approaching flow velocity w_∞ is considered instead of c_∞ for stationary blades. Further, let the resultant force acting on the blade because of the flow be **R**, whose components in the normal and parallel directions to the chord line are **L** and **D**, and the components in meridian and azimuthal planes are $\mathbf{F}_n$ and $\mathbf{F}_u$, respectively. The directions of **L** and **D** in the present case are, of course, somewhat different from the usual practice in aerodynamics, where these are normal and parallel to the approaching flow direction. Now for a two-dimensional cascade as that being considered here, all forces are in per unit length (that is, in N/m). Considering the flow between the radius r and $r + dr$, the power given to the blade element is $dP = H d\dot{m}$ where the element mass flow rate is

$$d\dot{m} = zt\rho_\infty w_\infty \sin(\beta_\infty - \alpha_\infty)$$

where z = number of blades, t = gap (pitch) between the blades, β_∞ is the blade chord angle, and α_∞ is the "angle of attack" of the approaching flow of the blade (with respect to the chord).

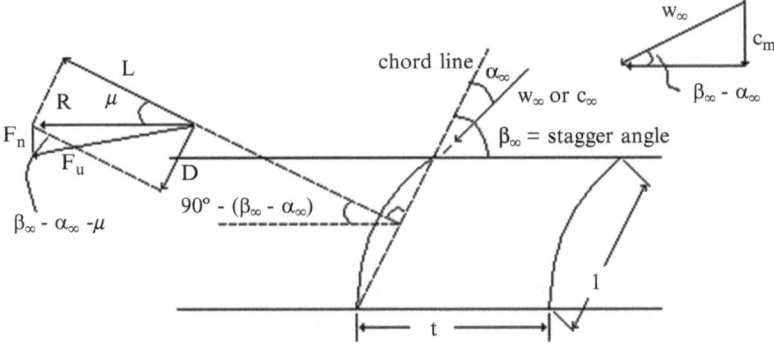

Fig. 7.1 Schematic forces acting on an axial blade cascade

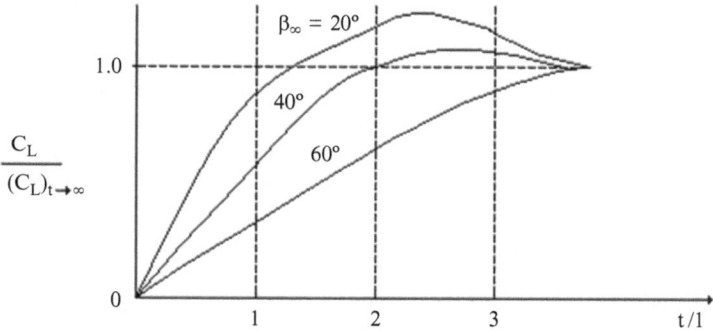

Fig. 7.2 Variation in lift coefficient c_L for different t/l and β_∞ as parameter

Further, the force in the direction of u is

$$dF_u = zR_u dr = zR\sin(\beta_\infty - \alpha_\infty \pm \mu)dr$$

where μ = angle between lift force direction and the normal to the chord.

Since power = force × velocity in the direction of motion, we have

$$dP = udF_u = zuR\sin(\beta_\infty - \alpha_\infty \pm \mu)dr = \frac{zLu}{\cos\mu}\sin(\beta_\infty - \alpha_\infty \pm \mu)dr$$

$$= Hd\dot{m} = zt\rho_\infty w_\infty u\Delta w_u \frac{\sin(\beta_\infty - \alpha_\infty \pm \mu)}{\cos\mu}dr$$

The *lift force* can now be written in turns of the lift coefficient, c_L, as

$$L = \frac{1}{2}\rho_\infty w_\infty^2 lc_L; l = \text{chord length}$$

and hence,

$$dP = \frac{1}{2}zu\rho_\infty w_\infty^2 lc_L \frac{\sin(\beta_\infty - \alpha_\infty \pm \mu)}{\cos\mu}dr = u\Delta w_u \rho_\infty w_\infty zt\sin(\beta_\infty - \alpha_\infty)dr$$

By a simple rearrangement, we further get (for μ small)

$$c_L\frac{l}{t} = \frac{2\Delta w_u}{w_\infty} = \frac{2H}{uw_\infty} \tag{7.5}$$

While (7.5) gives a simple expression to select a suitable c_L for a blade profile with proper chord length l and blade distant (pitch) t, unfortunately it is somewhat dependent on t/l, as c_L itself is somewhat dependent on t/l, as shown schematically in Fig. 7.2.

Furthermore, from (7.5), the following conclusions can be drawn:

1. For a given w_∞, a higher *turning angle* (larger $|\Delta w_u|$ or higher maximum *camber*) increases c_L, with a resultant higher P and higher $H = u\Delta w_u$.
2. A higher number of blades (z large and t small), if c_L is kept constant, gives higher power, but then H must be increased by a higher turning angle (higher c_L).
3. For high-speed machines (w_∞ large), $c_L(l/t)$ is small; that is, t/l is large [large blade separation (pitch) t or small chord length l], having a thinner profile near the root than the tip, which is also structurally convenient.

In older blades, the designer used to put a straight section near the trailing edge to force the flow at an exit angle required by the designer, which, however, causes additional friction. In the modern design practice of axial turbomachine blades, it has been realized that one can have an airfoil profile shape without making any special provision to force the flow in a particular direction. The problem of designing the blade may therefore be considered in either of two ways:

1. Given the entry and exit angles, find the profile shape and arrangement of blades (indirect method); or
2. Given are the entry angle and profile shape, find the exit angle (direct method).

We discuss these in the next section.

7.3   Calculating Potential Flow in a Blade Cascade

The present method of calculating the potential flow in a blade cascade is mostly computer-oriented, in which all effects, such as compressibility, can be taken into account. For this purpose, both finite-difference and finite-element methods have been proposed. However, for a better understanding of the physical flow phenomena, it is quite useful to look into some of the older analytical methods, developed mainly for incompressible flow.

7.3.1   Simple Vortex Theory

If Γ is the *vortex strength* (m^2/s) of a single *vortex*, there from the simple potential vortex theory of fluid dynamics in cylindrical coordinates (r, θ), the velocity potential and the stream function are given by the relations

$$\varphi = \frac{\Gamma\theta}{2\pi} \text{ and } \Psi = \frac{\Gamma}{2\pi} \ln r$$

Fig. 7.3 Representing a
blade cascade with a row of
discrete vortex filaments

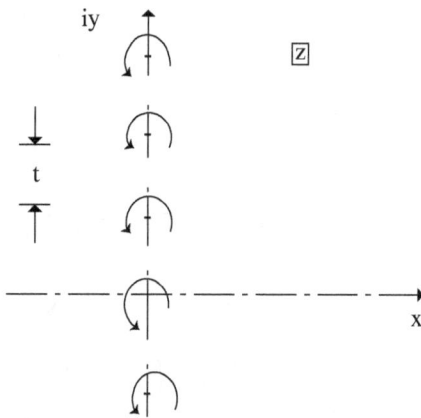

from which we get the relation for the *complex potential*

$$\chi = \varphi + i\Psi = -\frac{i\Gamma}{2\pi}\ln z$$

where $z = r\exp^{i\theta} = x + iy$ is the complex coordinate. Further, for a single vortex filament, the complex conjugate of velocity components, $\bar{\Omega} = u - iv$, is obtained by differentiating complex potential χ with respect to complex coordinate z to get

$$\bar{\Omega} = u - iv = \frac{d\chi}{dz} = -\frac{i\Gamma}{2\pi r\exp^{-i\theta}} = -\frac{i\Gamma}{2\pi r\exp^{-i\theta}}(\cos\theta - i\sin\theta)$$

$$= -\frac{x\Gamma}{2\pi r^2} - i\frac{y\Gamma}{2\pi r^2}$$

In the above equation, u and v are velocity components (m/s) in the x and y directions, respectively.

Therefore,

$$u = -\frac{x\Gamma}{2\pi r^2} \quad \text{and} \quad v = -\frac{y\Gamma}{2\pi r^2}$$

from which for absolute velocity $|V|$ we get

$$|V| = \sqrt{u^2 + v^2} = \frac{\Gamma\sqrt{x^2 + y^2}}{2\pi r} = \frac{\Gamma}{2\pi r}$$

Now, instead of a single vortex, we put a row of discrete vortex filaments of strength to represent the blade in a blade cascade (Fig. 7.3).

These are put at $(x = 0, y = nt)$, where n is an integer extending from $-\infty$ to $+\infty$. Therefore, the *complex potential* is

$$
\chi = -\frac{i\Gamma}{2\pi} \sum_{n=-\infty}^{+\infty} \ln(z - \text{i}nt) = -\frac{i\Gamma}{2\pi} \left[\ln z + \sum_{n=1}^{\infty} \ln\{(z + \text{i}nt)(z + \text{i}nt)\} \right]
$$

$$
= -\frac{i\Gamma}{2\pi} \left[\ln z + \sum_{n=1}^{\infty} (z^2 + n^2t^2) \right]
$$

$$
= -\frac{i\Gamma}{2\pi} \ln\left[z(z^2 + 4t^2)(z^2 + 9t^2) \ldots \right]
$$

$$
= -\frac{i\Gamma}{2\pi} \left[\ln\left\{ \pi\left(\frac{z}{t}\right)\left(1 + \frac{z^2}{t^2}\right)\left(1 + \frac{z^2}{2^2t^2}\right)\left(1 + \frac{z^2}{3t^2}\right) \ldots \right\} \right]
$$

$$
= -\frac{i\Gamma}{2\pi} \ln\left\{ \frac{tt^2(2^2t^2)(3^2t^2)}{\pi} \ldots \right\} \tag{7.6a}
$$

The second logarithmic term within the equation is not dependent on z and remains constant.

Noting that

$$
\pi x(1 + x^2)(1 + x^4) \ldots = \pi(x + x^3 + x^5 + x^7 + \ldots) = \sinh(\pi x)
$$

we write for the complex potential

$$
\chi = -\frac{i\Gamma}{2\pi} \ln\left[\sinh\left(\frac{\pi z}{t}\right) \right] \tag{7.6b}
$$

Further,

$$
\bar{\Omega} = u - iv = \frac{d\chi}{dz} = \frac{i\Gamma}{2t} \coth\left(\frac{\pi z}{t}\right) \tag{7.7}
$$

Noting further that

$$
\coth\left(\frac{\pi z}{t}\right) = \frac{1 + \coth\left(\frac{\pi x}{t}\right)\coth\left(\frac{i\pi y}{t}\right)}{\coth\left(\frac{\pi x}{t}\right) + \coth\left(\frac{i\pi y}{t}\right)} = \frac{1 - i\coth\left(\frac{\pi x}{t}\right)\coth\left(\frac{\pi y}{t}\right)}{\coth\left(\frac{\pi x}{t}\right) - i\coth\left(\frac{\pi y}{t}\right)}
$$

$$
= \coth\left(\frac{\pi x}{t}\right)\left[1 + \cot^2\left(\frac{\pi y}{t}\right)\right] - i\cot\left(\frac{\pi t}{t}\right)\frac{\left[\coth^2\left(\frac{\pi x}{t}\right) - 1\right]}{\coth^2\left(\frac{\pi x}{t}\right) + \cot^2\left(\frac{\pi y}{t}\right)}
$$

It follows that the velocity components are given by the relation

$$
u = \frac{\Gamma}{2t} \coth\left(\frac{\pi y}{t}\right)\frac{\left[1 - \cot^2\left(\frac{\pi x}{t}\right)\right]}{\coth^2\left(\frac{\pi x}{t}\right) + \cot^2\left(\frac{\pi y}{t}\right)} \tag{7.8a}
$$

and

$$v = \frac{\Gamma}{2t} \coth\left(\frac{\pi x}{t}\right) \frac{\left[1 - \cot^2\left(\frac{\pi y}{t}\right)\right]}{\coth^2\left(\frac{\pi x}{t}\right) + \cot^2\left(\frac{\pi y}{t}\right)} \tag{7.8b}$$

The limiting values of u and v obtained from (7.8a) and (7.8b) can, of course, be shown to be as follows:

$$y \to 0 : \cot\left(\frac{\pi y}{t}\right) \to \infty : u \to 0 \text{ and } v = \frac{\Gamma}{2t}\coth\left(\frac{\pi x}{t}\right)$$

and

$$x \to \pm\infty : \coth\left(\frac{\pi x}{t}\right) \to \pm 1, u \to 0 \text{ and } v = \pm\frac{\Gamma}{2t}.$$

$\bar{\Omega}_\infty = u_\infty - iv_\infty$ and for the velocity components given by (7.8a) and (7.8b) with u_∞ or v_∞ added to them, we get

$$u = u_\infty + \frac{\Gamma}{2t} \cot\left(\frac{\pi y}{t}\right) \frac{\left[1 - \coth^2\left(\frac{\pi x}{t}\right)\right]}{\left(\coth^2\left(\frac{\pi x}{t}\right) + \cot^2\left(\frac{\pi y}{t}\right)\right)} \tag{7.9a}$$

and

$$v = v_\infty + \frac{\Gamma}{2t} \cot\left(\frac{\pi x}{t}\right) \frac{\left[1 + \cot^2\left(\frac{\pi y}{t}\right)\right]}{\left(\coth^2\left(\frac{\pi x}{t}\right) + \cot^2\left(\frac{\pi y}{t}\right)\right)} \tag{7.9b}$$

Equations 7.9a and 7.9b have the limiting values

$$y \to 0 : u \to u_\infty, v = v_\infty + \frac{\Gamma}{2t}\coth\left(\frac{\pi x}{t}\right)$$

$$x \to \pm\infty : u \to u_\infty, v = v_\infty \pm \frac{\Gamma}{2t}$$

In earlier analysis, the meridian speed c_m was introduced, and it is evident that the present u_∞ is the same as c_m. Therefore, if the blade angles and one of the speeds were given, the value of α_i can be evaluated. With α_1 and α_e being the entry (inlet) and exit velocity, respectively, it is evident that at the leading edge

$$\sin \alpha_i = \frac{c_m}{w_i} = \frac{u_\infty}{v_\infty} + \frac{\Gamma}{2t}$$

and at the trailing edge

$$\sin \alpha_e = \frac{c_m}{w_\infty} = \frac{u_\infty}{v_\infty} - \frac{\Gamma}{2t}$$

Therefore,

$$u_\infty = c_m, v_\infty = \frac{c_m}{2}\left(\frac{1}{\sin \alpha_i} + \frac{1}{\sin \alpha_e}\right)$$

and

$$\Gamma = t c_m \left(\frac{1}{\sin \alpha_i} - \frac{1}{\sin \alpha_e}\right)$$

Since the lift generated by the blade profile is

$$L = \frac{1}{2}\rho_\infty V_\infty^2 l = \rho_\infty V_\infty \Gamma$$

where

$$V_\infty = \sqrt{u_\infty^2 + v_\infty^2} = c_m \sqrt{1 + \frac{1}{4}\left(\frac{1}{\sin \alpha_i} + \frac{1}{\sin \alpha_e}\right)}$$

we can write after some simple manipulation

$$c_L \frac{l}{t} = \frac{2\left(\frac{1}{\sin \alpha_i} - \frac{1}{\sin \alpha_e}\right)}{\sqrt{1 + \frac{1}{4}\left(\frac{1}{\sin \alpha_i} + \frac{1}{\sin \alpha_e}\right)^2}} = \frac{2|\Delta w_u|}{V_\infty} \tag{7.10}$$

While (7.10) gives an explicit equation to calculate $c_L(l/t)$, if the entry and exit angles are given, the method is too simple for practical use, since modern-day blades are like airfoils. It is therefore necessary to develop more complicated cascade theories.

7.3.2   Exact Calculation of Potential Flow

We are given the blade profile with entry angle α_i in an incompressible medium, and we would like to find the exit angle α_e and velocity distribution on the profile (Fig. 7.4).

We consider a line element of length ds and slope α on the blade at $\zeta = \xi + i\eta$, where there is a *vorticity strength* $\gamma(s)ds$. The effect of this line element is considered at a point $z = x + iy$. Earlier we had the relation

$$\bar{\Omega} = u - iv = -\frac{i\Gamma}{2t}\coth\left(\frac{\pi z}{t}\right)$$

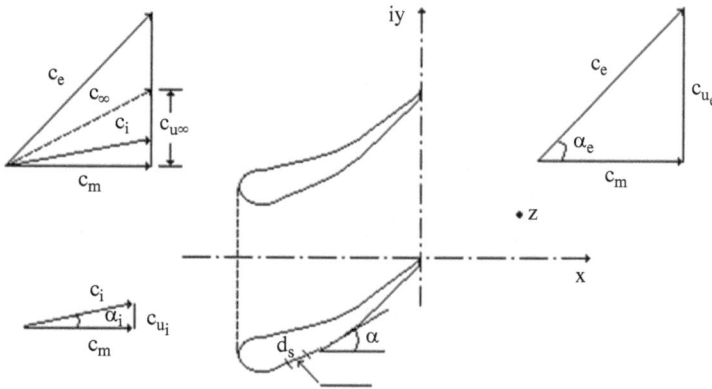

Fig. 7.4 To explain the theory of vorticity distribution

In an analogous manner, when the singularity due to vorticity is $\gamma(s)$ of length of the element ds is at z, its influence at z is given by the relation

$$d\bar{\Omega} = -i\gamma\left(\frac{s}{2t}\right) \coth\left[\frac{z - \zeta(s)}{t}\right] ds \qquad (7.11a)$$

Now we use the well-known mathematical expression

$$i \coth\left[\pi\frac{(z - \zeta(s))}{t}\right] = f_x - if_y$$

where

$$f_x = f_x(z, \zeta) = \frac{-\sin\left\{\frac{\pi(x-\xi)}{t}\right\} \cos\left\{\frac{\pi(x-\xi)}{t}\right\}}{\sinh^2\left\{\frac{\pi(x-\xi)}{t}\right\} + \sin^2\left\{\frac{\pi(y-\eta)}{t}\right\}} \qquad (7.12a)$$

and

$$f_y = f_y(z, \zeta) = \frac{-\sinh\left\{\frac{\pi(x-\xi)}{t}\right\} \cosh\left\{\frac{\pi(x-\xi)}{t}\right\}}{\sinh^2\left\{\frac{\pi(x-\xi)}{t}\right\} + \sin^2\left\{\frac{\pi(y-\eta)}{t}\right\}} \qquad (7.12b)$$

We note that in (7.12a) and (7.12b), if $(z \to \zeta)$, both f_x and f_y are indeterminate. On the other hand, if $(x \to \xi)$ only, then $f_x \to -\cot[\pi(y - \eta)/t]$ and $f_y \to 0$, but if $(y \to \eta)$ only, then $f_x \to 0$ and $f_y \to -\coth[\pi(x - \xi)/t]$. Therefore, (7.11a) becomes

$$d\bar{\Omega} = du - idv = \frac{\gamma(s)}{2t}[f_x(x, \zeta) - if_y(z, \zeta)]ds \qquad (7.11b)$$

or

$$u = \frac{1}{2t} \int \gamma(s) f_x \mathrm{d}s \tag{7.13a}$$

and

$$v = \frac{1}{2t} \int \gamma(s) f_y \mathrm{d}s \tag{7.13b}$$

We add to the expressions (7.13a) and (7.13b) the contribution of the approaching flow

$$\overline{\Omega}_\infty = c_\mathrm{m} - i c_\mathrm{m} \cot \alpha_\infty$$

and hence we therefore get

$$u = c_\mathrm{m} + \frac{1}{2t} \int \gamma(s) f_x \mathrm{d}s \tag{7.14a}$$

and

$$v = c_\mathrm{m} \cot \alpha_\infty + \frac{1}{2t} \int \gamma(s) f_y \mathrm{d}s \tag{7.14b}$$

The above expression is valid everywhere including on the blade surface Z, where the requirement for the tangency condition is

$$\frac{v(Z)}{u(Z)} = \cot \alpha_\infty(Z) \tag{7.15}$$

From (7.14a), (7.14b), and (7.15), we get

$$v(2t) = c_\mathrm{m}(2t) \cot \alpha_\infty + \int \gamma(s) f_y \mathrm{d}s = u(2t) \cot \alpha = c_\mathrm{m}(2t) \cot \alpha + \cot\alpha \int \gamma(s) f_x \mathrm{d}s$$

After rearranging, it follows that

$$2t c_\mathrm{m}(\cot \alpha - \cot \alpha_\infty) = \int \gamma(s) \left[f_y - f_x \cot \alpha \right] \mathrm{d}s \tag{7.15a}$$

Now, the circulation on each blade is

$$\Gamma = \int \gamma(s) \mathrm{d}s = t(c_\mathrm{ue} - c_\mathrm{ui}) = t c_\mathrm{m}(\cot \alpha_\mathrm{e} - \cot \alpha_\mathrm{i})$$

Since $c_{ui} = c_m \cot \alpha_i$, $c_{ue} = c_m \cot \alpha_e$, and $c_{u,ce} = c_m \cot \alpha_\infty$, and assuming also that

$$c_{u\infty} = \frac{1}{2}(c_{ui} + c_{ue}) = \frac{1}{2}c_m(\cot \alpha_i + \cot \alpha_e) = c_m \cot \alpha_\infty \rightarrow \cot \alpha_\infty$$

$$= \frac{1}{2}(\cot \alpha_i + \cot \alpha_e)$$

we therefore write

$$\Gamma = \int \gamma(s)ds = 2tc_m(\cot \alpha_\infty - \cot \alpha_i) \tag{7.15b}$$

In addition, on the point of singularity, when the *vorticity* is equal to the local surface speed, $\gamma \gamma(s) = c(s)$, we get from (7.15a) and (7.15b)

$$2tc_m(\cot \alpha - \cot \alpha_\infty) = \int c(s)\left[f_y - \cot \alpha f_x\right]ds = 2tc_m(\cot \alpha_\infty - \cot \alpha_i) = \int c(s)ds$$

Thus, we further find

$$2tc_m(\cot \alpha - \cot \alpha_i) = \int c(s)\left[f_y - f_x \cot \alpha + 1\right]ds$$

Therefore,

$$2t[\cot \alpha(Z) - \cot \alpha_i] = \int \frac{c(s)}{c_m}\left[f_y(Z,\zeta) - f_x(Z,\zeta)\cot \alpha(Z) + 1\right]ds \tag{7.16a}$$

Now we require a boundary condition, which is given by the well-known Kutta condition in fluid mechanics, at the trailing edge, $c(0) = 0$.
Further,

$$\Gamma = \int c(s)ds = tc_m(\cot \alpha_e - \cot \alpha_i)$$

from which it follows that

$$\cot \alpha_e = \cot \alpha_i + \frac{1}{t}\int \frac{c(s)}{c_m}ds \tag{7.16b}$$

Practical solution: Divide the surface into $(n + 1)$ equal parts; for example, in the figure $n + 1 = 19$, that is, $n = 18$.

Now, let $s_j = j$, where $j = 1$ to $(n + 1)$, $\Delta s = $ constant. Thus, from (7.16a), we find

$$2t\left[\cot\alpha(z_j) - \cot\alpha_i\right] = \sum \frac{c(s_k)}{c_m}\left[f_y - f_x\cot\alpha(Z_k) + 1\right]$$

which means that in the evaluation of the expression in a surface element, $(c(s_k)/c_m, k = 1$ to $n)$, Z_j is kept constant and Z_k is the coordinate for all other parts $(j = 1$ to $n)$.

Further from (7.16b), we have

$$\cot\alpha_e = \cot\alpha_i + \frac{\Delta s}{t}\sum \frac{c(s_k)}{c_m} \rightarrow \text{exit angle.}$$

Although for $j = k$, one gets f_x and f_y as being undetermined, one can put both f_x and f_y if $j = k$. Further calculations of u and v are from (7.14a) and (7.14b).

7.3.3 Numerical Calculation of Flow

The analytical methods discussed in the previous two subsections may be suitable utmost for liquid flow through blades and cannot reproduce results of *shocks* and other *gas-dynamic phenomenon* in a gas turbine.

Since Murman and Cole (1971) published their method of using type-dependent discretization of potential flow equation, many papers and books have been published, including one by this author (Bose 1997), and the flow equations have been described in many different forms. One such set of time-dependent equations in fully conservative form in vectorial form is as follows:

Eq. of continuity:

$$\frac{\partial\rho}{\partial t} + \nabla \bullet (\rho\mathbf{V}) = 0 \tag{7.17a}$$

Eq. of momentum:

$$\frac{\partial}{\partial t}(\rho\mathbf{V}) + [\{\nabla \bullet (\rho\mathbf{V})\}\mathbf{V}] = -\nabla p + \nabla \bullet \tau + \mathbf{F} \tag{7.17b}$$

Eq. of energy (in terms of stagnation specific enthalpy, h^o):

$$\frac{\partial}{\partial t}(\rho h^o - p) + \nabla \bullet (\rho h^o\mathbf{V}) = \nabla \bullet (k\nabla T) + \nabla \bullet (\tau \bullet \mathbf{V}) + \dot{Q} + \mathbf{F} \bullet \mathbf{V} \tag{7.17c}$$

Herein $\mathbf{V} = \{u, v, w\}$ is the *velocity vector* in the $\{x, y, z\}$ coordinate system, t is the time coordinate, ρ is the (mass) density of the fluid, p is the pressure, $\mathbf{F}$ is the volume force, $\dot{Q}$ is the volumetric energy input, $\tau = -\mu\nabla_n\mathbf{V}$, μ is the (dynamic) viscosity coefficient, and k is the heat conductivity coefficient.

Equations 7.17a, 7.17b, and 7.17c are written variously in finite-difference form, finite-volume form, finite-element form, and general coordinates system, and each of these forms has their own way of solution.

For blades, the solution has to be transformed first from regular to body-fitted coordinate system, for which various grids have been designed, for example, C-grids, O-grids, etc. (Thompson et al. 1985; Thompson and Weatherhill 1999). Sarkar (Ph.D. thesis, IIT Madras (Aero), 1997; see also his other papers in the "Bibliography") did a systematic analysis using Navier–Stokes equations and a two-equation turbulence model to compute numerically the flow and heat transfer over stationary turbine blade cascades. The two blade profiles, namely, a VKI turbine nozzle guide vane (Arts 1992) and a solid Rolls-Royce turbine rotor blade (Nicholson et al. 1984), were chosen to assess the performance of the flow solver for widely varying flow conditions. The effects of Mach number, Reynolds number, and free-stream turbulence intensity on the aerodynamics of blade and convective heat transfer were also determined. The film-cooling effectiveness was also calculated.

The time-dependent mass-averaged governing equations for mass, momentum, and energy were solved based on an explicit second-order accurate finite-volume formulation. Both the viscous and inviscid calculations were performed. An algebraically generated H-grid was used within the blade passage. For inviscid flow calculations, the wall boundary condition was that the flow velocity normal to the wall was zero, but for viscous flow calculations, the no-slip condition on the wall was maintained. The flow boundary condition before and after the blades was that the repeating velocity and the pressure difference were continuous. The bleeding from the surface for film cooling was imposed by boundary conditions. The predicted result illustrates aerodynamics and surface heat transfer with and without blade cooling. It also resolved the trailing-edge shock for some exit supersonic Mach numbers.

7.3.4 Design of Circular Arc Blades

Now, α_∞ is defined as $\alpha_\infty = (\alpha_1 + \alpha_2)/2$ Further, the *included angle* is

$$\theta = 180° - 2[\alpha_\infty + (90° - \alpha_2)] = 2\alpha_2 - \alpha_1 - \alpha_2 = \alpha_2 - \alpha_1$$

In addition, the radius of curvature $= R$, max. camber $= f_m$, and chord length $= l$.

Furthermore,

$$(R - f_m)^2 + \frac{l^2}{4} = R^2 \rightarrow \frac{R}{l} = \frac{f_m}{2l} + \frac{l}{8f_m}$$

The latter is a quadratic equation, whose solution is

$$\frac{f_m}{l} = \frac{R}{l} \pm \sqrt{\left(\frac{R}{l}\right)^2 - \frac{1}{4}}$$

Also, since $e = R(\cos \alpha_1 - \cos \alpha_2)$ and θ is the included angle,

$$\frac{l}{2R} = \sin\left(\frac{\theta}{2}\right) = \sin(\alpha_2 - \alpha_1)$$

Now, the following geometrical relations can be derived:

$$\frac{e}{l} = \sin \alpha_\infty = \frac{R}{l} |\cos \alpha_1 - \cos \alpha_2|$$

$$\frac{R}{l} = \frac{(e/l)}{|\cos \alpha_1 - \cos \alpha_2|} = \frac{\sin \alpha_\infty}{|\cos \alpha_1 - \cos \alpha_2|} = \frac{\sin\left(\frac{\alpha_1 + \alpha_2}{2}\right)}{|\cos \alpha_1 - \cos \alpha_2|}$$

$$\sin\left(\frac{\theta}{2}\right) = \sin\left(\frac{\alpha_2 - \alpha_1}{2}\right) = \frac{1}{2R}$$

$$\frac{d}{d_{max}} = x^* = \frac{x}{l}, x = \text{chord direction}$$

For moderate Mach numbers, Eckert (1953) gave the profile

$$f(x^*) = (2.975 - 2.667x^*)\sqrt{x^*(1 - x^*)}$$

with $f = 1$ at $x^* = 0.25$.
For high speeds, Eckert gave the NACA profile:

$$f(x^*) = 2.969\sqrt{x^*} - 1.260x^* - 3.516x^{*2} + 2.843x^{*3} - 1.0150x^{*4}$$
$$(f = 1 \text{ at } x^* = 0.33 \text{ and } f = 0.021 \text{ at } x* = 1).$$

7.4   Multiple Axial Compressors

Since for axial compressors the pressure increase per stage is small, multiple axial compressors are used in most of the gas-turbine engines in modern aircraft engines. They have a higher efficiency per stage than the centrifugal compressors and a higher air throughput.

For multiple axial compressors, designing blades with a constant average azimuthal speed is simpler to design, although the blade tip's diameter decreases continuously. One can keep the blade tip's diameter constant, which gives an aesthetically better design, but for a large number of stages, the blade length can become smaller faster than a constant-mean-diameter compressor. The third possibility, where the mean diameter reduces, can give a comparatively large blade height, even at a high pressure.

7.5 Two Examples

In the first example, we compute the overall thermodynamic cycle parameters, determine the rpm of the engine, and determine the number of stages of the compressor and of the turbine. We further compute the first stage of the compressor and the only turbine stage. In the second example, we compute somewhat in more detail the axial compressor only.

7.5.1 Example 1

Let's study the design of an aircraft gas-turbine engine with one axial compressor stage and one axial turbine stage in a straight-jet configuration with the following data (Fig. 7.5, shown schematically):

Mass flow rate $\dot{m} = 20$ kg/s, compression ratio $\pi_c^o = 5$, combustion chamber
  temperature $= 892°C \approx 1,165$ K, ambient state on ground: $p_\infty = 1.013$ bar,
  $T_\infty = 298$ K, and $M_\infty \approx 0$.

The calculation will be done in several steps, starting with overall cycle analysis, followed by preliminary design of the axial compressor.

Overall cycle analysis:
Assume

$$\eta_c^o = 0.84, \eta_t^o = 0.86, \eta_{c.c.}^o = 0.95, \eta_{noz} = 0.98$$
$$\Delta H_{p,fuel} = 4.27 \times 10^4 \text{ kJ/kg}, c_{pc} = 1.005 \text{ kJ/(kgK)}, \gamma_c = 1.4, c_{pt} = c_{p,noz.}$$
$$= 1.080 \text{ kJ/(kgK)},$$

$$\gamma_t = \gamma_{noz.} = 1.38, R = 287.5 \text{ kJ/(kgK)}$$

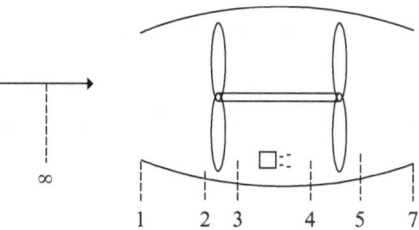

Fig. 7.5 Schematic of an
aircraft gas-turbine engine

All gas speeds inside the engine at all characteristic points are sufficiently small to warrant extra calculation of the compressibility effects.

Let $p_2^o \approx p_\infty = 1.013$ bar, $T_2^o \approx T_\infty = 298$ K, and thus, $\rho_2^o = \frac{p_2^o}{RT_2^o} = \frac{1.013 \times 10^5}{287.5 \times 298} = 1.1824$ kg/m^3

$$\frac{T_2^{*o}}{T_2^o} = \left(\frac{p_3^o}{p_2^o}\right)^{\frac{\gamma-1}{\gamma}} = (\pi_c)^{\frac{\gamma-1}{\gamma}} = 5^{0.286} = 1.5845 \rightarrow T_3^{*o} = 1.5845 \times 298 = 472.18 \text{ K}$$

from which it follows that $p_3^o = 5 \times 1.013 = 5.065$ bar.

Now, from stagnation efficiency we have

$$\eta_c^o = 0.84 = \frac{T_3^{*o} - T_2^o}{T_3^o - T_2^o} = \frac{472.18 - 298}{T_3^o - 298} \rightarrow T_3^o = 298 + \frac{174.18}{0.84} = 298 + 207.36$$

$$= 505.36 \text{ K}$$

and further,

$$\rho_3^o = \frac{5.065 \times 10^5}{287.5 \times 505.36} = 3.486 \text{ kg/m}^3$$

and so the state variables in position 4 (turbine inlet) are

$$T_4^o = 1,165 \text{ K}, \frac{(p_3^o - p_4^o)}{p_3^o} = 1 - 0.95 = 0.05 \rightarrow p_4^o = p_3^o(1 - 0.05)$$

$$= 5.065 \times 0.95 = 4.812 \text{ bar.}$$

For turbines: $R_t = \frac{\gamma_t - 1}{\gamma_t} c_{p,t} = \frac{0.38}{1.38} \times 1.080 \times 10^3 = 297.39 \frac{kJ}{kgK}$

Thus, the density at the turbine entry is $\rho_4^o = \frac{4.812 \times 10^5}{297.39 \times 1150} = 1.3889 \frac{kg}{m^3}$

Further,

$$c_{pc}(T_3^o - T_2^o) = c_{pt}(T_4^o - T_5^o):$$

$$T_5^o = T_4^o - \frac{c_{pc}}{c_{pt}}(T_3^o - T_2^o) = 1,165 - \frac{1.005}{1.080} \times 207.36$$

$$= 1,165 - 192.96 = 972.04 \text{ K}$$

Now, again from the definition of efficiency for turbine, we have

$$\eta_t^o = \frac{T_4^o - T_5^o}{T_4^o - T_5^{*o}} = 0.86 = \frac{192.96}{1,165 - T_5^{*o}} \rightarrow T_5^{*o} = 1,165 - \frac{192.96}{0.86} = 940.62 \text{ K}$$

Further, from the isentropic change-of-state relationship, we have

$$\frac{p_4^o}{p_4^o} = \left(\frac{T_4^o}{T_5^{*o}}\right)^{\frac{\gamma_t}{\gamma_t - 1}} = \left(\frac{1,165}{940.62}\right)^{\frac{1.38}{0.38}} = 1.2385^{3.63} = 2.175 \rightarrow p_5^o = 2.212 \text{ bar}$$

and the density at the turbine exit is $\rho_5^o = \frac{2.212 \times 10^5}{297.39 \times 972.03} = 1.652 \frac{\text{kg}}{\text{m}^3}$.

From the definition of the stagnation efficiency and adiabatic relation in nozzle, we have

$$\eta_{\text{Noz}} = \frac{T_5^o - T_7}{T_5^o - T_7^*} ; \frac{T_5^o}{T_7^*} = \left(\frac{p_5^o}{p_\infty}\right)^{\frac{\gamma-1}{\gamma}} = \left(\frac{2.212}{1.013}\right)^{0.2754} = 1.24 \rightarrow T_7^* = \frac{972.03}{1.24}$$

$$= 784.1 \text{ K}$$

Further results are

$$T_7 = T_7^o - \eta_{\text{Noz}}(T_5^o - T_7^*) = 972.03 - 0.95(972.03 - 784.1) = 793.5 \text{ K}$$

$$u_7 = \sqrt{2c_p(T_5^o - T_7)} = \sqrt{2 \times 1,080 \times (972.03 - 793.5)} = 621 \text{ m/s}$$

$$c_{p,cc} \approx \frac{1}{2}(c_{pc} + c_{pt}) = \frac{1}{2}(1.005 + 1.080) = 1.0425 \text{ kJ/kg}$$

From energy balance,

$$\dot{m}_a c_{p,cc}(T_4^o - T_3^o) = \dot{m}_f \Delta H_p$$

we get the fuel flow rate

$$\dot{m}_f = 20 \times 1.025 \times \frac{(1,165 - 505.37)}{4.27 \times 10^4} = 0.3221 \text{ kg/s} \rightarrow \dot{m}_f + \dot{m}_a = 20.3221 \text{ kg/s}$$

and the fuel–air ratio:

$$\frac{\dot{m}_f}{\dot{m}_a} = \frac{0.3221}{20} = \frac{1}{62.09} = 0.016.$$

Now, the thrust is given by

$$F = \dot{m}(u_7 - u_\infty) = 20.3221 \times 621 = 12.62 \text{ kN}$$

and thus the specific fuel consumption (per unit of thrust) is

$$\text{SFC} = \frac{\dot{m}_f}{F} = \frac{0.3221}{12.62 \times 10^3} = 25.25 \frac{\text{mg}}{\text{Ns}}$$

Further, the thermodynamic efficiency is

$$\eta_{th} = \frac{w}{q_{34}} = \frac{\left(u_7^2 - u_\infty^2\right)}{2c_{p,cc}\left(T_4^o - T_5^o\right)} = \frac{621^2}{2 \times 1,042.5 \times (1,165 - 505.37)} = 0.28$$

The jet exit Mach number is

$$M_7 = \frac{u_7}{\sqrt{\gamma R T_7}} = \frac{621}{\sqrt{1.38 \times 297.39 \times 793.5}} = \frac{621}{570.66} = 1.0882$$

Since the exit Mach number is marginally larger than 1, we now have the alternate option of having a conical nozzle (choked), which is easier to manufacture than a converging–diverging nozzle.

For a choked nozzle, we have

$$\frac{T_5^o}{T_7} = \frac{\gamma + 1}{2} = \frac{2.38}{2} = 1.19 \rightarrow T_7 = \frac{T_5^o}{1.19} = \frac{972.04}{1.19} = 816.83 \text{ K}$$

Further,

$$\frac{p_7^o}{p_7} = \left(\frac{T_7^o}{T_7}\right)^{\frac{\gamma}{\gamma-1}} = 1.19^{3.631} = 1.881 \rightarrow p_7 = \frac{2.212}{1.881} = 1.176 \text{ bar}$$

The exit velocity now is

$$u_7 = \sqrt{2c_{p,Noc}\left(T_5^o - T_7\right)} = \sqrt{2 \times 1.08 \times 10^{3 \times (972.03)}(972.04 - 816.83)}$$
$$= 579 \text{ m/s}$$

Now, the density is

$$\rho_7 = \frac{p_7}{R T_7} = \frac{1.176 \times 10^5}{297.39 \times 816.83} = 0.484 \text{ kg/m}^3$$

Thus, the cross-sectional area is

$$A_7 = \frac{\dot{m}_a + \dot{m}_f}{\rho_7 u_7} = \frac{20.3221}{0.484 \times 579} = 0.725 \text{ m}^2 \rightarrow D_7 = 0.3038 \text{ m}$$

The new thrust is now

$$F = \dot{m}(u_7 - u_\infty) = 20.3221 \times 579 = 11.776 \text{ kN}$$

and the specific fuel consumption

$$\text{SFC} = \frac{\dot{m}_f}{F} = \frac{0.3221}{11.776 \times 10^3} = 27.35 \; \frac{\text{mg}}{\text{Ns}}$$

The compressor's polytropic efficiency is

$$\eta_{\text{pol,c}} = \frac{\gamma - 1}{\gamma} \frac{\ln\left(p_3^o/p_2^o\right)}{\ln\left(T_3^o/T_2^o\right)} = 0.286 \frac{\ln 5}{\ln 1.696} = 0.871$$

and the turbine's polytropic efficiency is

$$\eta_{\text{pol,t}} = \frac{\gamma}{\gamma - 1} \frac{\ln\left(T_5^o/T_4^o\right)}{\ln\left(p_5^o/p_4^o\right)} = 3.63 \frac{\ln 0.8344}{\ln 0.46} = 0.8464$$

After having estimated the overall state points of the gas-turbine cycle, we now make a very preliminary design calculation.

Height of blade:
Let the tip and root diameters be $D_{\text{tip}} = D_m + b$ and $D_{\text{root}} = D_m - b$.
From area relations, we find

$$\frac{\pi}{4}\left(D_{\text{tip}}^2 - D_{\text{root}}^2\right) = 4\pi D_m b = 4\pi D_{\text{tip}}^2 \left(1 - \frac{b}{D_{\text{tip}}}\right)\frac{b}{D_{\text{tip}}}$$

leading to a quadratic equation for $\frac{b}{D_{\text{tip}}} = \frac{1}{2}\left(1 + \sqrt{1 - \frac{(1-v^2)}{4}}\right), v = \frac{D_{\text{root}}}{D_{\text{tip}}}$.

Preliminary design of axial compressor:
The average density is

$$\bar{\rho}_c = \frac{1}{2}\left(\rho_2^o + \rho_3^o\right) = \frac{1}{2}(1.1824 + 3.486) = 2.3342 \; \text{kg/m}^3$$

The average volume flow rate is

$$\dot{V}_c = \frac{20}{2.3342} = 8.568 \; \text{m}^3/\text{s}$$

The total work in the compressor is

$$H_c = c_{\text{pc}}\left(T_3^o - T_2^o\right) = 1.005(505.37 - 298) = 208.41 \; \text{kJ/kg}$$

The usual recommended value for an axial compressor is $\pi_c^{*o} \approx 1.2\text{--}1.5$, and for an axial turbine is $\left(\pi_t^{*o}\right)^{-1} \approx 1.73\text{--}2.17$. Since we have $\left(\pi_t^{*o}\right)^{-1} \approx 2.175$, we select for an axial turbine a single stage $z_t = 1$.

For an axial compressor, we select $\pi_{c,s}^o = 1.22$. Since

$$\left\{\pi_{c,s}^o\right\}^{z_c} = \pi_c^o \rightarrow z_c = \ln \pi_{c,s}^o = \ln(5)/\ln(1.22) = 0.69897/0.0863 = 8.09$$

the number of compressor stages we select is $z_c = 8$.

Let's also select the optimum values of nondimensional numbers for the compressor as follows:

$$\hat{\varphi}_s = 0.25, \hat{\Psi}_s = 0.4, \hat{K}_n = 0.7 \text{(subject to the condition that 1st stage}$$
$$\text{tip Mach number} < 1)$$

Now, for the compressor stage work,

$$\overline{H}_{c,s} = H_c/z_c = 208.41/8 = 26.0512 \text{ kJ/kg}.$$

and the heating factor for the compressor is

$$1 + f_c = \frac{\eta_{pol,c}}{\eta_{ad,c} - 1}\left(1 - \frac{1}{z_c}\right) + 1 = \frac{0.871}{0.84 - 1}\left(1 - \frac{1}{8}\right) + 1 = 1.0323$$

Therefore, the average isentropic work in the compressor is

$$\overline{H}_{cs}^* = \frac{H_c^*}{z_c}(1 + f) = \frac{c_{pc}\left(T_3^{*o} - T_2^o\right)}{z_c}(1 + f) = \frac{1.005(472.181 - 298)}{8} \times 1.0323$$
$$= 22.6 \text{ kJ/kg}$$

and the average compressor stage efficiency is

$$\eta_{cs}^o = \frac{\left\{\pi_{cs}^o\right\}^{\frac{\gamma - 1}{\gamma_e}} - 1}{\frac{\tau_c^o}{z_c} - 1} = \frac{5^{0.286/8} - 1}{\frac{505.87/298}{8} - 1} = 0.866$$

From the overall cycle analysis, we can now decide the distribution of the work load in each of the stages. For compressor stage I, we assume 15% less than the average work. Thus, let

$$H_I^* = 0.85\overline{H}_{cs}^* = 0.85 \times 22.6 = 19.21 \text{ kJ/kg}$$

Thus,

$$H_I = \frac{H_I^*}{\eta_{cI}^o} = \frac{19.21}{0.866} = 22.182 \text{ kJ/kg}$$

Further,

$$\Delta T_I^o = \frac{H_I}{c_p} = \frac{22.182}{1.005} = 22.072 \text{ K} \rightarrow T_{If}^o = 298 + 22.072 = 320.07 \text{ K}$$

is the stagnation temperature after the first stage (the subscript "f" denotes the final value of the particular stage).

$$\Delta T_I^{*o} = \frac{H_1^*}{c_p} = \frac{19.21}{1.005} = 19.11 \text{ K} \rightarrow T_{If}^{*o} = 298 + 19.11 = 317.11 \text{ K}$$

and the pressure ratio in the first stage

$$\pi_{cI}^o = \left(\frac{317.11}{298}\right)^{3.5} = 1.2431 \rightarrow p_{If}^o = 1.2431 \times 1.013 = 1.259 \text{ bar}$$

Therefore, the stagnation density at the end of the first stage is

$$\rho_{If}^o = \frac{1.259 \times 10^5}{287.5 \times 320.07} = 1.3684 \text{ kg/m}^3$$

and the average stagnation density in the first stage is $(1.3684 + 1.1824)/2 = 1.2754 \text{ kg/m}^3$. We assume the flow velocity to be small, and thus we consider the average stagnation density rather than the average static density. Therefore, the average volume flow rate in the first stage is

$$\dot{V}_I = \frac{20}{1.2754} = 15.681 \text{ m}^3/\text{s}.$$

We presume also that the first-stage optimum value for $\hat{K}_{nI}$ should be slightly more than the average. Let $\hat{K}_{nI} = 0.76$. In addition, let's select for the first stage

$$\hat{\varphi}_I = \frac{c_m}{u_{tip}} = \frac{4\dot{V}_I}{\pi D_{tip,I}^2(1 - v_I^2)u_{tip}} = \frac{4 \times 15.681}{\pi D_{tip,I}^2(1 - v_I^2)u_{tip}} = 0.25$$

$$\hat{\Psi}_I = 0.9\Psi_{aver.} = 0.9 \times 0.4 = 0.36 = \frac{2H_I}{u_{tip,I}^2} = \frac{2 \times 2.2182 \times 10^4}{u_{tip,I}^2} \rightarrow u_{tip,I}$$

$$= 351 \text{ m/s}$$

$$\hat{K}_{nI} = 0.76 = 0.03513\frac{n\dot{V}_I^{1/2}}{H_I^{3/4}} = 0.03513 \times \frac{n \times \sqrt{15.681}}{(2.2182 \times 10^4)^{3/4}} = 7.6537 \times 10^{-5}n$$

Therefore,

$$c_m = \hat{\varphi}_I u_{tip,I} = 0.25 \times 351 = 87.75 \text{ m/s},$$

$$w_{tip,I} = \sqrt{c_m^2 + u_{tip,I}^2} = \sqrt{87.78^2 + 351^2} = 351\sqrt{1 + \left(\frac{87.75}{351}\right)^2}$$

$$= 351 \times 1.0307 = 361.802 \text{ m/s}$$

Hence,

$$(1 - v_I^2) = \frac{4\dot{V}_I}{\pi D_{tip,I}^2 u_{tip}\hat{\varphi}_I} = \frac{4 \times 15.681}{\pi \times 0.675^2 \times 351 \times 0.25} = \frac{19.966}{0.25 \times 351 \times 0.455625}$$

$$= 0.4993 \rightarrow v_I = 0.7076$$

$$\frac{D_{I,root}}{D_{I,tip}} = 0.7076 \rightarrow D_{I,root} = 0.7076 \times 0.6775 = 0.4794 \text{ m}$$

Height of blade: $b_I = \frac{1}{2}(0.6775 - 0.4794) = 0.09905 \text{ m} = 9.9 \text{ cm}$

Now, the sonic speed at the entry for the first stage is $a_{s,I} = 20.05\sqrt{298} = 346.$
12 m/sand the tip Mach number of the first stage is given by

$$M_{tip,1I} = \frac{w_{tip,I}}{a_{s,I}} = \frac{361.802}{346.12} = 1.045$$

This is quite a modest supersonic value, and hence we put the condition of normal entry in the first stage of the compressor rotor. It can also be seen that the mean and root azimuthal speeds of the first stage rotor are $u_{m,I} = 325.52$ m/s and $u_{root,I} = 300.04$ m/s. By considering the entry meridian speed as 87.75 m/s, it is possible to compute the rotor blade entry angle at the root, mean diameter, and tip (in degrees) as 19.4°, 15.09°, and 14.04°, respectively. We'll do further calculations at the mean diameter.

Since

$$H_I = u_{m,I}(c_{um2,I} - c_{um1,I}) = u_{m,I}c_{um2,I} = 22.182 \times 10^3 \text{ J/kg}$$

thus at the mean diameter, $c_{um2,I} = 68.14$ m/s and $w_{u2m,I} = u_{2m,I} - c_{u2m,I} = 257.3$ 8 m/s.

Keeping the meridian speed at the trailing edge of the first-stage rotor again uniform and equal to the leading edge, we compute the velocities $w_2 = 271.93$ m/s and $c_2 = 226.37$ m/s, and the rotor trailing-edge angle = 18.82°. Further, if we keep the zero azimuthal component of the velocity at the trailing edge of the first-stage rotor, and keep the meridian speed again the same, we find $c_3 = c_{m1} = 87.75$ m/s. After we have gotten all the velocities at the characteristic points of the first stage and we have gotten the stagnation temperature and stagnation pressure from the overall analysis the work done, we can get easily the static temperatures: $T_1 = 294.17$ K, $T_2 = 313.93$ K, and $T_3 = 335.59$ m/s. We can also find the static pressures from knowledge of the efficiency at characteristic points. The method can no doubt be repeated for other compressor stages.

Now, from the overall cycle analysis for the single-turbine stage, the entry stagnation temperature $= 1,165$ K, the turbine exit stagnation temperature $= 972.03$ K, the entry stagnation pressure $= 4.88$ bar, and the exit stagnation pressure $= 2.212$ bar. This gives the work

$$H = c_{p,t}(T_4^o - T_5^o) = 208.41 \text{ kJ/kg}$$

Selecting reaction $= 0.5$, $\Psi_t = 2.0$ (equivalent to $c_{u1} = 0$), and an estimated meridian speed of 150 m/s, we get the azimuthal speed $u = 456.5$ m/s and the following blade angles: turbine nozzle and rotor entry angle $= 90°$, turbine rotor and nozzle exit angle $= 19.18°$. Keeping the mass flow rate and rpm in the turbine the same as those in compressor, we can now find some of the other data as follows: mean diameter $= 0.439$ m, blade height $= 0.071$ m.

7.5.2 Example 2

As a second example, we look into the case of a multistage aircraft axial compressor with inlet diffuser for operation of the aircraft at a designed altitude and speed. (For air, specific heat ratio, $\gamma = 1.4$, $c_p = 1.00$ kJ/(kgK), $R = 287.5$ kJ/(kgK)).

Given: $\dot{m} = 15.6$ kg/s, compression ratio $p_{ex}^o/p_{in}^o = 6$, flight speed: 700 km/h $= 194.5$ m/s, engine rpm $= 8,500$ min^{-1}.
For design altitude: 8 km with air state at $p_\infty = 0.3629$ bar, $T_\infty = 236$ K.
Flight Mach number: $M_\infty = 194.5/(20.05\sqrt{236}) = 194.5/308.0 = 0.631465$
Diffuser: Let $\eta_d = 0.85$.

Since

$$\frac{T_\infty^o}{T_\infty} = 1 + \frac{\gamma - 1}{2}M_\infty^2 = 1 + 0.2 \times 0.631465^2 = 1.07975 \rightarrow T_\infty^o = 254.821 \text{ K}$$

we further have

$$\frac{p_\infty^o}{p_\infty} = \left(\frac{T_\infty^o}{T_\infty}\right)^{\frac{\gamma}{\gamma-1}} = 1.07975^{3.5} = 1.30807 \rightarrow p_\infty^o = 0.474 \text{ bar}$$

Also,

$$\eta_d = 0.85 = \frac{(p_{in}^o/p_\infty) - 1}{[1 + \frac{\gamma-1}{2}M_\infty^2]^{\frac{\gamma}{\gamma-1}} - 1} \rightarrow \left(\frac{p_{in}^o}{p_{ki}}\right) = 1 + \eta_d\left[\left(\frac{p_\infty^o}{p_\infty}\right) - 1\right]$$

$$= 1 + 0.85 \times 0.30807 = 1.262$$

Hence, $p_{in}^o = 0.4565$ bar $\to p_{ex}^o = 6 \times p_{in}^o = 6 \times 0.4565 = 2.7393$ bar.
Inlet state of the compressor:

$$p_{in}^o = 0.4565 \text{ bar}, \quad T_{in}^o = 254.82 \text{ K}, \quad \rho_{in}^o = 0.6231168 \text{ kg/m}^3$$

$$T^{*o} = T_{in}^o \times 6^{0.286} = 425.39 \text{ K}$$

$$H^* = 1,005 \frac{\text{J}}{\text{kgH}} (425.39 - 254.82) = 1.714423 \times 10^5 \frac{\text{J}}{\text{kg}}$$

For aircraft application, the maximum velocity before combustion chamber <100 m/s, and degree of reaction $= 50\%$.
The entry volume flow rate for a compressor is

$$\dot{V}_{in} \approx \frac{15.6}{0.62312} = 25.035 \frac{\text{m}^3}{\text{s}}$$

Let the adiabatic efficiency for the compressor be 0.84:

$$H = \frac{H^*}{0.84} = \frac{1.714423 \times 10^5}{0.84} = 2.04098 \times 10^5 \frac{\text{J}}{\text{kg}}$$

Now for an axial compressor, the optimum values are $\varphi_{s,opt} = 0.45$ to $0.5, \Psi_{s,opt} = 0.2$ to 0.6.
Let's select: $\varphi_{s,opt} = 0.45, \Psi_{s,opt} = 0.45, v_{s,opt} = 0.7$

$$K_{ns} = 0.03513 \frac{n\dot{V}_s^{1/2}}{H_s^{3/4}} = \frac{\varphi_s^{1/2}}{\Psi_s^{3/4}} \sqrt{1 - v^2} = \frac{\sqrt{0.45}}{0.45^{3/4}} \sqrt{0.51} = \frac{0.71417}{0.8190} = 0.872$$

Now

$$T_{ex}^o = T_{in}^o + \frac{H}{c_p} = 254.82 + \frac{2.04098 \times 10^5}{1,005} = 457.90 \text{ K}$$

$$\rho_{ex}^o \approx \frac{2.7393 \times 10^5}{287.5 \times 457.9} = 2.08 \frac{\text{kg}}{\text{m}^3} \to \dot{V}_{ex} = \frac{15.6}{2.08} = 7.497 \frac{\text{m}^3}{\text{s}}$$

The average volume flow rate is

$$\dot{V}_{aver} = \frac{1}{2} \left(\dot{V}_{in} + \dot{V}_{ex} \right) = \frac{1}{2} (25.035 + 7.497) = 16.27 \frac{\text{m}^3}{\text{s}}$$

Now

$$K_n = 0.03513 \frac{n\sqrt{V_{\text{aver}}}}{H_s^{3/4}} = 0.03513 \frac{8,500\sqrt{16.27}}{H_s^{3/4}} = 0.872$$

$$H_s = \left(0.03513 \times \frac{8,500 \times \sqrt{16.27}}{0.872}\right)^{4/3} = 1.5383 \times 10^4 \ \frac{J}{kg}$$

Thus, the number of stages is

$$z = \frac{2.04 \times 10^5}{1.5385 \times 10^4} = 13.3 \rightarrow \text{select} : z = 13.\text{V}$$

$$\Psi_s = \frac{2H}{zu_{2m}^2} = 0.45 \rightarrow u_{2m} = \sqrt{\frac{2 \times 2.04 \times 10^5}{13 \times 0.45}} = 264 \text{ m/s}$$

$$u_{2m} = \frac{\pi n D_{\text{in}}}{60} \rightarrow D_m = \frac{60 \times 264}{\pi \times 8,500} = 0.593 \rightarrow \text{select } D_{2m} = 0.6 \text{ m}$$

$$\rightarrow u_{em} = \frac{\pi \times 0.6 \times 8,500}{60} = 267 \ \frac{m}{s}$$

$$\bar{\Psi}_s = \frac{2H}{zu_{2m}^2} = \frac{2 \times 2.04 \times 10^5}{13 \times 267^2} = 0.44$$

$$\varphi_s \approx \frac{\dot{V}_{\text{aver}}}{\pi \bar{D}_{2m}b_{2\infty}u} = 0.45 \rightarrow b_2 = \frac{16.27}{\pi \times 0.6 \times 0.45 \times 267} = 0.0718 \rightarrow \text{select} :$$
$$b_2 0.072 \text{ m}$$

Average inside diameter $= 0.6 + 0.072 = 0.672$ m; average inside diameter $= 0.6 - 0.072 = 0.528$ m:

$$\bar{v} = \frac{0.528}{0.672} = 0.787$$

Recalculate:

$$K_{ns} = \frac{\sqrt{\varphi_s}}{\Psi_s^{3/4}}\sqrt{1 - v^2} = \frac{\sqrt{0.45}}{0.44^{3/4}}\sqrt{1 - 0.787} = \frac{0.6708}{0.540} \times 0.213 = 0.240$$

$$H_s = \frac{H}{z} = \left(0.03513 \times 8,500 \times \frac{\sqrt{16.27}}{0.240}\right)^{4/3} = 8.592 \times 10^4 \ \frac{J}{kg}$$

No. of stages: $z = \frac{2.04 \times 10^5}{1.822 \times 10^4} = 1.12 \rightarrow$ select $z = 2$

$$u_{2m} = \sqrt{\frac{2 \times 2.04 \times 10^5}{11 \times 0.45}} = 287 \frac{m}{s}$$

$$D_m = \frac{60 \times 287}{\pi \times 8,500} = 0.6443 \rightarrow \text{select } D_{2m} = 0.644 \text{ m}$$

$$u_{2m} = \frac{\pi \times 0.644 \times 8,500}{60} = 286.6 \frac{m}{s}$$

$$\overline{\Psi}_s = \frac{2 \times 2.04 \times 10^5}{11 \times 286.6^2} = 0.451$$

$$\Psi_s = 0.45 = \frac{16.27}{\pi \times 0.644 \times 286.6 \times \overline{b}_2}$$

$$\overline{b}_2 = \frac{16.27}{0.45 \times \pi \times 0.64 \times 286.6} = 0.0627 \rightarrow \text{select; } \overline{b}_2 = 0.063 \text{ m}$$

Outside diameter $= 0.644 + 0.063 = 0.707$ m, root diameter $= 0.644 - 0.063 = 0.581$ m $m \rightarrow \overline{v} = 0.822$.

Final information:
Let

$$\overline{v} = 0.8 \rightarrow K_{ns} = \frac{\varphi_s^{1/2}}{\Psi_s^{3/4}} \sqrt{1 - v^2} = \frac{\sqrt{0.45}}{0.451^{0.75}} \sqrt{1 - 0.822^2} = 0.694$$

$$H_s = \frac{H}{z} = \frac{0.3513 \times 8,500 \times \sqrt{16.27}}{0.694^{4/3}} = 2.0856 \times 10^4 \frac{J}{kg}$$

$$z = \frac{2.04 \times 10^5}{2.0856 \times 10^4} = 9.78 \rightarrow \text{select } z = 10$$

$$u_{2m} = \sqrt{\frac{2 \times 2.04 \times 10^5}{10 \times 0.45}} = 301 \frac{m}{s} \rightarrow D_{2m} = 0.6768 \text{ m} \rightarrow \text{select } D_{2m} = 0.676 \text{ m}$$

$$\overline{b}_2 = 0.0566 \rightarrow \text{select: } \overline{b}_2 = 0.057 \text{ m} \rightarrow \overline{v} = 0.824$$

Preliminary design data:
No. of stages $= 10$

$$\overline{v} = 0.844, \varphi_s = 0.45, \overline{\Psi}_s = 0.45, K_{ns} = 0.694$$

$$\overline{D}_{2m} = 0.676 \text{ m}, \bar{b}_2 = 0.057 \text{ m}$$

$$\eta^o_{ad,c} = 0.84, \pi^o_c = 6 \rightarrow \pi^o_{cs} = 6^{1/10} = 1.1962$$

$$\eta_{pol} = \frac{\gamma - 1}{\gamma} \frac{\ln \pi^o_c}{\ln \left[1 + \frac{\pi_c^{\frac{\gamma-1}{\gamma}} - 1}{\eta^o_{ad,c}}\right]} = \frac{0.286 \times 1.7917}{0.586} = 0.874$$

$$\eta^o_{ad,cs} = \frac{\{\pi^o_{cs}\}^{\frac{\gamma-1}{\gamma}} - 1}{\tau^o_{cs} - 1} = 0.872$$

Heating factor:

$$1 + f_c = 1 + \left(\frac{\eta_{pol,c}}{\eta^o_{ad,c}} - 1\right)\left(1 - \frac{1}{z}\right) = 1 + \left(\frac{0.874}{0.840} - 1\right)\left(1 - \frac{1}{10}\right) = 1.03643$$

$$\overline{H}^*_s = \frac{H*}{10}(1 + f_c) = \frac{1.7763}{10} \times 1.03643 = 1.77688 \times 10^4 \frac{\text{J}}{\text{kg}}$$

Let us keep $D_{2m} = \overline{D}_{2m} = 0.676 \text{ m} \rightarrow u_{2m} = \frac{\pi \times 8,500 \times 0.6443}{60} = 286.7 \frac{\text{m}}{\text{s}}$
We now make estimates for the first and last stages.
First stage:
Let's set the work done in the first stage to be slightly smaller than the average. Thus, let

$$H^*_1 = 0.85 \times \overline{H}^*_s = 0.85 \times 1.77688 \times 10^4 = 1.51035 \times 10^4 \text{ J/kg}$$

$$H_1 = 1.51035 \times 10^4 / 0.84 = 1.798 \times 10^4 \text{ J/kg}$$

$$\Psi_1 = \frac{2H_1}{u^2_{2m,1}} = \frac{2 \times 1.798 \times 10^4}{301^2} = 0.397$$

Since $\dot{V}_1 = 25.035 \text{ m}^3/\text{s}$,

$$\varphi_1 \approx \overline{\varphi}_s = 0.45 = \frac{\dot{V}_1}{\pi D_{2m,1} b_{2m,1} u_{2m,1}} = \frac{25.035}{\pi \times 0.676 \times b_{2m,1} \times 301} \rightarrow b_{2m,1}$$

$$= 0.087 \text{ m}.$$

Thus, the first stage blade height is $b_{2m,1} = 0.087 \text{ m} \times 2 = 0.177 \text{ m}$.

Therefore,

$$v_1 = \frac{0.676 - 0.087}{0.676 + 0.087} = \frac{0.557}{0.731} = 0.762$$

Now,

$$c_{m1} = u_{m1}\overline{\varphi}_s = 301 \times 0.45 = 135.45\ \frac{m}{s}$$

$$u_{tip,1} = \frac{0.731}{0.676} \times 301 = 325.5\ \frac{m}{s}$$

Further,

$$T_1 = T_1^o - \frac{c_{m1}^2}{2,010} = 154.82 - \frac{135.45^2}{2,010} = 245.69\ K \rightarrow \text{sonic speed: } a_{s1}$$

$$= 314.2\ m/s.$$

Since there may be the problem of shock, we reduce the azimuthal speed to u_{m1} = 220 m/s, and therefore,

$$D_{m1} = \frac{60u_{m1}}{\pi n} = \frac{60 \times 220}{\pi \times 8,500} = 0.494 \rightarrow \text{select } D_{m1} = 0.5\ m.$$

We recalculate $u_{m1} = \pi \times 0.5 \times 8,300/60 = 222.53$ m/s, which means that

$$\Psi_s = \frac{2 \times 1.732 \times 10^4}{222.53^2} = 0.699$$

which is too high. Therefore, we may also reduce H_1^* and H_1. Now, select $\varphi_1 > \overline{\varphi}_s$ = 0.45.
Let

$$\varphi_1 = 0.6 = \frac{\dot{V}_1}{\pi D_{2m,1}b_{2m,1}u_{m,1}} = \frac{25.035}{\pi \times 0.5 \times 222.53 \times b_{2m,1}} \rightarrow b_{2m,1} = 0.119 \approx 0.12$$

Thus,

$$\varphi_1 = \frac{\dot{V}_1}{\pi D_{2m,1}b_{2m,1}u_{m,1}} = \frac{25.035}{\pi \times 0.5 \times 222.53 \times 0.12} = 0.596$$

$$v_1 = \frac{0.5 - 0.12}{0.5 + 0.12} = \frac{0.38}{0.62} = 0.612$$

$$u_{tip,1} = \frac{\pi \times 0.62 \times 8,500}{60} = 275.94 \; \frac{m}{s} \rightarrow M_{tip,1} = \frac{u_{tip,1}}{a_{s,1}} = \frac{275.94}{314.2} = 0.878$$

Xth stage:

Further, in the last stage, we increase the work per stage by 15% of the average value:

$$H_X^* = 1.15 \times H_s^* = 2.0434 \times 10^4 \; \frac{J}{kg}$$

Thus,

$$H_X = \frac{H_X^*}{\eta_{ad,X}^o} = \frac{2.0434 \times 10^4}{0.872} = 2.3434 \times 10^4 \; \frac{J}{kg}$$

In our design, we would like to have the mean diameter in the 10th stage be the same as in the first stage.

Thus,

$$u_{m,X} = 301 \; \frac{m}{s}, D_{m,X} = 0.676 \; m \rightarrow \Psi_X = \frac{2H}{u_{m,X}^2} = 0.517$$

With the volume flow rate in the 10th stage as $\dot{V}_X = 7.497 \; m^3/s$

$$\varphi_X \approx \varphi_s = \frac{7.497}{\pi \times 0.676 \times 301 \times b_{2m,X}} = 0.45 \rightarrow b_{2m,X} = 0.026 \; m$$

$$v_x = \frac{0.676 - 0.026}{0.676 + 0.026} = \frac{0.650}{0.702} = 0.926$$

We can now compute all intermediate stages. For this we would have to distribute the rest of the total H^* values to stages II to IX.

Considering for an axial compressor a typical ratio of compressor length to diameter of $L_{ax}/D_{tip,1} = 0.25 \; -0.35$, the length of the compressor $= 0.3 \times 10 \times 0.733 = 2.199 = 2.2 \; m$.

Chapter 8
Centrifugal Compressor

While single- and multiple-stage centrifugal compressors are used exclusively for terrestrial applications, since per stage, a centrifugal compressor gives a much higher compression ratio, they have not been used for aircraft applications, except in small engines, such as the old Goblin or the Rolls-Royce DART turboprop engine for the Avro-748 aircraft; in the latter, a two-stage centrifugal compressor is used. There are several reasons for the infrequent use of centrifugal compressors over axial compressors: (a) Centrifugal compressors have a slightly lower efficiency than axial compressors; (b) centrifugal compressors have a much smaller air throughput over axial compressors, and they also have a very large frontal area, giving a large drag. However, the centrifugal compressor rotors are structurally much more robust and can be built, especially for terrestrial applications, by simple welding of the blades and by using materials like plastics for applications with hazardous gases.

We begin our analysis with a small sketch of a centrifugal compressor (Fig. 8.1). The compressor consists of a rotor, whose blade entry diameter is D_1 and width is b_1, and blade exit diameter is D_2, and width b_2; it further consists of a stator with entry and exit diameters D_3 and D_4, respectively, with corresponding widths b_3 and b_4, and a spiral chamber as a diffuser.

As mentioned in Sect. 5.2, there will be losses at the entry to the rotor:

$$\Delta h_1 = \zeta_1 \frac{1}{2} c_1^2, \zeta_1 = 0.1 - 0.15 \tag{8.1a}$$

and further losses in the rotor due to turning of the flow and friction in rotor:

$$\Delta h_2 = \zeta_2 \frac{1}{2} w_1^2, \zeta_2 = 0.2 - 0.25 \tag{8.1b}$$

T. Bose, *Airbreathing Propulsion: An Introduction*,
Springer Aerospace Technology, DOI 10.1007/978-1-4614-3532-7_8,
© Springer Science+Business Media, LLC 2012

Fig. 8.1 Schematic sketch
of a centrifugal compressor

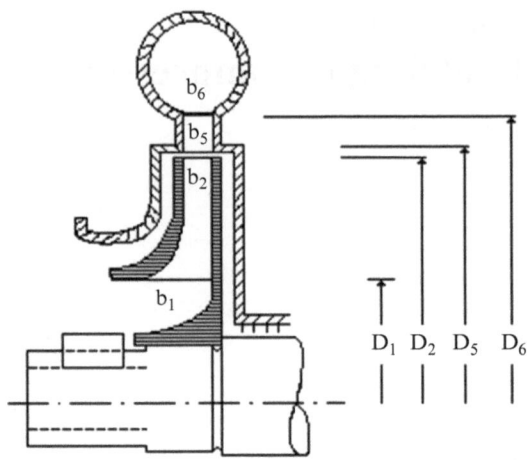

In the above equations, the ζs are *coefficients of losses*. In the case of no
azimuthal component of velocity at entry, $c_{u1} = 0$, that is, $c_{m1} = c_1$ and $w_1^2 = c_{m1}^2$
$+u_1^2 = c_1^2 + u_1^2$ we get for the total loss

$$\Delta h_{loss} = \Delta h_1 + \Delta h_2 = \frac{1}{2}\left(\zeta_1 c_1^2 + \zeta_2 w_1^2\right) = \frac{1}{2}\left[\zeta_1 c_{m1}^2 + \zeta_2\left(c_{m1}^2 + u_1^2\right)\right] \quad (8.2a)$$

For the *rotor's energy loss coefficient*, we therefore write

$$\Delta\Psi_{loss} = \frac{2\Delta h_{loss}}{u_2^2} = \left(\frac{c_{m1}}{u_2}\right)^2(\zeta_1 + \zeta_2) + \left(\frac{u_1}{u_2}\right)^2\zeta_2 \quad (8.2b)$$

Now, let

$$\chi = \frac{b_1}{D_1} \quad \text{and} \quad c_{m1} = \frac{\dot{V}_1}{\pi D_1 b_1}, b_1 = \text{channel width}$$

Therefore,

$$\frac{c_{m1}}{u_2} = \left(\frac{\dot{V}_1}{\pi\chi D_1^2 u_2}\right) = \frac{1}{4\chi\left(\frac{D_1}{D_2}\right)^2}\frac{\dot{V}_1}{\frac{\pi D_2^2}{4}} = \frac{\hat{\varphi}}{4\chi(D_1/D_2)^2} \quad (8.3)$$

where

$$\hat{\varphi} = \frac{4\dot{V}_1}{(\pi D_2^2)u_2}$$

and

$$\Delta \Psi_{\text{loss}} = \frac{(\zeta_1 + \zeta_2)\hat{\varphi}^2}{(4\chi)^2 (D_1/D_2)^4} + \zeta_2 \left(\frac{D_1}{D_2}\right)^2$$

To optimize the loss, we differentiate the previous equation with respect to (D_1/D_2) and set it equal to zero to get

$$-\frac{4(\zeta_1 + \zeta_2)\hat{\varphi}^2}{(4\chi)^2 (D_1/D_2)^5} + 2\zeta_2 \left(\frac{D_1}{D_2}\right) = 0$$

Therefore,

$$\left(\frac{D_1}{D_2}\right)_{\text{opt}}^6 = \frac{2(\zeta_1 + \zeta_2)\hat{\varphi}^2}{(4\chi)^2 \zeta_2} H$$

Let's take the following reasonable values for the constants: $\zeta_1 = 0.1$, $\zeta_2 = 0.2$, and $\chi = 0.2$.

Therefore,

$$\left(\frac{D_1}{D_2}\right)_{\text{opt}} \approx 1.3\sqrt[3]{\hat{\varphi}} \tag{8.4a}$$

While we have taken the value of χ to be 0.2, we will now try to justify it. Let's assume that for the sake of continuity, the cross-sectional ring area at the compressor entry, $\pi(D_1^2 - d_H^2)/4$, where d_H is the hub diameter, must be larger than the rotor entry area, $\pi D_1 b_1$. Let's further assume that the former needs to be larger than the latter by only 5%. Therefore,

$$\frac{\pi}{4}(D_1^2 - d_H^2) = 1.05\pi b_1 D_1 \rightarrow \frac{b_1}{D_1} = \frac{1}{4.2}\left[1 - \left(\frac{d_H}{D_1}\right)^2\right]$$

Now d_H can be 0 (frontal entry without any shaft in the front) to a maximum of about 0.5, beyond which there will not be enough passage area. Therefore, $\chi = b_1/D_1 = 1/4.2$ to $1/5.6$, and on average, $\chi = 0.2$ is fully justified. Thus, we further have

$$\left(\frac{c_{m1}}{u_2}\right)_{\text{opt}} = \frac{\hat{\varphi}}{4\chi(D_1/D_2)_{\text{opt}}^2} = \frac{\hat{\varphi}}{0.8 \times (1.3\sqrt[3]{\hat{\varphi}})^2} = 0.739\hat{\varphi}^{1/3} \tag{8.4b}$$

Now,

$$\tan \beta_1 = \frac{c_{m1}}{u_1} = \frac{c_{m1}}{u_2} \cdot \frac{u_2}{u_1} = \frac{c_{m1}}{u_2} \cdot \frac{D_2}{D_1} = \frac{0.739\hat{\varphi}^{1/3}}{1.3 \times \hat{\varphi}^{1/3}} = 0.4372 \qquad (8.4c)$$

and therefore strictly for $c_{u1} = 0, \beta_{1,\text{opt}} = 30°$.

Further, from the principal equation of turbomachinery, (4.17a), for $c_{u1} = 0, H = u_2 c_{u2}$, and therefore the work coefficient (for $c_{u1} = 0$) is

$$\Psi = \frac{2H}{u_2^2} = \frac{2c_{u2}}{u_2} .$$

From the velocity triangle, we have

$$\tan \beta_2 = \frac{c_{m1}}{u_2 - c_{u2}} = \frac{c_{m2}}{u_2 \cdot \frac{1}{1 - \Psi/2}}$$

which means that for $c_{u1} = 0$,

$$\Psi = 2 \left[1 - \frac{c_{m2}}{u_2 \tan \beta_2} \right]$$

which is a straight-line equation in terms of (c_{m2}/u_2). Further, the degree of reaction is

$$\hat{r} = \frac{H_{\text{rotor}}}{H} = \frac{\left[(u_2^2 - u_1^2) + (w_1^2 - w_2^2) \right]}{2u_2 c_{u2}}$$

Once again by assuming $c_{u1} = 0$ and $c_{m1} = c_{m2} = c_m$, we write

$$c_1^2 = c_{m1}^2 = w_1^2 - u_1^2 \text{ and } w_2^2 - c_m^2 = (u_2 - c_{u2})^2$$

and therefore,

$$w_1^2 - w_2^3 - u_1^2 = -(u_2 - c_{u2})^2$$

The degree of reaction thus becomes (for $c_{u1} = 0$)

$$\hat{r} = \frac{u_2^2 - (u_2 - c_{u2})^2}{2u_2 c_{u2}} = 1 - \frac{c_{u2}}{2u_2} = 1 - \frac{\Psi}{4} \qquad (8.5a)$$

If one defines a *rotor energy coefficient*

$$\Psi_{\text{rotor}} = \hat{r}\Psi = \frac{2H_{\text{rotor}}}{u_2^2} \qquad (8.5b)$$

Fig. 8.2 Relations among ψ, ψ_{rotor}, $\hat{r}$, and c_{m2}/u_{2m}

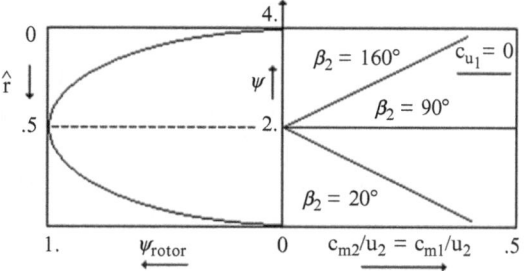

Fig. 8.3 Schematic sketch of forward-moving, radial exit, and rearward-moving rotors

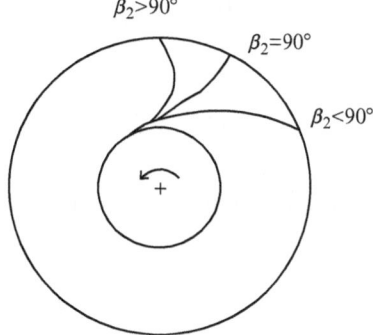

then

$$\Psi_{rotor} = \Psi\left(1 - \frac{\Psi}{4}\right) \tag{8.5c}$$

Equations 8.5a, 8.5b, and 8.5c are now shown in Fig. 8.2, strictly for $c_{u1} = 0$.

We can see that for $\beta_2 > 90°$, the reaction $\hat{r} < 0$, and there will be very little static pressure increase, although the overall pressure increase, mostly in the stator and diffuser, can be substantial. Therefore, these so-called forward-moving blades (Fig. 8.3) are good for a large mass flow rate and a large overall compression ratio per stage, but with little pressure increase in the rotor. On the other hand for $\beta_2 < 90°$, with little curvature, the compressors have good efficiency and are chosen for stationary units. For aircrafts, $\beta_2 = 90°$ is the standard practice since this is the best form from a structural point of view.

While for $c_{u1} = 0$ we have discussed the relationship among $(\hat{r}, \Psi, \beta_2)$, we have not discussed any optimum values for these. We will this do now by bringing in from (8.4b) the result that

$$\left(\frac{c_{m1}}{u_2}\right)_{opt} \approx \left(\frac{c_{m2}}{u_2}\right)_{opt} \approx 0.739\hat{\varphi}^{1/3}$$

Therefore, after substituting in (8.5a) and (8.5c), we get

$$\tan \beta_{2,\text{opt}} = \frac{c_{m2}}{u_2} \frac{1}{1 - \frac{\Psi}{2}} = \frac{0.739 \hat{\varphi}^{1/3}}{1 - 2(1 - \hat{r})} = \frac{0.739 \hat{\varphi}^{1/3}}{2\hat{r} - 1}$$

For a typical range of values of $\hat{\varphi}$ from 0.03 to 0.075 (average, say, 0.05), we therefore get $(c_{m1}/u_2)_{\text{opt}} = 0.3684$ and $\beta_{2,\text{opt}} = \tan^{-1}[0.3684/(2\hat{r} - 1)]$. Centrifugal compressors for aircraft applications may not, however, have the condition $c_{u1} = 0$ always satisfied. Instead, one could assume $w_{u2} = 0$. Hence, from (4.17a), we have

$$\Psi = \frac{2H}{u_2^2} = \frac{2}{u_2^2}[u_2^2 - u_1 c_{u1}] = 2\left[1 - \frac{u_1}{u_2}\frac{c_{u1}}{u_2}\right] = 2\left[1 - \left(\frac{u_1}{u_2}\right)^2 \frac{c_{u1}}{u_1}\right]$$

$$= 2\left[1 - \left(\frac{u_1}{u_2}\right)^2 \left(1 - \frac{w_{u1}}{u_1}\right)\right] = 2\left[1 - \left(\frac{u_1}{u_2}\right)^2 \left(1 - \frac{w_{u1}}{c_{m1}}\right) \cdot \frac{c_{m1}}{u_2} \cdot \frac{u_2}{u_1}\right]$$

Taking the following optimum values:

$$\left(\frac{c_{m1}}{u_2}\right)_{\text{opt}} = 0.739 \hat{\varphi}^{1/3}; \left(\frac{u_1}{u_2}\right)_{\text{opt}} = 1.3\sqrt[3]{\hat{\varphi}} \text{ and } \left(\frac{w_{u1}}{c_{m1}}\right)_{\text{opt}} = \frac{1}{\tan \beta_{1,\text{opt}}} = 1.732$$

we get further

$$\Psi = 2\left[1 - 1.69\hat{\varphi}^{2/3}\left\{1 - \frac{1.732 \times 0.739}{1.3} \frac{\hat{\varphi}^{1/3}}{\hat{\varphi}^{1/3}}\right\}\right]$$

$$= 2[1 - 1.69\hat{\varphi}^{2/3}\{1 - 0.986\}] = 2[1 - 0.024\hat{\varphi}^{2/3}]$$

Taking the value of $\hat{\varphi} = 0.03 - -0.075$, the corresponding work coefficient range is 1.9–2.0.

The work done in a centrifugal stage is smaller when the finite number of blades is H, whereas for the infinite number of blades, it is H_∞. The ratio of the two is the *power-lowering factor*, also called the *slip factor*, and for centrifugal compressors is given by the expression

$$\mu = \frac{H}{H_\infty} = \frac{1}{1 + \frac{\pi}{2} \frac{\sin \beta_2}{z\left(1 - \frac{r_1}{r_2}\right)}}$$

where z is the number of blades. It can be seen that for $z \to 0 : \mu \to 0$ (no blade) and $z \to \infty : \mu \to 1$ (infinite number of blades). However, if there are an infinite number of blades, there cannot be any flow, and if z is large only, then the flow would be better, but there would be more friction losses. On the other hand, if z is small, there can be secondary flow losses. An optimum number of blades may be obtained if the blade channel can be considered a diffuser, and from the limit of the diffuser opening angle, one can limit the value of t/l to within 0.35–0.45, where t is the gap between blades (pitch) and l is the chord length. Further, from geometrical considerations we may write for the number of blades

$$z = \frac{2\pi \sin\left(\frac{\beta_1 + \beta_2}{2}\right)}{\frac{t}{l} \ln\left(\frac{D_2}{D_1}\right)}$$

Now, the rotor exit's width is obtained from the continuity equation

$$b_2 = \frac{\dot{V}_2}{\pi D_2 c_{m2}}$$

For the design of blades for given entry and exit angles, one could do the design in various ways. One of the simplest ways is to consider a *circular arc blade*, whose *radius of curvature* is given by the relation

$$R = \frac{r_2^2 - r_1^2}{2(r_2 \cos \beta_2 - r_1 \cos \beta_1)}$$

An alternate method is to prescribe the distribution of the local blade angle $(\beta(r), \varphi)$ to be linear, for example. In such a case, the azimuthal angle φ from the inner radius $(r_1 = D_1/2)$ is given by the relation

$$\varphi^\circ = \frac{180}{\pi} \int_{r_1}^{r} \frac{dr}{r \tan \beta(r)}$$

On exit from the rotor, one has generally an area 3–4 without vanes, followed by stator vanes 5–6. The width of the entire channel is generally given by

$$b_6 = b_5 = b_4 = b_2 + (1 \text{ to } 2 \text{ mm})$$

Further, let $c_{m3} \approx c_{m2}$ and $c_{u3} \approx c_{u2}$. Therefore, in the space without guide vanes and without viscous action, it follows that

$$c = \sqrt{c_m^2 + c_u^2} = \sqrt{c_{m2}^2 \left(\frac{r_2 b_2}{rb}\right)^2 + \left(\frac{r_2}{r}\right)^2 c_{u2}^2} = \left(\frac{r_2}{r}\right)\sqrt{c_{m2}^2 \left(\frac{b_2}{v}\right)^2 + c_{u2}^2} \approx \left(\frac{r_2}{r}\right)c_2$$

However, for a viscous flow,

$$\text{Friction coefficient}\, \frac{1}{rc_\text{u}} = \frac{1}{r_2 c_\text{u2}} + \frac{c_\text{f} \pi (r - r_2)}{2\dot{V}}$$

in which the $c_\text{f} \approx 0.02 - 0.04$.

The local slope can be taken approximately for the inviscid case as

$$\tan \alpha = \frac{c_\text{m}}{c_\text{u}} = \left(\frac{r_2 b_2}{rb} \right) \frac{r}{r_2} \frac{c_\text{m2}}{c_\text{u2}} = \frac{b_2}{b} \tan \alpha_2 \approx \tan \alpha_2$$

where α_2 is the angle of the absolute velocity at the rotor exit. For stator vanes, the following values were recommended by Eckert (1961):

$$\frac{D_5}{D_2} = 1 + \left(\frac{1}{8} \text{ to } \frac{1}{14} \right)$$

$$\tan \alpha_5 = \tan \alpha_4 = \tan \alpha_3 + c_\text{f} \frac{(r_5 - r_2)}{4 b_2}$$

where again the friction coefficient $c_\text{f} \approx 0.02 - 0.04$.

Further, $D_6/D_5 = 1.25 - 1.35$, and $C = $ opening ratio of stator blade as diffuser $= \frac{D_6 \sin \alpha_6}{D_5 \sin \alpha_5} \approx 2.5$.

8.1 Calculation of a Centrifugal Compressor

The numerical example is based on the one given by Eckert (1961). The values given are *mass flow rate* $\dot{m} = 5$ kg/s, *compression ratio* $\pi_\text{c} = 4.0$, *ambient pressure* $p_\infty = 1$ bar, and *temperature* $T_\infty = 303$ K, under static conditions.

Now, the *ambient density* is

$$\rho_\infty = \frac{p_\infty}{RT_\infty} = \frac{10^5}{287.5 \times 303} - 1.1479 \, \frac{\text{kg}}{\text{m}^3} = \rho_1$$

The volume flow rate is

$$\dot{V}_1 = \frac{\dot{m}}{\rho_1} = \frac{5}{1.1479} = 4.355625 \, \frac{\text{m}^3}{\text{s}}$$

The exit stagnation pressure is

$$p_\text{ex}^o = \pi_\text{c} p_\infty^o - 4 \text{ bar}$$

and the corresponding isentropic stagnation temperature is

$$T_{\text{ex}}^{*o} = T_\infty^o \pi_{\text{c}}^{\frac{\gamma-1}{\gamma}} = 303 \times 4^{0.286} = 450.43 \text{ K}$$

The isentropic specific work is

$$H^* = c_{\text{p}}\left(T_{\infty,\text{ex}}^{*o} - T_\infty^o\right) = 1,005 \times (450.43 - 303) = 1.4871 \times 10^5 \text{ J/kg}$$

Let $\eta_{\text{c}}^o = 0.984$. Therefore,

$$H - \frac{H^*}{\eta_{\text{c}}^o} = \frac{1.4871 \times 10^5}{0.84} = 1.7693 \times 10^5 \, \frac{\text{J}}{\text{kg}}$$

Now,

$$\hat{K}_{\text{n}} = 0.03513 \times n\{\text{min}^{-1}\}\left[\dot{V}_1\{\text{m/s}\}\right]^{1/2}\left[H\{\text{J/kg}\}\right]^{-3/4}$$
$$= 0.03513 \times n \times \sqrt{4.3556}\left[1.7369 \times 10^5\right]^{-3/4} = 8.518066 \times 10^6 \times n$$

Further, taking optimized values gives us

$$\hat{\varphi} = 0.05, \; \Psi^* = 1.4 \rightarrow \Psi = \frac{\Psi*}{\eta_{\text{c}}^o} = \frac{1.4}{0.84} = 1.67$$

Since

$$\hat{K}_{\text{n}} = \sqrt{\hat{\varphi}}\Psi^{-3/4} = \frac{\sqrt{0.05}}{1.67^{0.75}} = \frac{0.2236}{1.469} = 1.1522$$
$$= 8.518066 \times 10^{-6} \times n \rightarrow n = 17869.36, \text{ select}: n = 17,000$$

Note that the rpm for a centrifugal compressor is much larger than for an axial compressor.

Now,

$$\Psi = 1.67 = \frac{2H}{u_2^2} \rightarrow u_2 = \sqrt{\frac{2H}{\Psi}} = \sqrt{\frac{2 \times 1.7639 \times 10^5}{1.67}} = 459.6 \, \frac{\text{m}}{\text{s}}$$

$$u_2 = \frac{\pi D_2 n}{60} \rightarrow D_2 = \frac{60 \times u_2}{\pi n} = \frac{60 \times 459.6}{\pi \times 17000} = 0.5163 \text{ m}$$

Again from

$$\hat{\varphi} = \frac{4\dot{V}_1}{\pi D_2^2 u_2} \rightarrow D_2 = \sqrt{\frac{4\dot{V}_1}{\pi u_2 \hat{\varphi}}} = \sqrt{\frac{4 \times 4.3556}{\pi \times 459.6 \times 0.05}} = 0.491 \text{ m} \rightarrow \text{select} : D_2 = 0.500 \text{ m}$$

$$u_2 = \frac{\pi D_2 n}{60} = \frac{\pi \times 0.5 \times 17,000}{60} = 445 \frac{\text{m}}{\text{s}}$$

$$\Psi = \frac{2H}{u_2^2} = \frac{2 \times 1.76393 \times 10^5}{445^2} = 1.781$$

$$\hat{\varphi} = \frac{4\dot{V}_1}{\pi D_2^2 u_2} = \frac{4 \times 4.3556}{\pi \times 0.25 \times 445} = 0.0498$$

Select: Exit angle $\beta_2 = 90°$ and entry angle $\beta_1 = 30°$. Further,

$$\left(\frac{D_1}{D_2}\right)_{opt} = 1.30\sqrt[3]{\hat{\varphi}} = 1.30 \times 0.0498^{1/3}$$

$$= 0.4782 \rightarrow D_1 = 0.500 \times 0.4782 = 0.239 \text{ m}$$

Select: $D_1 = 0.24$ m, and thus, $b_1 = 0.2 D_1 = 0.048$ m $= 48$ mm.

Now, $(\beta_1 + \beta_2)/2 = (90 + 30)/2 = 60°$. Further, $t/l = 0.35\text{–}0.45$, *select*: $(t/l) = 0.4$. Therefore, the number of blades is

$$z = \frac{2\pi \sin\left(\frac{\beta_1+\beta_2}{2}\right)}{\frac{t}{l} \ln\left(\frac{D_2}{D_1}\right)} = \frac{2\pi \sin 60°}{0.4 \ln(0.5/0.24)} = \frac{2\pi \times 0.866}{0.4 \times 0.734} = 18.53$$

Select: number of blades $z = 18$. We assume that the cross-sectional area in the inlet region is slightly bigger than the flow area at the entry to the blade row. By "slightly bigger," we mean the former is 1.05 times the latter. Thus, let

$$\frac{\pi}{4}(D_1^2 - d_H^2) = 1.05\pi D_1 b_1 \rightarrow \left[1 - \left(\frac{d_H}{D_1}\right)^2\right] = 4.20\frac{b_1}{D_1} \rightarrow \frac{d_H}{D_1}$$

$$= \sqrt{1 - 4.20\left(\frac{b_1}{D_1}\right)}$$

Now we assume $b_1/D_1 = 0.2$. Thus,

$$\frac{d_H}{D_1} = \sqrt{1 - 0.84} = 0.4 \rightarrow d_H = 0.4 \times 0.24 = 0.096 \text{ m}$$

Now,

$$c_{m1} = \frac{\dot{V}_1}{\pi D_1 b_1} = \frac{4.3557}{\pi \times 0.24 \times 0.048} = 120.35 \ \frac{m}{s}$$

Let $c_{u1} = 0$. It follows that

$$c_1 = c_{m1} = u_1 \tan \beta_1 = u_2 \frac{D_1}{D_2} \tan \beta_1 = 445 \times \frac{0.24}{0.5} \tan 30° = 123.32 \ m/s$$

Further,

$$u_1 = u_2 \left(\frac{D_1}{D_2}\right) = 445 \times \frac{0.24}{0.5} = 213.6 \ m/s$$

and

$$w_{u1} = \frac{c_{m1}}{\tan \beta_1} = \frac{u_1 \tan \beta_1}{\tan \beta_1} = u_1 = 213.6 \ m/s.$$

Thus,

$$w_1 = \sqrt{c_{m1}^2 + u_1^2} = u_1 \sqrt{1 + \tan^2 \beta_1} = \frac{u_1}{\cos \beta_1} = \frac{213.6}{0.866} = 246.65 \ m/s.$$

Now let $c_{m2} = c_{m1} = 123.32 \ m/s$. Then

$$H = u_2 c_{u2} - u_1 c_{u1} = u_2 c_{u2}$$

and for an infinite number of blades, the power-lowering factor is

$$\mu = \frac{H}{H_\infty} = \frac{c_{u2}}{c_{u2,\infty}} = 1 + \frac{\pi \sin \beta_2}{2z \left[1 - \left(\frac{r_1}{r_2}\right)\right]} = 1 + \frac{\pi}{2 \times 18 \times \left(1 - \frac{0.24}{0.9}\right)} = 0.8563$$

Since $\beta_2 = 90° \rightarrow c_{u2,\infty} = u_2 = 445 \ m/s \rightarrow c_{u2} = 445 \times 0.8563 = 381.0 \ m/s$
Therefore,

$$w_{u2} = u_2 - c_{u2} = 445 - 381 = 64 \ m/s.$$

$$w_2 = \sqrt{w_{u2}^2 + c_{m2}^2} = \sqrt{64^2 + 123.32^2} = 138.94 \ m/s$$

$$c_2 = \sqrt{c_{u2}^2 + c_{m2}^2} = \sqrt{381^2 + 123.32^2} = 400.46 \ m/s.$$

Losses:

$$\Delta h_1 = \frac{1}{2}\zeta_1 c_1^2 = \frac{0.1}{2} \times 123.32^2 = 760.39 \ \frac{m^2}{s^2}$$

$$\Delta h_2 = \frac{1}{2}\zeta_2 w_1^2 = \frac{0.2}{2} \times 246.65^2 = 6,083.13 \ \frac{m^2}{s^2}$$

Let the velocity at the end of the diffuser/stator section be $c_{ex} = 120$ m/s.

$$\Delta h_3 = \frac{1}{2}\zeta_3 \left(c_2^2 - c_{ex}^2\right) = \frac{0.25}{2}(400.46^2 - 120^2) = 18,246.03 \ \frac{m^2}{s^2}$$

Therefore, the total losses are

$$\Delta h_{total} = \Delta h_1 + \Delta h_2 + \Delta h_3 = 25,089.55 \ m^2/s^2$$

The efficiency is recalculated as

$$\eta_c^o = \frac{H^*}{H^* + \Delta h_{total}} = \frac{1.48171 \times 10^5}{(14.817 + 2.5089) \times 10^4} = \frac{14.817}{17.326} = 0.8552$$

and thus,

$$\frac{T_{ex}^{*o} - T_{in}^o}{T_{ex}^o - T_{in}^o} = \frac{450 - 303}{T_{ex}^o - 303} = \eta_c^o = 0.855 \rightarrow T_{ex}^o = 303 + \frac{147}{0.855} = 475.0 \ K$$

State at rotor exit point 2:

$$T_2^o = T_{ex}^o = 475.0 \ K$$

$$p_2^o = \frac{p_{ex}^o}{\left(1 - \frac{\Delta h_3}{c_p T_{ex}^o}\right)^{\frac{\gamma}{\gamma-1}}} = \frac{4}{(1 - 0.039)^{3.5}} = 4.6 \ bar$$

$$T_2 = T_2^o - \frac{c_2^2}{2c_p} = 475.0 - \frac{400.46^2}{2010} = 395.2 \ K$$

$$M_2 = \frac{c}{\sqrt{\gamma R T_2}} = \frac{400.46}{20.05\sqrt{395.2}} = \frac{400.46}{398.59} = 1.0047$$

which shows that there would be weak shock at the stator leading edge if the stator immediately follows the rotor, but the problem may not be serious and could be ignored.

Further,

$$p_2 = p_2^o \left(\frac{T_2}{T_2^o}\right)^{\frac{\gamma}{\gamma-1}} = 4.6 \times \left(\frac{395.2}{475.0}\right)^{3.5} = 2.416 \text{ bar}$$

$$\rho_2 = \frac{p_2}{RT_2} = \frac{2.416 \times 10^5}{287.5 \times 395.2} = 2.126 \frac{\text{kg}}{\text{m}^3}$$

$$\dot{V}_2 = \frac{\dot{m}}{\rho_2} = \frac{5}{2.13} = 2.347 \frac{\text{m}^3}{\text{s}}$$

Thus, the rotor channel width is

$$b_2 = \frac{\dot{V}_2}{\pi D_2 c_{m2}} = \frac{2.347}{\pi \times 0.5 \times 123.32} = 0.0121 \text{ m} \rightarrow \text{select}: b_2 = 0.012 \text{ m} = 12 \text{ mm}$$

The state at the air suction point "in" is
Let $c_{in} \approx c_1 = c_{m1} = 123.32$ m/s
Thus, the inlet Mach number is

$$M_{in} = \frac{c_{in}}{20.05\sqrt{T_{in}}} = \frac{123.32}{20.05\sqrt{T_{in}}} \quad \text{and} \quad T_{in} = \frac{T_{in}^o}{1 + \frac{\gamma-1}{2}M_{in}^2} = \frac{303}{1 + 0.2M_{in}^2}$$

Solving the two equations by alternate substitution of the two variables, we get $M_{in} = 0.348$ and $T_{in} = 295.83$ K.
Thus,

$$p_{in} = p_{in}^o \left(\frac{295.8}{303}\right)^{3.5} = 0.919 \text{ bar}$$

State at rotor entry point 1:

$$T_1^o = T_i^o = T_\infty = 303 \text{ K}$$

$$p_1^o = p_{in}\left[1 + \frac{c_{in}^2 - \zeta_1 c_{in}^2}{2c_p T_{in}}\right]^{3.5} = 0.919\left[1 + \frac{120.35^2}{2,010 \times 295.8}\right] = 0.991 \text{ bar}$$

$$p_1 = p_1^o\left[1 - \frac{c_1^2}{2c_p T_1^o}\right]^{\frac{\gamma}{\gamma-1}} = 0.991\left[1 - \left(\frac{120.35^2}{2,010 \times 303}\right)\right]^{3.5} = 0.911 \text{ bar}$$

$$T_1 = T_1^o\left[1 - \frac{c_1^2}{2c_p T_1^o}\right] = 303\left(\frac{0.911}{0.991}\right)^{0.286} = 295.8 \text{ K}$$

$$M_1 = \frac{c_1}{\sqrt{\gamma R T_1}} = \frac{120.35}{20.05\sqrt{295.81}} = 0.349 \approx M_{in}$$

$$\rho_1 = \frac{p_1}{R T_1} = \frac{0.911 \times 10^5}{287.5 \times 295.8} = 1.07123 \; \frac{kg}{m^3}$$

$$\dot{V}_1 = \frac{\dot{m}}{\rho_1} = \frac{5}{1.07123} = 4.667 \; \frac{m^3}{s} = c_{m1}\frac{1}{2}\pi D_1 b_1 = \frac{1}{2} \times 120.35 \times \pi \times b_1 \rightarrow b_1$$

$$= \frac{4.667}{\pi \times 120.35 \times 24} = 0.051$$

R-select: $b_1 = 0.051$ m $= 51$ mm $\rightarrow \frac{b_1}{D_1} = \frac{0.051}{0.24} = 0.2125$

$$d_H = D_1\sqrt{1 - 4.2 \times \frac{b_1}{D_1}} = 0.24\sqrt{1 - 4.2 \times \frac{0.2125}{0.8925}} = 0.079 \; m$$

Reselect: $d_H = 0.079$ m $= 79$ mm.

Labyrinth loss: Let's select labyrinth no. $z_{lab} = 3$
  While $p_1 = 0.911$ bar, $p_2 \times \frac{0.87}{\sqrt{0.68+z_{lab}}} = \frac{2.427 \times 0.87}{\sqrt{3.68}} \times 1.1$ bar
  Since

$$p_1 < p_2 \times \frac{0.87}{\sqrt{0.68 + z_{lab.}}},$$

we take the subcritical solution, (5.18b).
   Now, let the mean diameter of the labyrinth seal on the front side be

$$D_m = \sqrt{d_H^2 + \frac{4\dot{V}_1}{\pi c_{m1}}} = \sqrt{0.079^2 + \frac{4 \times 4.667}{\pi \times 120.35}} = 0.2353 \approx D_1 = 0.24 \; m$$

Select: $D_m = 0.24$ m, labyrinth gap $= s_m = 0.1$ mm.
   Further,

$$\rho_1 = \rho_2 = 2.13 \; \frac{kg}{m^3}, \quad p_1 = p_2 = 2.427 \; bar.$$

The labyrinth air mass flow rate is

$$\dot{m}_L = \pi D_m s_m \sqrt{\frac{p_i \rho_i}{0.68 + z_{lab}}} = \pi \times 0.24 \times 10^{-4}\sqrt{\frac{2.427 \times 10^5 \times 2.13}{3.68}} = 0.028 \; \frac{kg}{s}$$

Thus, $\frac{\dot{m}_L}{\dot{m}} = \frac{0.028}{5} = 45.65 \times 10^{-3} = 0.565\%$.

The radius of curvature of the circular arc rotor blade is

$$R = \frac{|r_2^2 - r_1^2|}{|r_2 \cos \beta_2 - r_1 \cos \beta_1|} = \frac{|D_2^2 - D_1^2|}{2|D_2 \cos \beta_2 - D_1 \cos \beta_1|} = \frac{0.5^2 - 0.24^2}{0.25} = 0.802 \text{ m}$$

Thus, the final data are $D_1 = 0.34$ m; $D_2 = 0.5$ m; $b_1 = 0.051$ m; $b_2 = 0.013$ m. Now, $D_3 = D_2 = 0.5$ m, $D_4 = D_5 = D_2\left[1 + \frac{1}{8} \text{ to } \frac{1}{14}\right] \rightarrow D_2 \times 1.1 = 0.55$ m. Further, let $D_6 = D_5 \times 1.32 = 0.726$ m. Select: $D_5 = 0.55$ m, $D_6 = 0.726$ m.

$$b_5 = b_6 = b_2 + (1 \text{ to } 2) \text{ mm} = 0.014 \text{ m}.$$

Diffuser without vanes (space 3–4): $b_3 = b_2 + 1$ mm $= 0.014$ m.

$$c_{m3} = \frac{b_2}{b_3} c_{m2} = \frac{0.013}{0.014} \times 120.35 = 111.75 \; \frac{\text{m}}{\text{s}}, c_{u3} = c_{u2} = 381 \; \frac{\text{m}}{\text{s}}$$

$$c_3 = \sqrt{c_{m3}^2 + c_{u3}^2} = \sqrt{381^2 + 111.75^2} = 397.0 \text{ m/s}$$

The slope of the radial streamline is

$$\tan \alpha_3 = \frac{c_{m3}}{c_{u3}} = \frac{111.75}{381} = 0.2933 \rightarrow \alpha_3 = 16.35°$$

The hydraulic diameter of the gap is $d_{\text{hydr}} = 2b_3 = 0.028$ m. For the diffuser without vanes (inviscid),

$$\rho_3 = \rho_4 \rightarrow c_{m4} = \frac{D_3}{D_4} c_{m3} = \frac{0.50}{0.55} \times 111.75 = 101.59 \; \frac{\text{m}}{\text{s}}$$

$$c_{u4} = \frac{D_3}{D_4} c_{u3} = \frac{0.50}{0.55} \times 381 = 363.36 \; \frac{\text{m}}{\text{s}}$$

$$c_4 = \sqrt{c_{m4}^2 + c_{u4}^2} = \sqrt{101.59^2 + 363.36^2} = 377.29 \text{ m/s}$$

$$T_4 = T_3 + \frac{c_3^2 - c_4^2}{2c_p} = T_2 + \frac{c_3^2 - c_4^2}{2c_p} = 397.41 + \frac{397^2 - 377^2}{2,010} = 404.0 \text{ K}$$

$$\frac{p_4^o}{p_d} = \frac{p_4}{p_2} = \left[1 + \frac{c_3^2 - c_4^2}{2c_p T_3}\right]^{\frac{\gamma}{\gamma-1}} = \left[1 + \frac{7.5927}{396.41}\right]^{3.5} = 1.06866 \rightarrow p_4$$

$$= 1.06866 \times 2.427 \text{ bar} = 2.5963 \text{ bar}.$$

Further,

$$\rho_4 = \frac{p_4}{RT_4} = \frac{2.59363 \times 10^5}{287.5 \times 404} = 2.13 \ \frac{kg}{m^3}$$

(Note that $\rho_2 = \rho_3 = 2.13 \ kg/m^3$.)

$$\dot{V}_4 = \frac{\dot{m}}{\rho_4} = \frac{5}{2.239} = 2.239 \ \frac{m^3}{s}$$

The calculation may be repeated as follows:

$$c_{m4} = \frac{D_3}{D_4} c_{m3} = \frac{0.50}{0.55} \times \frac{2.13}{2.233} \times 111.75 = 96.90 \ \frac{m}{s} \ ; c_{u4} = 363.34 \ \frac{m}{s}$$

Thus,

$$c_4 = \sqrt{c_{m4}^2 + c_{u4}^2} = \sqrt{96.90^2 + 363.36^2} = 376.06 \ m/s$$

Further,

$$\frac{c_3^2 - c_4^2}{2c_p} = \frac{397^2 - 376.06^2}{2,010} = 8.045 \ K$$

Therefore,

$$T_4 = T_3 + \frac{c_3^2 - c_4^2}{2c_p} = 396.41 + 8.054 = 404.46 \ K$$

Now

$$\frac{p_4}{p_2} = \left[1 + \frac{c_3^2 - c_4^2}{2c_pT_3}\right]^{\frac{\gamma}{\gamma-1}} = \left[1 + \frac{8.0454}{396.41}\right]^{3.5} = 1.0729 \rightarrow p_4 = 2.604 \ bar.$$

$$\rho_4 = \frac{p_4}{RT_4} = \frac{2.604 \times 10^5}{287.5 \times 404.46} = 2.2394 \ \frac{kg}{m^3} \rightarrow \dot{V}_4 = \frac{5}{2.2394} = 2.233 \ \frac{m^3}{s}$$

$$c_{m4} = \frac{\dot{V}_4}{\pi D_4 b_4} = \frac{2.233}{\pi \times 0.55 \times 0.014} = 92.3 \ \frac{m}{s}$$

The calculation may again be repeated!
For a diffuser without vanes (viscous flow):
The viscosity coefficient for air is $\mu = 2.1 \times 10^{-5} \left(\frac{T}{300}\right)^{0.6} \frac{kg}{ms}$

With
$p_2 = 2.427$ bar, $T_2 = 396.41$, $\rho_2 = 2.13$ kg/m^3 and $\mu_2 = 2.4822 \times 10^{-5}$ kg/ms.

$$\mathrm{Re}_3 = \frac{\rho_3 c_3 d_{\text{hydr.}}}{\mu_3} = \frac{2.13 \times 397.0 \times 0.028}{2.4822 \times 10^{-5}} = 9.54 \times 10^5 \rightarrow \text{turbulent!}$$

For a channel flow, the friction formulas from the literature are

Laminar: $c_f = \frac{24}{\mathrm{Re}}$

Turbulent (smooth): $c_f = 0.3164\mathrm{Re}^{-1/4}$

Turbulent (rough): $c_f = \dfrac{1}{2\log_{10}\left(\frac{d}{k}+1.74\right)^2}$, where d = diameter, k = surface roughness.
Let

$$\frac{d}{k} = 100 \rightarrow c_f = 0.03$$

Therefore,

$$\frac{1}{r_r c_{u4}} = \frac{1}{r_2 c_{u2}} + \frac{c_f \pi (r_4 - r_2)}{\sqrt{2}} \rightarrow \frac{1}{D_4 c_{u4}} = \frac{1}{D_2 c_{u2}} + \frac{c_f \pi (D_4 - D_2)}{8\dot{V}_2}$$

from which it follows that

$$\frac{1}{0.55 c_{u4}} = \frac{1}{0.5 \times 381} + \frac{0.03 \times \pi \times 0.05}{8 \times 2.347} = 5.2492 \times 10^{-3} + 2.5098 \times 10^{-4}$$
$$= 5.5003 \times 10^{-3}$$

Thus, $c_{u4} = 330.58$ m/s, $c_{m4} = 92.3$ m/s.
Further,

$$c_4 = \sqrt{330.58^2 + 92.3^2} = 343.22 \text{ m/s} \quad (\text{inviscid}: c_4 = 376.06 \text{ m/s})$$

Slope: $\tan \alpha_4 = 92.3/330.58 = 0.279 \rightarrow \alpha_4 = 15.6°$.

$$\Delta h_{34} = \frac{0.25}{2} \times (c_3^2 - c_4^2) = 0.125 \times (397^2 - 343.22^2) = 4976.13 \text{ m}^2/\text{s}^2$$

Since

$$\frac{c_3^2 - c_4^2}{2c_p} = \frac{397^2 - 343.22^2}{2010} = 19.80 \text{ K}$$

therefore,

$$T_4 = T_2 + \frac{c_3^2 - c_4^2}{2c_p} = 396.41 + 19.80 \text{ K} = 416.21 \text{ K}.$$

$$p_4 = p_2 \left[1 + \frac{c_3^2 - c_4^2}{2c_p T_3}\right]^{\frac{\gamma}{\gamma-1}} = 2.427 \times \left[1 + \frac{19.80}{396.41}\right]^{3.5} = 2.878 \text{ bar}$$

Thus,

$$\rho_4 = \frac{p_4}{RT_4} = \frac{2.878 \times 10^5}{287.5 \times 416.21} = 2.405 \frac{\text{kg}}{\text{m}^3}$$

Further,

$$\dot{V}_4 = \frac{5}{2.405} = 2.079 \frac{\text{m}^3}{\text{s}}$$

$$c_{m4} = \frac{2.079}{\pi \times 0.55 \times 0.014} = 85.94 \frac{\text{m}}{\text{s}}$$

and

$$c_4 = \sqrt{85.94^2 + 330.58^2} = 341.57 \frac{\text{m}}{\text{s}}$$

The calculation may be repeated!

$$\frac{c_3^2 - c_4^2}{2c_p} = \frac{397^2 - 341.57^2}{2,010} = 20.37 \text{ K}$$

$$T_4 = T_2 + \frac{c_3^2 - c_4^2}{2,010} = 396.41 + 20.37 = 416.78$$

$$p_4 = p_2 \left[1 + \frac{c_3^2 - c_4^2}{2c_p T_3}\right]^{\frac{\gamma}{\gamma-1}} = 2.427 \left[1 + \frac{20.37}{396.41}\right]^{3.5} = 2.892 \text{ bar}$$

$$\rho_4 = \frac{p_4}{RT_4} = \frac{2.892 \times 10^5}{287.5 \times 416.78} = 2.413 \frac{\text{kg}}{\text{m}^3}$$

$$\dot{V}_4 = \frac{5}{2.413} = 2.072 \frac{\text{m}^3}{\text{s}}$$

$$c_{m4} = \frac{2.072}{\pi \times 0.55 \times 0.0114} = 85.65 \; \frac{m}{s}$$

$$c_4 = \sqrt{85.65^2 + 330.58^2} = 341.5 \; \frac{m}{s}$$

$$\tan \alpha_4 = \frac{85.65}{330.58} = 0.259 \rightarrow \alpha_4 = 14.52°$$

Bladed diffuser:

Let the opening ratio be

$$C = 2.5 = \frac{D_6 \sin \alpha_6}{D_5 \sin \alpha_5} = \frac{0.726}{0.55} \times \frac{\sin \alpha_6}{\sin 14.73°} = 5.263 \sin \alpha_6 \rightarrow \sin \alpha_6 = 0.475 \rightarrow \alpha_6$$
$$= 28.36°$$

$$D_5 = D_4 = 0.55 \; m \rightarrow D_6 = 1.32 D_5 = 0.726 \; m$$

$$c_{m6} = c_{m5} \frac{D_5}{D_6} = 85.94 \times \frac{0.55}{0.726} = 65.10 \; \frac{m}{s}$$

$$c_6 = \frac{c_{m6}}{\sin \alpha_6} = \frac{65.10}{\sin 28.36°} = 137.05 \; \frac{m}{s}$$

$$\frac{c_5^2 - c_6^2}{2c_p} = \frac{341.5^2 - 137.05^2}{2,010} = 48.68 \; K$$

$$T_6 = T_5 + \frac{c_5^2 - c_6^2}{2c_p} = 416.78 + 48.68 \; K = 465.46 \; K$$

$$p_6 = p_5 \left[1 + \frac{c_5^2 - c_6^2}{2c_p T_5}\right]^{\frac{\gamma}{\gamma-1}} = 2.878 \left[1 + \frac{48.68}{416.78}\right]^{3.5} = 4.24 \; bar$$

Further decrease the speed to $c_{ex} = 100 \; \frac{m}{s} \rightarrow p_{ex}^o = 4.3 \; bar$

Chapter 9
Off-Design Running of Aircraft Gas Turbines

While the aircraft gas turbines are designed to run efficiently in a given environment and at given operating conditions, it is evident that, except under cruise conditions, the engine is off the design conditions. In this chapter, we therefore consider operation of the gas turbine with respect to operation of the design conditions.

For a single-stage machine, since

$$\frac{T^o}{T} = 1 + \frac{\gamma - 1}{2}M^2 = \left(\frac{p^o}{p}\right)^{\frac{\gamma-1}{\gamma}}$$

we can write for any stagnation pressure ratio with respect to the same under on-design condition

$$\frac{p^o}{\overline{p^o}} \sim \frac{p_\infty^o}{\overline{p_\infty^o}} \sim \frac{p_\infty \left[1 + \frac{\gamma-1}{2}M_\infty^2\right]^{\frac{\gamma-1}{\gamma}}}{\overline{p_\infty}\left[1 + \frac{\gamma-1}{2}\overline{M}_\infty^2\right]^{\frac{\gamma-1}{\gamma}}}$$

provided the efficiencies are the same as under the design conditions, designated with a bar $(^-)$ (Fig. 9.1).

Further,

$$\dot{m} = \rho u A = \frac{p}{RT}Ma_s \sim \frac{p_\infty}{T_\infty^{1/2}}M_\infty \rightarrow \frac{\dot{m}}{\overline{\dot{m}}} \sim \frac{p_\infty}{\overline{p}_\infty}\sqrt{\frac{\overline{T}_\infty}{T_\infty}} \cdot \frac{M_\infty}{\overline{M}_\infty}$$

Also,

$$\frac{\dot{m}}{\overline{\dot{m}}} \sim \frac{\rho c_m}{\overline{\rho c_m}} \sim \frac{p}{T}\cdot\frac{\overline{T}}{\overline{p}}\frac{c_m}{\overline{c}_m} \sim \frac{p_\infty}{\overline{p}_\infty}\cdot\frac{\overline{T}_\infty}{T_\infty}\cdot\frac{c_m}{\overline{c}_m} \sim \frac{p_\infty}{\overline{p}_\infty}\sqrt{\frac{\overline{T}_\infty}{T_\infty}}\frac{M_\infty}{\overline{M}_\infty}$$

T. Bose, *Airbreathing Propulsion: An Introduction*,
Springer Aerospace Technology, DOI 10.1007/978-1-4614-3532-7_9,
© Springer Science+Business Media, LLC 2012

Fig. 9.1 Velocity components under design and off-design conditions

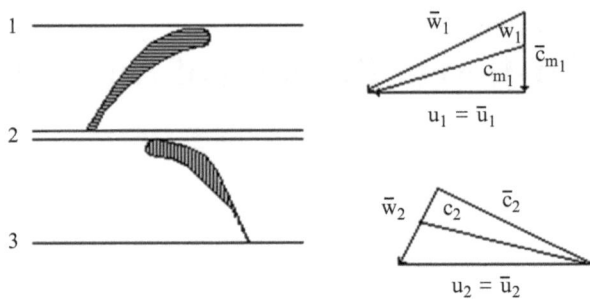

Therefore,

$$\frac{c_m}{\bar{c}_m} = \frac{\dot{V}}{\bar{\dot{V}}} \sim \frac{M_\infty}{\bar{M}_\infty} \sqrt{\frac{\bar{T}_\infty}{T_\infty}}$$

Now assuming rpm $=$ *constant*, that is, $n = \bar{n}$ and $c_{u1} = \hat{c}_{u1} = 0$, we get under an off-design condition a positive angle of attack if $c_{m1} < \bar{c}_{m1}$. Therefore, even for $n = \bar{n}$, there is a shift in the direction of the flow, which is equivalent to an angle of attack, which can cause stalling of the blades. On the other hand, if $n \neq \bar{n}$ there is one direction for a zero angle of attack, when $\frac{c_m}{\bar{c}_m} = \frac{u}{\bar{u}}$ (for best efficiency).

Now, we'll first consider when $n = \bar{n}$ but there is a shift in the direction of the flow. If at the design point, $c_{u1} = 0$, then

$$\bar{H} = u_2 c_{u2} \quad \text{and} \quad H = u_2 c_{u2} - u_1 c_{u1}.$$

Now,

$$c_{u2} = u_2 - w_{u2} = u_2 - c_{m2} \cot \beta_2 = u_2 - \frac{\dot{V}}{\bar{\dot{V}}} \bar{c}_{m2} \cot \beta_2$$

and

$$c_{u1} = u_1 - \frac{\dot{V}}{\bar{\dot{V}}} \bar{c}_{m1} \cot \beta_1$$

Therefore,

$$H = \bar{u}_2 \left[\bar{u}_2 - \frac{\dot{V}}{\bar{\dot{V}}} \bar{c}_{m2} \cot \beta_2 \right] - \bar{u}_1 \left[\bar{u}_1 - \frac{\dot{V}}{\bar{\dot{V}}} \bar{c}_{m1} \cot \beta_1 \right]$$

$$= (\bar{u}_2^2 - \bar{u}_1^2) - \frac{\dot{V}}{\bar{\dot{V}}} (\bar{u}_2 \bar{c}_{m2} \cot \beta_2 - \bar{u}_1 \bar{c}_{m1} \cot \beta_1) \approx \frac{\dot{V}}{\bar{\dot{V}}} u \bar{c}_{m1} \cot \beta_1$$

[for axial machines, $\bar{u}_2 \approx \bar{u}_1$, $\cot \beta_1 > \cot \beta_2$].

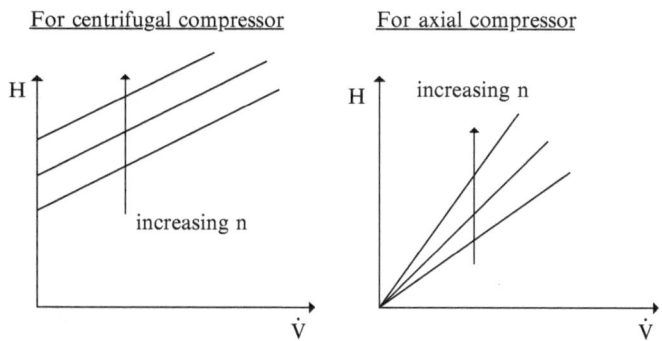

Fig. 9.2 Theoretical Performance Map for Compressor and Turbine

This gives a linear relationship between H and $((\dot{V}/\bar{\bar{V}}))$; schematic results have been plotted in Fig. 9.2 for centrifugal and axial compressors.

We now consider various losses that will occur under off-design conditions. First, we'll deal with *friction losses*. From (3.28), we write

$$\frac{\mathrm{d}p^o}{p^o} = -\gamma M^2 \left(\frac{2}{D} c_f \mathrm{d}x\right)$$

and from the friction coefficient for the turbulent boundary layer, we know that

$$c_f \sim \mathrm{Re}^{0.8} \sim u^{0.8} \quad \text{and} \quad M \sim u$$

Therefore,

$$\Delta H_\mathrm{m} \sim \frac{\mathrm{d}p^o}{p^o} \sim \frac{U^2}{U^{0.8}} \sim U^{1.2}.$$

Further, we consider *collision* or *shock loss*, which has a parallel with an increase in drag for single airfoils at different angles of attack. Actually, the name is somewhat unfortunate since it may very well be confused with the nonisentropic shock loss that can occur in a supersonic flow. However, here we are considering basically a viscous effect increasing the "form drag" of the blade when the flow direction deviates from the minimum drag direction (Fig. 9.3). This is considered from the component normal to the $\overline{w}_\infty$ (for the rotor) or $\bar{c}_\infty$ (for the stator).

The normal component is

$$w_n \sim \left(\overline{\overline{V}} - \dot{V}\right) = \overline{\overline{V}}\left(1 - \frac{\dot{V}}{\overline{\overline{V}}}\right) \sim \left(1 - \frac{\dot{V}}{\overline{\overline{V}}}\right)$$

Fig. 9.3 Blade flow for minimum drag

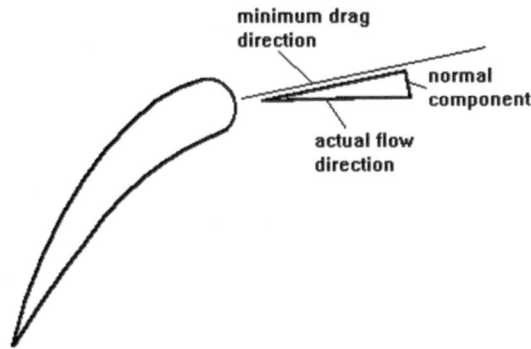

Fig. 9.4 Characteristic performance curve of a single-stage machine (rpm = const.)

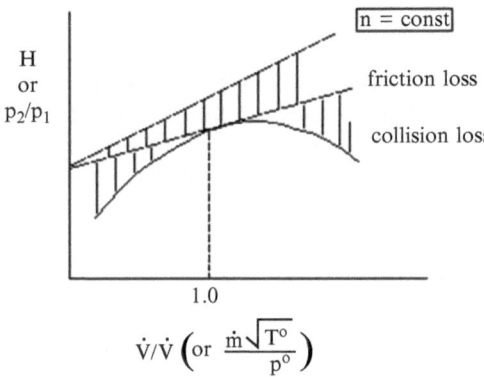

Collision loss (for changing the direction) is taken into account by introducing a loss coefficient, and we write for the loss in enthalpy as

$$\Delta h_{\text{coll}} = k_{\text{coll}} \frac{1}{2} \left(1 - \frac{\dot{V}}{\bar{\bar{V}}} \right)^2$$

by combining the contribution from the rotor and the stator,

$$\Delta h_{\text{coll}} = \frac{1}{2} \left(1 - \frac{\dot{V}}{\bar{\bar{V}}} \right)^2 \left[u_1^2 + u_2^2 \left(\frac{D_2}{D_1} \right)^2 \right], k_{\text{coll}} = 0.5 \text{ to } 0.7$$

where D_1 is the entry diameter of the stator (the formula is from Pfleiderer 1964, p. 183). For an optimum operation of an engine at any given rpm, the optimum value will be the one at maximum H, designated as H_{opt}, corresponding to the minimum drag, which is under shock-free flow of a blade (Fig. 9.3).

Therefore, for a single-stage machine, we add various contributions schematically and show these in Fig. 9.4.

Basically, the above is true for both the turbine and the compressor with some modification. In addition, we discuss the rpm effect.

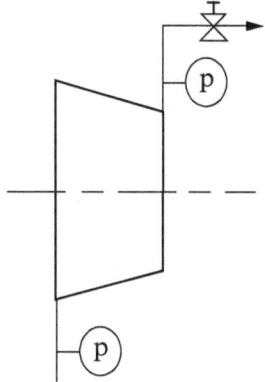

Fig. 9.5 Schematic arrangement of a compressor with closing valve

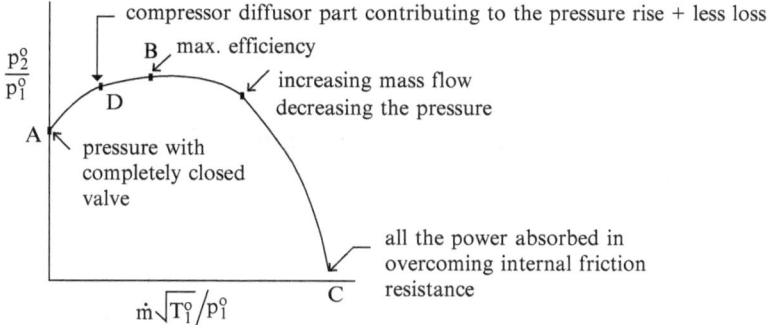

Fig. 9.6 Schematic performance curve of a turbo compressor ($n = \text{const.}$)

Now, $H_{\text{opt}} \sim u^2 \sim n^2$ and $c_{\text{m,opt}} \sim u$. Hence, $\dot{V}_{\text{opt}} \sim n$. Therefore, $H_{\text{opt}} \sim \dot{V}_{\text{opt}}^2$.

However, before describing a typical set of characteristics at various rpms, it will be better to consider what might be expected to occur for a single-stage compressor when a valve, placed in the *delivery line* of the compressor running at constant speed, is slowly operated. The entire arrangement for this compressor is shown schematically in Fig. 9.5.

Initially, when the valve is shut and the mass flow is zero, the pressure head is produced by the action of the impeller on the air trapped between the vanes. As the valve is opened and the flow commences, the diffuser begins to contribute its quota of pressure rise, and the pressure ratio increases (Fig. 9.6). At point B, the efficiency reaches a maximum and the pressure ratio is also at a maximum, and any further increase in mass flow will result in a decrease in the pressure ratio. At C, when the value is fully opened, all of the pressure increase is offset with the pressure loss in overcoming the internal frictional resistance. Around B is the surge point; to the left of B surge occurs and there is violent aerodynamic pulsation, which is transmitted through the whole machine.

There is a slight difference in the characteristic curves of axial and centrifugal compressors. For a centrifugal compressor, if we let the compressor operate at a point D on the part of the characteristic curve having a positive slope, then a decrease in the mass flow is accompanied with full delivery pressure. If the pressure downstream of the compressor has fallen also, the flow repeats the cycle (*surging*). Surging for centrifugal compressors need not be immediately at the left side of point B since the actual surge point depends on the fall in pressure downstream of the compressor. For an axial compressor, the characteristic curves take a form similar to the centrifugal type, but the characteristics for fixed values of $n/\sqrt{T_1^o}$ cover a much narrower range of mass flow than in the case of the centrifugal compressor, and may be close to the maximum of the characteristic curve. Along this connection, we need to make special mention of what is known as the rotating stall. Because of the nonuniformity in the flow or geometry of the channels between the vanes and blades, let there be a breakdown in one channel, say B. Thus, C receives fluid at a reduced angle of incidence and channel A at an increased incidence. Therefore, channel A stalls, but channel B recovers.

9.1 Analytical Calculation of Performance Map of a Multistage Axial Compressor

Generally, the calculation of off-design conditions in a multistage axial compressor is somewhat amenable to analysis. However, for axial and radial turbines, as well as for radial compressors, one has to depend more on the experimental performance maps. For the axial compressor, we assume that the multistage axial compressor consists of similar individual axial compressor stages, which under normal operating conditions all work at the same point of the individual performance map. Further, we examine the flow for multiple stages until a choking behavior occurs. The basic idea is that if the total pressure change is done in small stages, then one can think about differential change of state in each stage.

Let's start with the performance characteristic of a single axial compressor stage. As an example, we consider one given in Fig. 9.7.

In addition, we assume the compressor as given, for which a typical analytical expression is

$$f = 1.2 - |g - 0.75|^{0.161}$$

Herein again the super-bar ($^-$) denotes the design value. From the above figure, we may now write a relation

$$\frac{\Psi_s}{\overline{\Psi}_s} = f(g), g = \frac{\varphi_s}{\overline{\varphi}_s}$$

Fig. 9.7 Relative characteristc map of an axial compressor stage (Courtesy of Eckert (1953), Fig. 279)

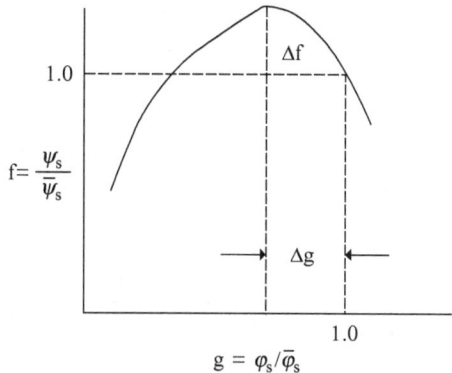

Now, the increase in pressure in each stage is given by the relation

$$\Delta p_s = \frac{\rho \overline{\Psi}_s}{2} u_2 \frac{\varphi_2}{\overline{\varphi}_s}$$

If there are x stages, then the pressure increase per stage ($\Delta x = 1$) is

$$\Delta p_s = \frac{\Delta p}{\Delta x} = \frac{dp}{dx} = \frac{\rho \overline{\Psi}_s}{2} u^2 f \left(\frac{\varphi_s}{\overline{\varphi}_s} \right)$$

Now, we'll define the nondimensional variables:

$$\dot{m}^* = \frac{\dot{m}}{\overline{\dot{m}}}; p^* = \frac{p}{\overline{p}}; n^* = \frac{n}{\overline{n}} = \frac{u}{\overline{u}}; \rho^* = \frac{\rho}{\overline{\rho}}$$

where p and π are dependent on the number of stages passed, and the change of state within the compressor can be thought to be a part of a polytropic change of state pv^ν constant, where ν is the polytropic exponent. By guessing a compressor's efficiency, we can calculate the polytropic exponent, as given later in this section.

Further, we consider the following two ratios:

$$\frac{c_m}{\overline{c}_m} \approx \frac{\dot{V}}{\overline{\dot{V}}} = \frac{\dot{m} v}{\overline{\dot{m}} \overline{v}} = \frac{\dot{m}}{\overline{\dot{m}}} \left(\frac{\overline{p}}{p} \right)^{1/\nu} = \dot{m}^* p^{*1/\nu}$$

and

$$\frac{\varphi_s}{\overline{\varphi}_s} = \frac{c_m/u}{\overline{c}_m/\overline{u}} = \frac{c_m}{\overline{c}_m} \frac{\overline{u}}{u} = \frac{\dot{m}^*}{n^* p^{*1/\nu}}$$

Further, the density and pressure are given by

$$\rho = \overline{\rho} \left(\frac{p}{\overline{p}} \right)^{1/\nu} = \overline{\rho} p^{*1/\nu} \quad \text{and} \quad p = \overline{p} p^*$$

Let's assume that the initial pressure under both design and off-design conditions are the same; that is,

$$p_i v_i^v = p v^v \rightarrow \frac{\bar{\rho}}{\rho} = \frac{v}{v} = \left(\frac{\bar{p}}{p}\right)^{1/v} = \frac{1}{p*^{1/v}} \rightarrow \rho* = p * 1/v$$

Now

$$\frac{dp}{dx} = \frac{d}{dx}(\bar{p}p^*) = \bar{p}\frac{dp^*}{dx} + p^*\frac{d\bar{p}}{dx} = \frac{1}{2}\bar{\rho}p*^{1/v}\bar{u}^2 n^{*2}\varphi_s f\left(\frac{\dot{m}^*}{p*^{1/v}n*}\right)$$

$$= \left(\bar{p}\frac{dp^*}{d\bar{p}} + p^*\right)\frac{d\bar{p}}{dx}$$

Now, at the *design point* $\Psi/\bar{\Psi} = f(\varphi/\bar{\varphi}) = 1$ and

$$\frac{d\bar{p}}{dx} = \frac{1}{2}\bar{\rho}\bar{u}^2\bar{\Psi}_s \rightarrow dx = \frac{d\bar{p}}{\frac{1}{2}\bar{\rho}\bar{u}^2\bar{\Psi}_s}$$

Therefore,

$$\left(\bar{p}\frac{dp^*}{dp} + p^*\right)\frac{d\bar{p}}{dx} = \left(\bar{p}\frac{dp^*}{dp} + p^*\right)\cdot\frac{1}{2}\bar{\rho}\bar{u}^2\bar{\Psi}_s = \frac{1}{2}\bar{\rho}p*^{1/v}\bar{u}^2 n^{*2}\bar{\Psi}_s f\left(\frac{\dot{m}^*}{p*^{1/v}n*}\right)$$

from which it follows that

$$\bar{p}\frac{dp*}{d\bar{p}} + p^* = p*^{1/v}n*^2 f\left(\frac{\dot{m}*}{p*^{1/v}n*}\right)$$

By simple manipulation, we write

$$\frac{d\bar{p}}{\bar{p}} = \frac{dp^*}{p*^{1/v}n*^2}f\left(\frac{\dot{m}*}{p*^{1/v}n*}\right) - p^*$$

By integrating from $\bar{p} = p_i$ to $\bar{p} = p_f$, we get

$$\int_{p_i}^{p_f}\frac{d\bar{p}}{\bar{p}} = \ln\bar{\pi}_c = \int_{p*=1}^{\bar{p}_f}\frac{dp^*}{p*^{1/v}n*^2}f\left(\frac{\dot{m}*}{n*p*^{1/v}}\right) - p^* = \int_{p*=1}^{\bar{p}_f}\frac{dp^*}{y}$$

It is obvious that the second expression is positive, and therefore, the numerator and denominator within the integral have the same sign. Since $\bar{\pi}_c$ is known for a particular compressor, we have to integrate from $p* = 1$ to p_f^* for a given $\dot{m}*$ and $n*$.

Afterward, we get $p_f/p_i = p_f^*/\bar{\pi}_c$. The required polytropic exponent v can be determined from the (guessed) stage efficiency of the compressor. Since

$$\eta_{cs}^o = \frac{\pi_{cs}^{o(\gamma-1)/\gamma} - 1}{\tau_{cs}^o - 1} = \frac{\pi_{cs}^{o(\gamma-1)/\gamma} - 1}{\tau_{cs}^{o(v-1)/v} - 1}$$

we thus get

$$\frac{v}{v-1} = \frac{\log_{10}/\pi_{cs}^o}{\log_{10}\left[1 - \pi_{cs}^{o(\gamma-1)/\gamma} - \frac{1}{\eta_{cs}^o}\right]} \approx \frac{\gamma}{(\gamma-1)}\,\eta_{cs}^o$$

The above equations can be evaluated for different n^*, p^*, and $\dot{m}^*$ for given π_c^o and *polytropic exponent* and plotted in a chart. However, several aspects of the chart can be pointed out. First, the chart gives no information about the compression efficiency or surge limits. Second, it is much better and more accurate to measure experimentally the performance map of the entire compressor, rather than just for a stage; in fact, the number of stages does not go into the calculations. Third, a gas turbine consists of not only a compressor part, but there are other parts, including combustion chamber, turbine, etc. Therefore, in the next section, we discuss running the gas-turbine engines under any specific operating condition.

9.2   Equilibrium Running Condition

In the previous section, we discussed the method to get a *performance map* of a compressor. Unfortunately, such maps are not available for all the components of a gas-turbine engine; they have to be obtained for individual components from experiments, and then they have to be considered together for all components. While such consideration for all types of gas-turbine engines with single or multiple spools will be out of the scope of this book, we consider only the performance characteristic of a typical straight-jet turbojet with a single spool and conical nozzle. Such an engine is typical of a subsonic or transonic flying aircraft, in which the jet exhaust speed, under cruise conditions, should be as close to the flying condition as possible to obtain high propulsive efficiency of the engine.

For this purpose, we have typical performance maps for the compressor, turbine, and nozzle, as given in Figs. 9.8 and 9.9.

Note that from one-dimensional gas dynamics for a choking nozzle exit $(A_7 = A_t)$, we have

$$\frac{\dot{m}\sqrt{T^o}}{p^o A_t} = \sqrt{\gamma}\left(\frac{2}{\gamma+1}\right)^{\frac{\gamma+1}{2(\gamma+1)}}$$

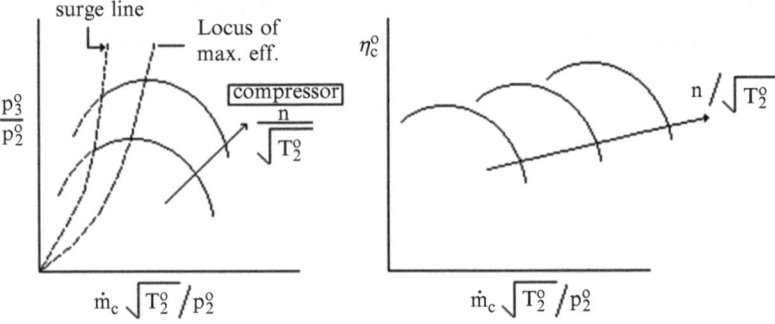

Fig. 9.8 Characteristic maps of a typical compressor

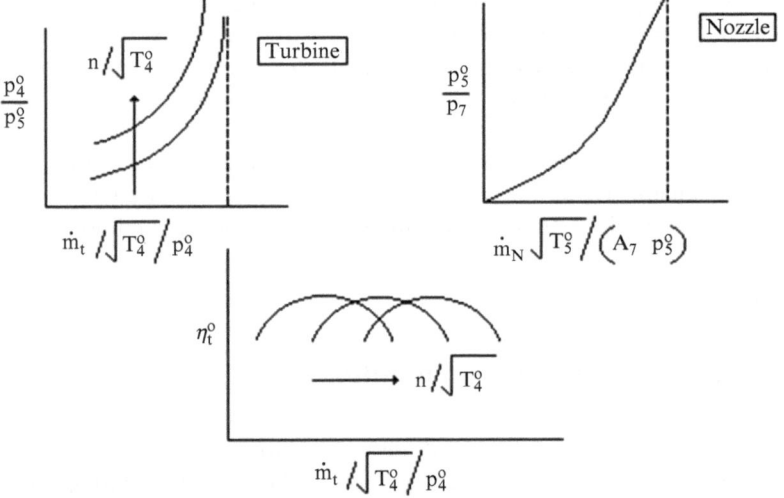

Fig. 9.9 Characteristic maps of typical turbine and nozzle

For various specific heat ratios γ, the left-hand side of the above equation can be evaluated easily, and they are given in the following table:

$\gamma =$	1.40	1.33	1.30
$\frac{m\sqrt{T^o}}{p^o A_t}$	0.6847	0.6726	0.6623

Under equilibrium running condition (between the compressor and the turbine and other components, such as the nozzle), we have

$$\dot{m}_c \approx \dot{m}_t \approx \dot{m}_{Noz}, \; p_3^o \approx p_4^o, \; \frac{T_2^o}{T_2} = \frac{T_\infty^o}{T_\infty} = 1 + \frac{\gamma - 1}{2} M_\infty^2; \; \frac{p_2^o}{p_2} = \frac{p_\infty^o}{p_\infty} = \left(\frac{T_\infty^o}{T_\infty}\right)^{\frac{\gamma}{\gamma - 1}}$$

Either by guessing or by assuming efficiency data, η_c^o and η_t^o, we consider the following equations:

$$\frac{T_3^o}{T_2^o} = 1 + \frac{1}{\eta_c^o}\left[\left(\frac{p_3^o}{p_2^o}\right)^{\frac{\gamma_c-1}{\gamma_c}} - 1\right] \tag{9.1a}$$

$$\frac{T_5^0}{T_4^o} = 1 - \frac{1}{\eta_t^o}\left[1 - \left(\frac{p_5^o}{p_4^o}\right)^{\frac{\gamma_t-1}{\gamma_t}}\right] \tag{9.1b}$$

$$\frac{\dot{m}_t\sqrt{T_4^o}}{p_4^o} \approx \frac{\dot{m}_c\sqrt{T_2^o}}{p_2^o}\cdot\frac{p_2^o}{p_3^o}\cdot\sqrt{\frac{T_4^o}{T_2^o}} \tag{9.2}$$

$$\frac{n}{\sqrt{T_4^o}} = \frac{n}{\sqrt{T_2^o}}\cdot\sqrt{\frac{T_2^o}{T_4^o}} \tag{9.3}$$

The *torque moments* absorbed in the compressor and developed in the turbine are

$$\mu_c = \dot{m}_c c_{pc}\left(T_3^o - T_2^o\right)\frac{30}{n\pi} \tag{9.3a}$$

$$\mu_t = \dot{m}_c c_{pc}\left(T_3^o - T_2^o\right)\frac{30}{n\pi} \tag{9.3b}$$

In a general case,

$$\frac{dn}{dt} = \frac{30}{\pi I}(\mu_t - \mu_c) \tag{9.4}$$

where I = *polar mass moment of inertia* and n is the rpm. Depending on the sign of the bracketed quantity, the rpm must change. However, for equilibrium conditions in a single-spool engine, both torque moments must balance each other and we get a relationship connecting the difference in the stagnation temperature of the compressor and that of the turbine.

Nozzle:

$$\frac{\dot{m}_{Noz}\sqrt{T_5^o}}{A_7 p_5^o} \approx \frac{\dot{m}_c\sqrt{T_2^o}}{A_7 p_2^o}\cdot\frac{p_2^o}{p_3^o}\cdot\frac{p_4^o}{p_5^o}\sqrt{\frac{T_5^o}{T_2^o}} \tag{9.5}$$

Also,

$$\frac{p_5^o}{p_7} = \frac{p_5^o}{p_4^o}\cdot\frac{p_3^o}{p_2^o}\cdot\frac{p_\infty^o}{p_7} = \frac{p_5^o}{p_4^o}\cdot\frac{p_3^o}{p_2^o}\cdot\frac{p_\infty}{p_\infty}\cdot\frac{p_\infty}{p_7} \tag{9.6}$$

For a convergent nozzle, the nozzle exit flow Mach number may be maximum sonic. If the flow is subsonic, then $p_7 = p_\infty$, but if there is just a choking (critical) condition at the throat, then $p_7 = p_{\text{crit}}$. On the other hand, for a convergent–divergent nozzle, the optimum condition in the nozzle exit plane is $p_7 = p_\infty$, and in general for a choked nozzle, $p_7 \geq p_\infty$.

In some cases, we can additionally have an equation for the intercomponent volume, V, for example, in the combustion chamber. Here

$$p^o = \rho^o R T^o = \frac{M}{V} R T^o \rightarrow \frac{dM}{dt} = \frac{V}{RT^o} \frac{dp^o}{dt}$$

where M is the equivalent mass of the gas volume V. Therefore, in a combustion chamber, we may write

$$\frac{dp_3^o}{dt} \approx \frac{RT_3^o}{V} \left(\dot{m}_c - \dot{m}_t \right) \tag{9.7}$$

The quantity within the parentheses in the above equation may be approximately equal to zero, since the mass flow rate of the fuel is only a small fraction of the air mass flow rate. Thus, for equilibrium running condition,

$$\frac{dn}{dt} = 0 \rightarrow \mu_t = \mu_c$$

and let $\dot{m}_c = \dot{m}_t$. Therefore, from energy balance,

$$\dot{m}_t c_{pt} \left(T_4^o - T_5^o \right) = \dot{m}_c c_{pc} \left(T_3^o - T_2^o \right)$$

which leads to

$$\frac{T_5^o}{T_2^o} = \frac{T_4^o}{T_2^o} - \frac{c_{pc}}{c_{pt}} \left(\frac{T_3^o}{T_2^o} - 1 \right)$$

and further

$$\frac{T_5^o}{T_2^o} = \frac{T_5^o}{T_4^o} \cdot \frac{T_4^o}{T_2^o} = \frac{T_4^o}{T_2^o} - \frac{c_{pc}}{c_{pt}} \left(\frac{T_3^o}{T_2^o} - 1 \right)$$

We get, therefore, the final expression

$$\frac{T_5^o}{T_4^o} = 1 - \frac{c_{pc}}{c_{pt}} \frac{T_\infty^o}{T_4^o} \left(\frac{T_4^o}{T_2^o} - 1 \right)$$

If the pilot suddenly allows more fuel, T_4^o is increased. Obviously, from the turbine map, $n/\sqrt{T_4^o}$ decreases and thus $\left(p_4^o/p_5^o \right)$ decreases, $\left(T_4^o/T_5^o \right)$ decreases, and

Fig. 9.10 Typical thrust map of a gas-turbine engine

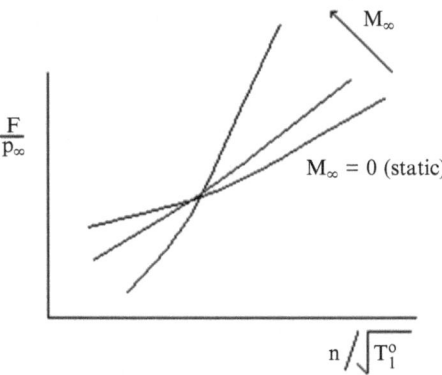

$\left(T_5^o, p_5^o\right)$ increase. Further, from (9.3a) and (9.3b), $\mu_t > \mu_c$ with a consequent increase in the rpm resulting in an increase in $\dot{m}_t$ and $\dot{m}_{\text{Noz}}$, and a change in the value of p_5^o. Now, the general expression for the net thrust is

$$F = \dot{m}(u_7 - u_\infty) + (p_7 - p_\infty)A_7$$

The above equation can be written immediately in the form

$$\frac{F}{\infty} = \frac{\dot{m}\sqrt{T_2^o}}{p_2^o} \cdot \frac{p_2^o}{p_\infty} \left[\frac{u_7}{\sqrt{T_5^o}} \sqrt{\frac{T_5^o}{T_4^o} \cdot \frac{T_4^o}{T_2^o}} - \frac{u_\infty}{\sqrt{T_2^o}} \right] + \left(\frac{p_7}{p_\infty} - 1 \right) A_7$$

Obviously, the above expression has some use in engineering practice, and the above and similar expressions are always mentioned as nondimensional numbers in the English-language literature. Since the true nondimensional number is $F/(p_\infty A)$ for a given engine, A may be kept constant and we get the above expression. A typical variation of thrust is given in Fig. 9.10.

The method, as described above, is similar even for multispool engines, except that it is also necessary to satisfy compatibility conditions of flow between the spools. Thus, the compatibility requirement gives rise to the phenomenon of aerodynamic coupling, which determines the ratio of the rotor speeds even though the rotors are mechanically independent of each other.

9.3 Surge and Stall

Figure 9.11 gives a schematic picture of the compressor performance map in which the constant-efficiency lines are approximately oblique oval circles and constant lines are constant-rpm lines of parabolic shape shifting upward at increasing rpm, while the constant-surge line gives the limitation in stable compressor performance at a given rpm.

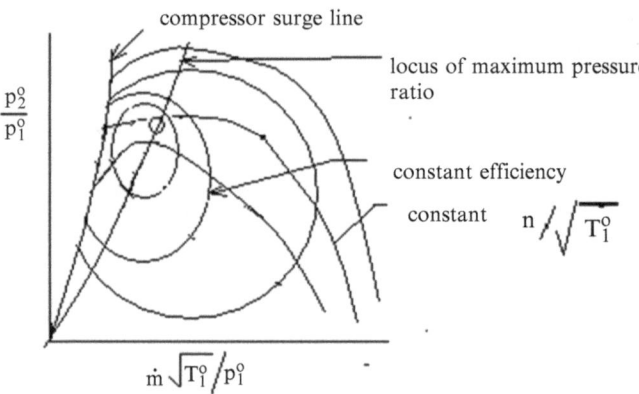

Fig. 9.11 Schematic sketch of a compressor map

Earlier in this chapter we discussed the reasons for a *surge*, where

> The surge line gives the locus of unstable operation of the compressor. .... The surge can lead to a large (violent), periodically steady oscillation in pressure rise for a single-stage compressor. Designers can take certain steps to influence the location of the surge line. One of them is to make sure the flow at inlet is as uniform as possible. Other and more subtle measures have to do with blading configurations at the back end of the compressor and even with the texture of the casing surface adjacent to the tips of the blades. (Hill and Peterson 1992)

On the other hand, the tendency of the boundary-layer separation causes reduction in the flow in the channel. Not all stages in an axial compressor may exhibit the flow separation at the same time. As a single blade results in flow diversion, then among the adjacent blades, one is overloaded and the other is unloaded (*rotating stall*).

> The overloaded blade then stalls, and the resulting flow diversion unloads the other blade. The process is repeated and it progresses in the direction of the suction surface of the blade. The alternate loading and unloading of the blades sets up an alternating stress on each blade. This stress in itself is not large unless the forcing frequency happens to match a blade vibrational frequency. In this case large stresses are incited and *fatigue failure* can occur, resulting in complete destruction of an entire blade row. (Hill and Peterson 1992)

9.3.1   Casing Treatment for Rotors

Enhancing the *stall margin* without degrading engine performance in the case of the fan of a fanjet engine is done by treating the casing by incorporating a passageway to remove flow from the main flow stream (*case treatment*). The location and amount of the low-momentum flow relative to the rotor being removed are critical. The flow is removed at a judicious location downstream of the leading edge of the rotor blade and returned at a judicious location upstream of the point of removal. The quantity of the removed flow must not be greater than 8% of the total flow in the fan rotor. Antiswirl vanes in the passageway and discreetly sized inlet and outlet passages of the passageway return the flow to the main stream at an increased velocity (Koff et al. 1994; Bock et al. 2010).

9.4 Matching of Running of Turbine and Compressor

The designer has to try to match the turbine and the compressor, so that the compressor runs near its peak efficiency through its entire range of operation, especially during acceleration of the engine. If the fuel flow is suddenly increased, it causes an increase in the turbine inlet temperature and increases the shaft speed without increasing the air flow immediately, which would result in moving toward the surge line. Thus, the fuel control system must limit the rate of the additional flow and the operation moves from one equilibrium point to another somewhat parallel to the surge line. However, since there would be imbalance in the work output in the turbine and the compressor, the rpm would increase, although the direction of change in the equilibrium power point would be approximately parallel to the surge curve.

In large gas turbines, because of difficulties in matching all the compressor stages with all the turbine stages, there are *multiple-spool engines*, where each spool or shaft connects a block of multiple compressor stages with a corresponding block of multiple turbine stages running at different rpm.

9.5 Fuel and Engine Controls

An intercontinental transport aircraft flying at an altitude of about 12–15 km will have a pressure and density ratio of 0.18 with respect to these variables on the ground. In addition, most commercial passenger aircraft fly at a transonic range with a free-stream Mach number of 0.8. Since the *fuel–air ratio* at all altitudes is very much the same, it is obvious that the actual fuel flow rate must change in substantial quantity. Therefore, the fuel supply system must have barometric pressure information and the pilot's throttle setting to decide about the actual fuel flow rate. There would be a servo-controlled hydraulic fuel system integrating all this information and adjusting the fuel supply according to the changing requirements. There can also be *electronic engine control*, which is essentially a hydromechanical fuel control but added electronic components to prevent overheating or overspeeding of the engine. If the electronic part of the control fails, then the control would revert back to a standard hydraulic control. There can also be a *full authority digital engine control* (FADEC), which is a digital computer controlling a servo-operated valve, and the power-level adjustment by the pilot can be connected to the computer and fuel control only electrically.

A gas-turbine engine control applies nonlinear control and system management. To do this, in addition to the engine modeling and simulation, one has to design set-point controllers as well as transient and limit controllers.

Besides the fuel control, a gas-turbine engine must provide predictable thrust performance over the entire operating envelope of the engine, covering altitudes ranging from sea-level to many kilometers, and operating flight speeds from takeoff through very high speeds, including supersonic Mach numbers.

Strong nonlinearities are present in aircraft gas-turbine engines due to the large range of operating conditions and power levels during a typical mission. Also, turbine operation is restricted due to mechanical, aerodynamic, thermal, and flow limitations.

In a gas-turbine engine driving an electric generator, for example, as an *auxiliary power unit*, the speed must be kept constant regardless of the electric load. A decrease in the load from the design maximum can be matched by burning less fuel while keeping the engine speed constant. Fuel flow reduction would lower the turbine exit temperature of the combustion chamber. Although this reduces the turbine efficiency, it does not affect the compressor, which still handles the same quantity of air.

9.6   Exercises

1. Given a performance map of the compressor, turbine, and nozzle, explain how the equilibrium running lines can be determined. [Hint: Select the pair of data points $\left(\dot{m}_c \sqrt{T_2^o}/p_2^o, n/\sqrt{T_2^o}\right)$.]
2. Explain surge and stall.
3. Readers are encouraged to use the Internet to study the actual design of the *case treatment*.
4. While flying at an equilibrium running point, the pilot opens the throttle suddenly, in order to increase the aircraft speed and/or to fly to a higher altitude. What happens to the engine?

Chapter 10
Propeller Aerodynamics

Before the jet-propelled, engine-driven aircrafts came into existence, propellers were used in aircrafts with piston and turboprop engines. In addition, they have been used for multiple applications, such as in cooling towers, cooling of car radiators, windmill generators, etc. As an example of the last application, Fig. 10.1 is a photograph of a series of windmills set up on a hill in a California desert.

During the latter half of the nineteenth century, ship propulsion very often occurred with paddles attached to wheels, but some development took place with propellers. Rankine (1865) and Froude (1869) developed the *axial momentum theory*, for which knowledge about the blade shape was not required. During the beginning of the twentieth century, Gustaf Eiffel, a French structural engineer who built the tower in Paris bearing his name, had experimented with twisted propellers, such that each local airfoil section of the propeller was at the same angle of attack with respect to the local relative wind. This was to keep each airfoil section as close to the angle of maximum efficiency as possible, which also meant operation of each airfoil section at the minimum drag-to-lift ratio. Eiffel clearly showed that the propeller efficiency varied with the parameter $V/(nD)$, where n was the propeller's rpm, D was its diameter, and V was the free-stream velocity; today the parameter is called the *advance ratio*. However, designing propellers with aerodynamic shape came only when the lifting line and lifting surface theories were developed before World War I in Germany by Ludwig Prandtl and his coworkers. By 1911, Prandtl was trying out the idea of modeling the effects of airflow over a finite wing by simply replacing the wing with a single line of vortex that would run from one wing tip to the other and then would run downstream from the two tips. By 1919, Prandtl's coworker Albert Betz developed a more general theory to compute the distribution of the lift when the wing shape and angle of attack were specified; applied to a propeller, this was the *blade element theory*. Both these theories are for constant-density fluid.

T. Bose, *Airbreathing Propulsion: An Introduction*,
Springer Aerospace Technology, DOI 10.1007/978-1-4614-3532-7_10,
© Springer Science+Business Media, LLC 2012

Fig. 10.1 Windmills atop a hill in a California desert

10.1   Axial Momentum Theory

Let's consider the fluid density ρ to be constant and consider a stream tube along the stream line touching the tip of the propeller (Fig. 10.2). Viscosity effects are neglected. Then the continuity equation, applied to three cross sections, is

$$v_1 A_1 = v_p A_p = v_2 A_2 \tag{10.1}$$

where v is the (axial) velocity in the flow direction (assumed constant) in the lateral direction inside the stream tube and A is the cross-sectional area.

The thrust developed by the propeller can now be computed by considering the *momentum flux*, and is given by the relation

$$F = \rho A_2 v_2 (v_2 - v_1) \tag{10.2}$$

Since the flow speed with respect to the propeller is v_p, the power required to drive the propeller (equated with the increase in the flow kinetic energy) is

$$P = v_p F = \rho A_2 v_2 v_p (v_2 - v_1) = \frac{1}{2} \rho A_2 v_2 (v_2^2 - v_1^2) = \Delta E_{\text{kin}} \tag{10.3}$$

Now, the useful work done due to thrust for a moving propeller is

$$W = F v_1 = \rho A_2 v_1 v_2 (v_2 - v_1) \tag{10.4}$$

Therefore, the *propeller efficiency* is

$$\eta_p = \frac{W}{P} = \frac{\rho A_2 v_1 v_2 (v_2 - v_1)}{\frac{1}{2} \rho A_2 v_2 (v_2^2 - v_1^2)} = \frac{2 v_1}{(v_1 + v_2)} = \frac{2}{\left(1 + \frac{v_2}{v_1}\right)} \tag{10.5}$$

Fig. 10.2 Schematic stream tube of a propellar

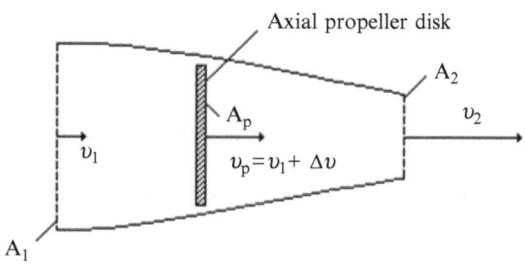

Fig. 10.3 Schematic of increase in pressure across propeller disk

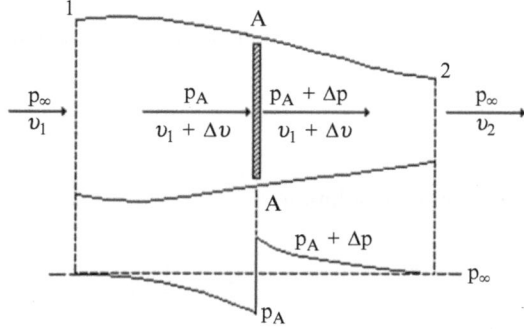

Since $v_2/v_1 > 1$, obviously $\eta_p \leq 1$, which goes to show that only a fraction of the kinetic energy change is converted to the work done by the thruster. Now, in order to compute pressure across the propeller, we consider *Bernoulli's equation* (Fig. 10.3).

We note that $p_A \leq p_\infty$ and $p_A + \Delta p \geq p_\infty$. Using the equation (for low-speed flow, $\rho = $ const.), we write for the total pressure (with superscript "o") as follows:

Upstream of the disk:

$$p_1^o = p_\infty + \frac{1}{2}\rho v_1^2 = p_A + \frac{1}{2}(v_1 + \Delta v)^2 \tag{10.6a}$$

Downstream of the disk:

$$p_2^o = p_\infty + \frac{1}{2}\rho v_2^2 = p_A + \Delta p + \frac{1}{2}(v_1 + \Delta v)^2 = p_\infty + \frac{1}{2}\rho v_1^2 + \Delta p$$
$$= p_1^o + \Delta p \tag{10.6b}$$

Thus,

$$\Delta p = p_2 - p_1 = \frac{1}{2}\rho(v_2^2 - v_1^2) \tag{10.6c}$$

Further, let $v_p = v_1 + \Delta v = (v_1 + v_2)/2$, where Δv is the increment in flow velocity, and we get the following expressions:

$$\Delta v = \frac{1}{2}(v_1 + v_2) - v_1 = \frac{1}{2}(v_2 - v_1) \tag{10.6d}$$

The thrust depends on the product of the blade disk area and the difference in pressure across the blade disk and is given by the relation

$$F = A_p \Delta p = A_p \frac{1}{2} \rho (v_2^2 - v_1^2) = \rho A_p v_p (v_2 - v_1)$$
$$= \rho A_2 v_2 (v_2 - v_1) = 2\rho A_p v_p \Delta v = 2\rho A_p (v_1 + \Delta v)\Delta v \tag{10.7}$$

After slight modification, from the above equation, we get the quadratic equation

$$(\Delta v)^2 + v_1 \Delta v - \frac{F}{2\rho A_p} = 0$$

which has the solution

$$\Delta v = \frac{1}{2}\left[-v_1 + \sqrt{v_1^2 + \frac{2F}{\rho A_p}} \right] \tag{10.8}$$

The term $\lfloor F/\rho A_p \rfloor$ is the *disk loading* of the propeller, and it has been shown that the increment in the flow speed is related to the square root of the disk loading. Applying the energy theorem to the system, we get the power P, added to the flow, by the propeller as

$$P = v_p F = F(v_1 + \Delta v) = Fv_1 + F\Delta v = W + F\Delta v$$

Herein P is related to the engine power to drive the propeller, while W is the useful work; the difference between the two may be considered as the induced (loss in) power.

Now, the propeller efficiency is given by the relation

$$\eta_p = \frac{W}{P} = \frac{1}{1 + \frac{\Delta v}{v_1}} = \frac{1}{1 + \frac{1}{2}\left(-1 + \sqrt{\frac{2F}{\rho A_p v_1^2}}\right)} = \frac{2}{1 + \sqrt{\frac{2F}{\rho A_p v_1^2}}} \tag{10.9}$$

Obviously, $F \to 0 : \Delta v \to 0, P \to W \to 0, \eta_p \to 1$; that is, the propeller efficiency is maximum when the thrust goes to zero. For the last, obviously

$$\frac{2F}{\rho A_p v_1^2} = 1, \text{ that is } F \to 0, v_1 \to 0.$$

Let's now define an *axial inflow factor*:

$$a = \frac{\Delta v}{v_1} = \frac{v_2 - v_1}{2v_1} \tag{10.10}$$

which tends to zero as $\Delta v \to 0$.

Thus, in terms of this, the following equations can be derived:

$$\frac{v_2}{v_1} = \frac{A_1}{A_2} = 1 + 2a \tag{10.11a}$$

$$\frac{v_p}{v_1} = \frac{A_1}{A_p} = 1 + \alpha \tag{10.11b}$$

$$\eta_p = \frac{2}{1 + (v_2/v_1)} = \frac{2}{1 + 2a} \tag{10.11c}$$

$$F = \rho A_2 v_2 (v_2 - v_1) = 2\rho A_2 v_1 v_2 a = 2\rho A_1 v_1^2 a(1 + a) \to \frac{F}{\rho A_1 v_1^2}$$

$$= 2a(1 + a) \tag{10.11d}$$

$$P = \frac{1}{2}\rho A_2 v_2 (v_2^2 - v_1^2) = 2\rho A_p v_1^3 a(1 + a)^2 \to \frac{P}{\rho A_p v_1^3} = 2a(1 + a)^2 \tag{10.11e}$$

An estimate of a can be done by noting that η_p must be less than 1, which means that $a \geq 1/2$.

For the stationary case of a propeller ($v_1 = 0$), written with the subscript "o," (10.11a), (10.11b), (10.11c), (10.11d), and (10.11e) are, of course, meaningless. For this, however, from (10.9), we have

$$\Delta v_o = \sqrt{\frac{F_o}{2\rho A_p}} \tag{10.12a}$$

and

$$P_o = F_o \Delta v_o = \frac{F_o^{3/2}}{\sqrt{2\rho A_p}} \tag{10.12b}$$

Examples:

1. A 2-m-diameter propeller is driven by a 150-kW (200-HP) engine. Under standard sea-level density, $\rho = 1.226$ kg/m^3. Thus,

$$F_o = P_o^{2/3}(2\rho A_p)^{1/3} = (150 \times 10^3 \text{ W})^{2/3}(2 \times 1.226 \times \pi)^{1/3} = 5,600 \text{ N}$$

This is the upper limit and is not attainable in practice because of neglecting the profile drag and azimuthal velocity component.

The increase in velocity across the disk in this case ($v_1 = 0$) is

$$\Delta v_o = \sqrt{\frac{F_o}{2\rho A_p}} = \sqrt{\frac{5,600}{2 \times 1.226 \times \pi}} = 26.96 \left[\frac{m}{s}\right]$$

Therefore,

$$v_2 = 2\Delta v_o = 53.92 \left[\frac{m}{s}\right] \qquad \text{and}$$

$$\Delta p_o = \frac{1}{2}\rho v_2^2 = 1,782.5 \text{ N}$$

2. A Cherokee 180 aircraft is in flight with a cruising speed 60.4 m/s, propeller diameter = 1.88 m, aircraft drag = 1,390 N at sea level.
   Therefore,

$$\frac{2F}{\rho A_p v_1^2} = \frac{2 \times 1,390 \times 0.224}{1.226 \times \pi \times 0.94^2 \times 60.4^2} \rightarrow \eta_p = 0.9497$$

The actual efficiency in flight is more like 0.83 since

$$\eta_p = \frac{1}{1 + \Delta v/v_1} \rightarrow \frac{\Delta v}{v_1} = \frac{1}{\eta_p} - 1 = 0.05296$$

and the relative increase in the flow speed is only 5.3%. Obviously, an increase in the relative speed reduces the propeller efficiency to an unacceptable level. Actually, $\Delta v = \left(\frac{1}{0.83} - 1\right)v_1 = 0.20482 \times 60.4 = 12.37$ m/s

$$v_p = v_1 + \Delta v = 60.4 + 12.37 = 72.77 \text{ m/s}$$

$$v_2 = v_1 + 2\Delta v = 60.4 + 24.74 = 85.14 \text{ m/s}$$

$$\Delta p = \frac{1}{2}\rho(v_2^2 - v_1^2) = \frac{1}{2} \times 1.226 \times (85.14^2 - 60.4^2) = 2,207.2 \text{ [N/m}^2\text{]}$$

Engine power in flight:

$$P = Fv_p = 1,390 \times 72.77 \left[\frac{Nm}{s}\right] = 101.15 \text{ kW}$$

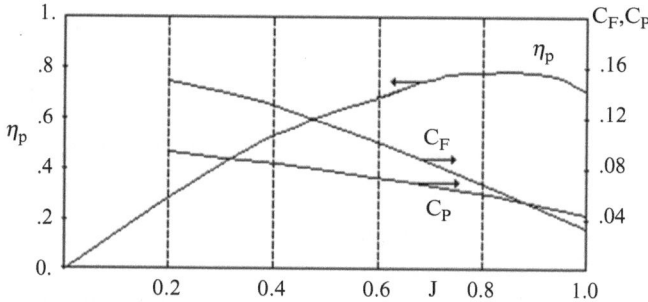

Fig. 10.4 Typical propeller coefficients

We'll now examine a few other propeller characteristic parameters, by writing them down as follows:

Propeller diameter $= D$

rpm $= n$

Revolutions per second $= n_s = n/60$

Angular speed of blades $\omega = \pi n/30 = 20\pi n_s$

Flight speed $= v_1$

Tip speed $= \frac{D\omega}{2}$

Torque moment [Nm] $= M$

Power of the engine $= P = M\omega$

Useful work done $= Fv_1$

Propeller efficiency: $\eta_p = \frac{Fv_1}{P}$

Propeller speed ratio: $\lambda =$ (forward speed)/(tip speed) $= \frac{2v_1}{\omega D}$

Advance ratio: $J = \frac{v_1}{n_s D} = \pi\lambda$

Propeller torque coefficient: $C_M = \frac{M}{\rho n_s^2 D^5}$

Propeller power coefficient: $C_P = \frac{P}{\rho n_s^3 D^5} = \frac{2\pi M}{\rho n_s^2 D^5} = 2\pi C_M$

Propeller thrust coefficient: $C_F = \frac{F}{\rho n_s^2 D^4}$

Propeller efficiency: $\eta_p = \frac{Fv_1}{M\omega} = \frac{JC_F}{C_P}$

Solidity: $\sigma = \frac{\pi D}{2\Delta x N}$, $=$ number of blades, $\Delta x =$ width of each blade

Older propellers had a solidity of typically 200–400, with a very small number of tip vortices. However, for modern propellers, $\sigma \to 1$.

Typical propeller parameters: The propeller performance parameters such as power or thrust coefficient or the efficiency are typically given for a particular propeller as a function of the advance ratio, as has been given in Fig. 10.4. Such performance curves also depend on the Reynolds number (viscosity effect, critical for small thrust), tip speed/sound speed, geometrical shape of the blade (pitch, turning angle, chord length), solidity, etc.

Pitch of the blade: As mentioned already, the propeller incident angle needs to be kept at the maximum propeller efficiency. The term "pitch" comes from the "pitch" of a screw, which is the advance of the screw for one turn. Now, for a propeller

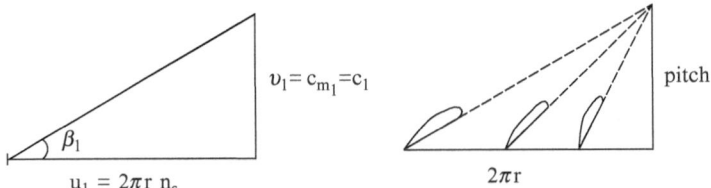

Fig. 10.5 Velocity triangles in a propeller blade

rotating n_s times per second, the time for one revolution is $1/n_s$, during which the propeller advances by v_1/n_s, which is defined as the *pitch of the propeller*. Now the tangent of the angle made by the propeller advancing speed to propeller azimuthal speed is

$$\tan \beta = \frac{v_1}{u_1} = \frac{2v_1}{\omega D} = \frac{2v_1}{2\pi n_s D} = \frac{\text{pitch}}{\pi D} = \lambda$$

which is the propeller speed ratio. If the pitch at the leading edge is kept constant in the span-wise direction of the blade, then we can find the tangent of the angle at the leading edge of the blade inversely proportional to the radius and the leading-edge blade angle becomes flatter from the root to the tip of the blade (Fig. 10.5). At this stage, the propeller blade angle at the trailing edge has no effective role to play, since the trailing-edge angle is a part of the propeller blade theory. However, from the axial momentum theory, the first method was to keep the propeller pressure surface parallel to the chord of the propeller. The performance of the propeller was, therefore, evaluated through experiments. However, at the trailing edge of the blade, from our discussions in the earlier chapters, we need to have the relation $c_{u2}r = $ constant to have c_{m2} uniform (generally, $c_{m2} \geq c_{m1}$). Let $k_1 = 2\pi n_s$ and $k_2 = c_{u2}r$. Then at the trailing edge of the blade,

$$\tan \beta_2 = \frac{c_{m2}}{w_{u2}} = \frac{c_{m2}}{u_2 - c_{u2}} = \frac{c_{m2}}{2\pi r_2 n_s - k_2/r_2}$$

This is all we can find from the axial momentum theory. But since a propeller blade section can be considered an airfoil section, we will consider the flow over a propeller blade from our knowledge of aerodynamics in the next section.

10.2   Blade Element Theory

We consider a propeller of diameter D, which is rotating around its axis (Fig. 10.6). In this propeller we cut a small airfoil section of radius r of thickness dr. From Fig. 10.6, let's first define a few variables as follows:

V_e = effective velocity; V_r = relative velocity; v_1 = flight speed
u = azimuthal speed; Δw = induced speed (note: $\Delta v = \Delta w \cos \varphi$)

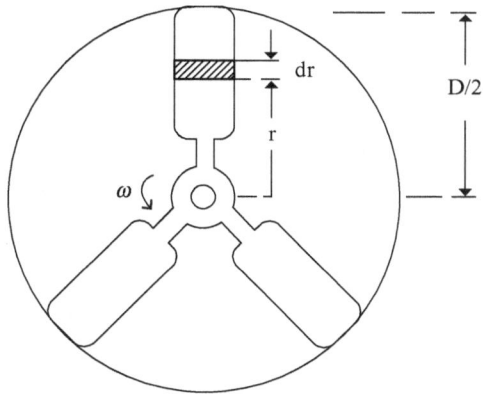

Fig. 10.6 Blade elemental theory of a propeller blade

α = angle of attack; α_i = induced angle of attack

$\theta = \alpha + \gamma$ = blade angle; $\gamma = \theta - \alpha = \varphi + \alpha_i$ = effective pitch angle

$\omega = 2\pi n_s$ = radian frequency of blade; $r^* = r/R = 2r/D$

Further, the advance ratio is defined as

$$J = \frac{v_1}{n_s D} = \frac{v_1}{2\pi n_s r} \cdot \frac{\pi r}{R} = \pi r * \tan \varphi$$

We consider now propeller blades with a given distribution of the chord length $l \sim r^*$ and the blade angle $\theta \sim r^* \theta$ is smaller from hub to tip, as for a pitch setting. We select the advance ratio J. Thus, in principle, we select $\tan \varphi = J/(\pi r^*)$ and, therefore, $\varphi = \varphi(r^*)$.

From Fig. 10.7, we can now write the following transformation matrices:

$$r_{\text{pile}} \{xy\} = l_{\text{pile}} \begin{Bmatrix} \cos \gamma - \sin \gamma \\ \sin \gamma \cos \gamma \end{Bmatrix} \cdot r_{\text{pile}} \begin{Bmatrix} x' \\ y' \end{Bmatrix}; r_{\text{pile}} \begin{Bmatrix} x' \\ y' \end{Bmatrix}$$

$$= l_{\text{pile}} \begin{Bmatrix} \cos \gamma - \sin \gamma \\ \sin \gamma \cos \gamma \end{Bmatrix} r_{\text{pile}} \{xy\}$$

In addition, the relationship between the forces in two coordinates is

$$r_{\text{pile}} \begin{Bmatrix} -dF_x \\ dF_y \end{Bmatrix} = l_{\text{pile}} \begin{Bmatrix} \cos \gamma - \sin \gamma \\ \sin \gamma \cos \gamma \end{Bmatrix} \cdot r_{\text{pile}} \begin{Bmatrix} -dD \\ dL \end{Bmatrix}$$

where L and D are the lift and drag force, respectively. The last two are taken from aerodynamics, except that the convention in aerodynamics is to represent these forces as normal and parallel to the approaching flow direction, but here they are

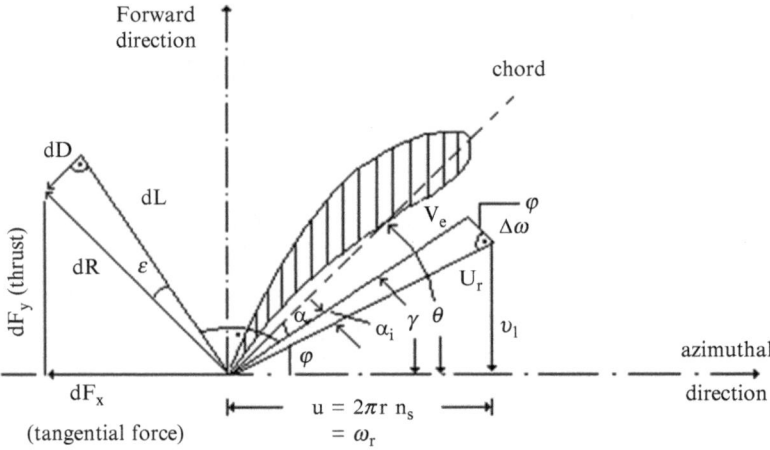

Fig. 10.7 Forces acting on a propeller blade

with respect to the chord. These differences are necessitated because of the rotary motion of the propeller blade. Therefore,

$$dF = dF_y = dL \cos \gamma - dD \sin \gamma \quad \text{(towards thrust)}$$

Similarly, the torque moment is given as

$$dM = r[dL \sin \gamma + dD \cos \gamma]$$

Now we write the lift and drag in terms of the respective coefficients as

$$dL = \frac{1}{2} \rho V_e^2 l c_L dr; \, dD = \frac{1}{2} \rho V_e^2 l c_D dr$$

Also, let $c_L = a\alpha = a(\theta - \varphi - \alpha_i)$, where $\theta = \varphi + \alpha_i + \alpha$. Thus,

$$\varphi = \arctan \left[\frac{v_1}{2\pi r n_s} \right] = \arctan \left[\frac{J}{\pi r*} \right]$$

Obviously, as $r_{hub}^* < 1, \varphi_{hub} > \varphi_{tip}$.

Further, let α_i small and (dD/dL) be small, $V_e \approx V_r$, and with N_B = number of blades, and we get the relation

$$dF = N_B \frac{\rho}{2} V_e^2 l [c_L \cos \gamma - c_D \sin \gamma] dr \approx N_B \frac{\rho}{2} V_e^2 l c_L \cos \gamma dr$$

$$\approx N_B \frac{\rho}{2} V_r^2 l c_L \cos \gamma dr = N_B \frac{\rho}{2} V_r^2 l a (\theta - \varphi - \alpha_i) \cos(\varphi + \alpha_1) dr$$

$$\approx N_B \frac{\rho}{2} V_r^2 l a (\theta - \varphi - \alpha_i) \cos \varphi dr$$

Further, by applying the momentum principle, we also write

$$dF = \rho(2\pi r dr)(v_1 + V_r \alpha_i \cos \varphi)2V_r \alpha_i \cos \varphi$$

Equating the two equations, we get

$$8\pi r(v_1 + V_r \alpha_i \cos \varphi)V_r \alpha_i \cos \varphi = N_B V_r^2 la(\theta - \varphi - \alpha_i) \cos \gamma$$

and further with some manipulation, we get a quadratic equation for α_i:

$$\alpha_i^2(8\pi r \cos \varphi) + \alpha_i\left[N_B la \cos \gamma + 8\pi r \frac{v_1}{V_r} \cos \varphi\right] - N_B la(\theta - \varphi) \cos \gamma = 0$$

Now, noting the relations

$$\frac{v_1}{V_r} = \sin \varphi, \frac{v_1}{V_r} \cos \varphi = \frac{1}{2 \sin 2\varphi} \text{ and } \cos\gamma \approx \cos\varphi$$

the above equation, being a quadratic equation, has the solution

$$\alpha_i = \frac{1}{16\pi r \cos^2\varphi}\left[\frac{-(N_B la \cos \varphi + 4\pi r \sin 2\varphi)+}{\sqrt{(N_B la \cos \varphi + 4\pi r \sin 2\varphi)^2 + 32\pi r \cos^2\varphi N_B la(\theta - \varphi) \cos \varphi}}\right]$$

Note that if $\theta = \varphi$, $\alpha_i = 0$, and since $\theta - \alpha = \varphi + \alpha$, $\alpha = 0$. This means that without the slip condition $\theta = \varphi$, the induced velocity for a symmetric blade is zero, giving zero thrust and maximum propulsive efficiency. To get good thrust, however, one must set $\alpha > 0$. Now further,

$$dF_x = dR \sin(\gamma + \varepsilon); dF_y = dR \cos(\gamma + \varepsilon)$$

from which it follows that

$$\frac{dF_y}{dF_x} = \frac{\cos(\gamma + \varepsilon)}{\sin(\gamma + \varepsilon)}; \frac{v_1}{u} = \tan \varphi$$

and the propeller element's efficiency is

$$\eta_e = \frac{v_1 dF_y}{u dF_x} = \frac{v_1 \cos(\gamma + \varepsilon)}{u \sin(\gamma + \varepsilon)} = \frac{\tan \varphi}{\tan(\gamma + \varepsilon)} = \frac{\tan(\gamma - \alpha_i)}{\tan(\gamma + \alpha_i)}$$

Now let's consider the calculation procedure for a given symmetric blade, for which the given data are the chord length $l \sim r*$, the blade angle $\theta \sim r^*$, the propeller diameter D, the number of propellers N_B, and $a = c_{L\alpha}$. Let θ be smaller from hub to tip, and at the tip it is given by the relation

$$\theta_{tip} = \arctan\left(\frac{v_1}{2\pi r_{tip} n_s}\right)$$

Now $J = v_1/(n_s D) = \pi r * \tan\varphi$ and J has a fixed value for given $v_1, n_s,$ and D. From the expression $\varphi = \arctan\{J/(\pi r*)\}$, we compute $\varphi = \varphi(r*)$ and, further, α_i. Therefore, from the formula $\gamma = \varphi + \alpha_i = \theta - \alpha$, we get $\alpha = \alpha(r*)$, and further

$$c_L(r*), c_D(r*), \eta = \arctan(c_D/c_L) = \varepsilon(r*)$$

$$\eta_e = \frac{\tan\varphi}{\tan(\varphi + \varepsilon)} = \eta_e(r*)$$

$$\frac{dF_y}{dr} = N_B \frac{\rho}{2} V_r^2 c_L \cos\gamma \, dr; V_r = \frac{v_1}{\cos\varphi}$$

$$\frac{dF_x}{dr} = \frac{\sin(\gamma + \varepsilon)}{\cos(\gamma + \varepsilon)} \frac{dF_y}{dr}$$

We can now get distributions of $\frac{dL}{dr}, \frac{dD}{dr}$, and $\frac{dM}{dr}$ and further, after integration, L, D, M, and F. Further, we can determine some of the overall propeller performance coefficients:

$$C_F = \frac{F}{\rho n_s^2 D^4} ; C_M = \frac{M}{\rho n_s^3 D^5} ; C_P = 2\pi C_M = \frac{P}{\rho n_s^3 D^5}$$

10.3   Exercise

For a propeller with two blades of diameter 2.4 m and rotating at 400 min^{-1} with an approaching flow speed of 120 m/s, use the axial momentum theory to compute first the disk loading and then iteratively the inflow and outflow velocities (initially take the inflow velocity $= 0$), the propeller efficiency, thrust, advance ratio, pressure change across the blade disk, and other relevant coefficients.

Chapter 11
Materials and Structural Problems

11.1 Materials Problems

Whittle's W-1 engine developed around 1940 had austenitic steel turbine blades, but by September 1942, he had already turned to high-nickel alloys. Whittle's engine had a turbine inlet temperature of 718°C, but today's engine run at about 1,500°C. As the years passed and the temperature climbed, engines had to adopt superalloys mostly of nickel and cobalt. The first superalloys had about 20% Cr, enough to protect against high-temperature oxidation (corrosion), and later the chromium was cut back by adding aluminum ("aluminium" in British literature) and titanium. Single-crystal alloys (*monocrystal alloy*), where the entire blade is made of a single crystal, have appeared also, with very good effect.

The fans and compressors have blades made of lightweight aluminum or titanium alloys and exhibit temperature stability. Since they do not have a uniform temperature distribution and therefore must have sufficient strength at the temperature of operation, they must be able to withstand large thermal stresses. In order to fabricate these out of metal sheets, they must have good low-temperature forming capacity, since these have in parts to be bent, drawn, or pressed. A flame tube is made out of point welding of individual parts, even by using an automatic welding procedure.

For turbine blades, many companies in various countries have developed high-temperature steels. The Austrian company Boehler developed various ferrite-type steel containing small quantities of carbon and chromium and large quantities of molybdenum and vanadium. The same company's austenite steels, known as Turbotherm, have very little carbon, about 20% each of chromium and nickel, about 10–42% of cobalt, and the rest a small quantity of tungsten, molybdenum, and niobium. Some German companies, including Krupp and others, have again the main composition of nickel and chromium, and also small quantities of carbon, silicon, manganese, molybdenum, tungsten, and other metals. In the twentieth century, when aircraft was invented Brown-Firth Laboratories did pioneering

T. Bose, *Airbreathing Propulsion: An Introduction*,
Springer Aerospace Technology, DOI 10.1007/978-1-4614-3532-7_11,
© Springer Science+Business Media, LLC 2012

work in this field, and companies such as William Jessop & Sons, and Mond Nickel Co. developed many different types of steel. Mond Nickel developed the Nimonic type of mainly nickel- and chromium-based steel. Similarly, American companies, including Crucible Steel, Westinghouse Electric Corp., Universal Cyclops Steel Corp., Allegheny Ludlum Steel Corp., Union Carbide, and Carbon Co., among others, developed high-temperature steels with extensive applications of many different alloy components. A complete description of all these materials would take us away from the main topic of airbreathing propulsion, and we suggest that readers could look into the extensive tables given in Kruschik (1960).

The physical and mechanical properties of various high-temperature steels depend, of course, on the composition. For the Nimonic type of steels, the density is about 7,970 (Nimonic DS) to 8,350 (Nimonic 75), the maximum allowable strength between 74 (Nimonic DS) to 129 kgf/mm^2 (Nimonic 95), 0.1% expansion limit from 35 (Nimonic 75) to 83 kgf/mm^2 (Nimonic 100), coefficient of linear expansion around 15×10^{-7} m/m°C, elasticity module around 20×10^{-3} kgf/mm^2 at 20°C to about half the value at 900°C. Creep and fatigue strengths are two very important properties of these materials depending on the temperature and the loading duration. Data about these, although somewhat old, can be seen in the extensive tables given in Kruschik (1960).

11.2   Normal Stress Analysis

Let's consider a blade of material density ρ, height b in the span-wise direction, and section area $A(r)$ going from root to tip as A_{root} and A_{tip}. In a particular section of the blade, the *centrifugal force* is

$$dF_{\text{C}} = (\rho A dr)\omega^2 r \text{ N}$$

for which

$$R = r_{\text{tip}} - r, x = 0 : r = r_{\text{tip}}; x = b = r_{\text{tip}} - r_{\text{root}}$$

Integrating the above equation from any radius to the tip, we get

$$F_{\text{C}} = \rho\omega^2 \int_{r}^{r_{\text{tip}}} Ar \, dr$$

and further dividing by the local cross section, we get the *normal stress* at that section due to the centrifugal force as

$$\sigma_{\text{c}} = \frac{F_{\text{c}}}{A} = \frac{\rho\omega^2}{A} \int_{r}^{r_{\text{tip}}} Ar dr \ \frac{\text{N}}{\text{m}^2}$$

Fig. 11.1 Area distribution
in a blade

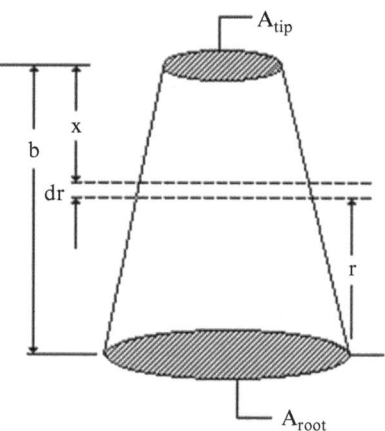

Thus, it is evident that the maximum stress is at the root and zero stress is at the tip of the blade. For a blade with a constant cross section (A = constant), the equation can be integrated easily, and we can get the stress at the root section as

$$\sigma_{c,root} = \frac{\rho\omega^2}{A}\left[r_{tip}^2 - r_{root}^2\right] = \frac{\rho\omega^2}{2}bD_m = \frac{1}{2}\rho\left(\frac{\pi n}{30}\right)^2 bD_m$$

$$= 17.4533\rho\left(\frac{n}{100}\right)^2 \pi b D_m \left[\frac{N}{m^2}\right] \qquad (11.1)$$

If the density is given in kg/m³, n is the rpm, and the lengths are in meters. It is interesting to note that the root stress is directly proportional to the cross section of the gas flowing through the turbine blade row. On the other hand, for A not equal to a constant, one can do a similar analysis by introducing a correction factor k_v, which we can write formally as

$$\sigma_c = 17.4533\rho k_v\left(\frac{n}{100}\right)^2 \pi b D_m \left[\frac{N}{m^2}\right] \qquad (11.2)$$

where $k_v = k_v\left(x/b, A_{tip}/A_{root}\right)$ and the distribution of A/A_{root} is defined as the ratio of the normal tensile stress to the tensile stress of a blade of constant cross section at the root (Fig. 11.1). For a linear distribution of A/A_{root}, it can be shown that $k_v = \left(1 + A_{tip}/A_{root}\right)/2$ at the tip ($x/b = 1$). In addition, let

$$A^* = A/A_{root}, r^* = (2r - D_{root})/\left(D_{tip} - D_{root}\right)$$

which means

$$r = \frac{1}{2}\left[D_{root} + \left(D_{tip} - D_{root}\right)r^*\right] \rightarrow dr = b dr^*$$

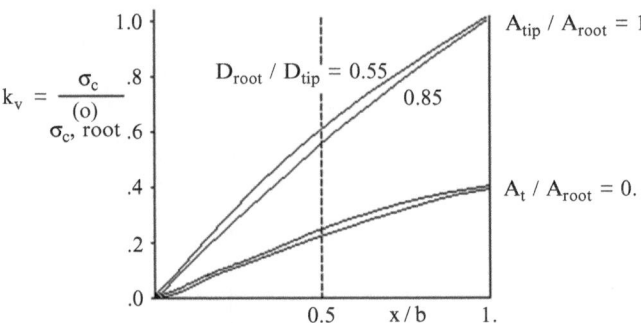

Fig. 11.2 Blade stress distribution correction factor for different area distributions

Therefore, upon integration, we get

$$\sigma_c(r^*) = \frac{\rho\omega^2}{2A^*}\left[bD_{root}\int_0^1 A^* dr^* + 2b^2\int_{r^*}^1 A^* dr^*\right]$$

At the root, we therefore get

$$\sigma_c(0) = \frac{\rho\omega^2}{2}\left[bD_{root}\int_0^1 A^* dr^* + \frac{2}{3}b^2\right]\qquad(11.2a)$$

For cylindrical blades, $A^* = 1$, we get further the root stress

$$\sigma_c = \frac{\rho\omega^2 b}{2}\left[D_{root} + \frac{2}{3}b\right],\ \left[\frac{N}{m^2}\right]\qquad(11.2b)$$

In the general case, it is possible to compute k_v for various A distributions, and a schematic sketch is given in Fig. 11.2.

11.3   Bending Stress (for Axial Blades)

The force (in N) acting on blades in the azimuthal (tangential) direction is

$$dF_t = \frac{2\pi r dr}{z}(\rho_1 c_{m1} c_{u1} - \rho_2 c_{m2} c_{u2})\qquad(11.3)$$

and the force acting in the axial direction is

$$dF_a = \frac{2\pi r dr}{z}(p_1 - p_2) + \rho_1 c_{m1}^2 - \rho_2 c_{m2}^2\qquad(11.4)$$

where z is the number of blades. Thus, the above expressions become

$$\mathrm{d}F_t = \frac{2\pi r\mathrm{d}r}{z}\rho c_m \Delta c_u; \quad \mathrm{d}F_\alpha = \frac{2\pi r\mathrm{d}r}{z}\Delta p$$

where

$$\Delta c_u = c_{u1} - c_{u2} \quad \text{and} \quad \Delta p = p_1 - p_2$$

Now, the *tangential* and *axial bending moments* (in Nm) are

$$\mathrm{d}M_{bt} = (r - r_{root})\mathrm{d}F_t = \frac{2\pi r(r - r_{root})}{z}(\rho_1 c_{m1} c_{u1} - \rho_2 c_{m2} c_{u2})\mathrm{d}r$$

and

$$\mathrm{d}M_{ba} = (r - r_{root})\mathrm{d}F_a = \frac{2\pi r(r - r_{root})}{z}(p_1 - p_1 + \rho_1 c_{m1}^2 - \rho_2 c_{m2}^2)\mathrm{d}r$$

Integration has to be done between $r = r_{root}$ to r_{tip}, where $r_{tip} - r_{root} = b$. Once the tangential and axial moments are known, the stresses can be given from the known cross-sectional dimension. The largest stress will be at the leading and trailing edges and the convex side of the blades (at root of the blades).

For the bending stress calculation, one has to determine the *principal axes* and the *mass moment of inertia;* the origin of the axes is the *center of gravity*, which is determined from the relations

$$\int x\mathrm{d}m = 0 \quad \text{and} \quad \int y\mathrm{d}m = 0$$

Now if (x', y') are the directions of the principal axes, then by definition of the *moment of inertia*,

$$I_{xy} = \frac{1}{\rho}\int\frac{xy}{r}\mathrm{d}m, \left[m^4\right], \quad \text{but} \quad I_{x'y'} = \frac{1}{\rho}\int\frac{x'y'}{r}\mathrm{d}m = 0$$

where the mass

$$\mathrm{d}m = \rho\mathrm{d}A\mathrm{d}r \ [\mathrm{kg}]$$

The relations between the (x, y) and (x', y') coordinates (Fig. 11.3) are

$$\begin{Bmatrix} x \\ y \end{Bmatrix} = \begin{bmatrix} \cos\theta & -\sin\theta \\ \sin\theta & \cos\theta \end{bmatrix}\begin{Bmatrix} x' \\ y' \end{Bmatrix} \rightarrow \begin{Bmatrix} x' \\ y' \end{Bmatrix} = \begin{bmatrix} \cos\theta & \sin\theta \\ -\sin\theta & \cos\theta \end{bmatrix}\begin{Bmatrix} x \\ y \end{Bmatrix}$$

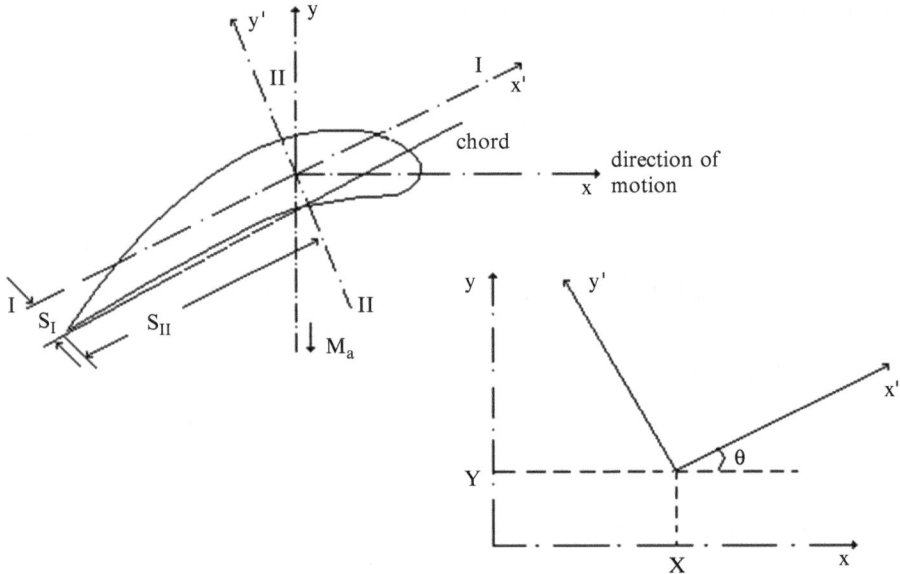

Fig. 11.3   To explain bending moments in a blade

From (Fig. 11.3), we have

$$I_{x'y'} = \frac{1}{\rho} \int \frac{x'y'}{r} dm = 0 = \frac{1}{\rho} \int \frac{1}{r} [(x - X) \cos \theta + (y - Y) \sin \theta - (x - X) \sin \theta$$
$$+ (y - Y) \cos \theta] dm$$

$$= \frac{1}{\rho} \int \frac{1}{r} \left[\frac{1}{2} \left\{ -(x - X)^2 + (y - Y)^2 \right\} \sin 2\theta + \{(x - X)(y - Y)\} \cos 2\theta \right] dm$$

For $X = Y = 0$, one can easily find the value of θ from the relation

$$\tan 2\theta = \frac{\int xy dm}{\int (x^2 - y^2) dm} = \frac{I_{xy}}{I_{xx} - I_{yy}}$$

where

$$I_{xy} = \frac{1}{\rho} \int \frac{xy}{r} dm, I_{xx} = \frac{1}{\rho} \int \frac{x^2}{r} dm, \text{ and } I_{yy} = \frac{1}{\rho} \int \frac{y^2}{r} dm$$

Now if I–I and II–II are the principal axes of the mass moment of inertia, then the *bending moment* (Nm) along each axis will be

$$M_{I-I} = M_t \cos \theta - M_a \sin \theta; \quad M_{II-II} = M_t \sin \theta - M_a \cos \theta$$

and the *bending stress* at point S will be

$$\sigma_b \left[\frac{N}{m^2}\right] = \left[\frac{M_{I-I}s_I}{I_{I-I}} + \frac{M_{II-II}s_{II}}{I_{II-II}}\right] \tag{11.5}$$

where

$$I_{I-I} = \frac{1}{\rho} \int \frac{y'^2}{r} dm \quad \text{and} \quad I_{II-II} = \frac{1}{\rho} \int \frac{x'^2}{r} dm$$

For compressor blades, generally it will be a very small mistake if the principal axes I–I are put parallel to the chord profile.

For the selection of bending blades, it is usual to select a blade profile for a given chord length and then the bending stress is calculated. If σ_b is the calculated bending stress and σ_b' is the desired stress, then the chord must be changed in the ratio σ_b/σ_b', but t/l is not changed.

11.4   Torsional Stress

Rotating blades for turbomachines with an axial flow can have a torsional moment due to the centrifugal force if the blade sections do not fall on each other in the radial direction. Considering a differential centrifugal force due to a differential mass $dm = \rho dAdr$, we have

$$dF_c = r\omega^2 dm = \rho r\omega^2 dAdr \ [N]$$

If (x, y) are coordinates in the plane parallel to a blade section, then the torsion moment is

$$M[Nm] = \int \frac{xy}{r} dF_c = \rho\omega^2 \iint xy dy dr = \rho\omega^2 \int J_{xy} dr \tag{11.6}$$

where $J_{xy} = \int xy dA \ [m^4]$ is the *area moment of inertia*.

The *torsional stress* (Fig. 11.4) is then given by the equation

$$\sigma_\tau = \frac{Mr}{J_{xy}}, \ \left[\frac{N}{m^2}\right] \tag{11.7}$$

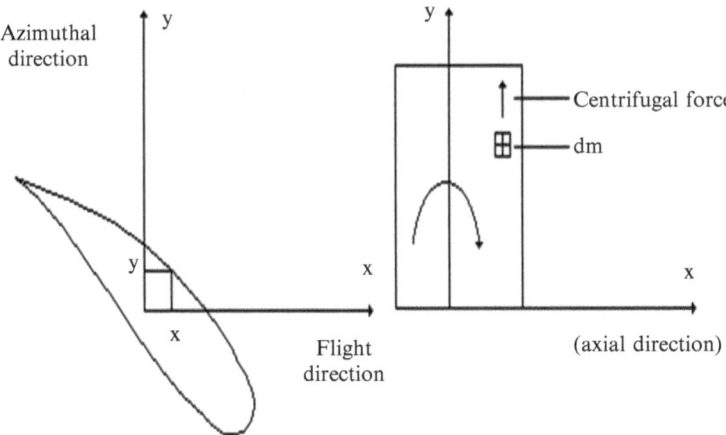

Fig. 11.4 To explain torsional stress

11.5 Oscillation of Blades

Axial compressor and turbine blades without a tip band oscillate like a cantilever beam under load, although the actual determination of the oscillation frequency of the blades is not very simple. Therefore, approximate methods are used for design purposes.

11.5.1 Bending Oscillation

Let's consider a beam with uniform load $q(x)$ [N/m] in the transverse direction as

$$q(x) = \frac{d^2 M}{dx^2}, M = \text{bending moment [Nm]}$$

The bending moment is given by

$$M \approx EI \frac{d^2 w}{dx^2}$$

where E = *elasticity module* [N/m^2] and I = area moment of second order = *area moment of inertia* [m^4]

$$= \int z^2 dA \, [\text{m}^4], \quad (\text{if on } y \text{ - axis})$$

The maximum stress is now $\sigma_x = \frac{M}{I} \cdot z \left[\frac{N}{m^2} \right]$ and the *radius of moment of inertia* is $i = \sqrt{\frac{I}{A}} \, [m]$.

For a uniform beam, $M = EIw''$, $q = M''$ and therefore, $q = EIw''''$. The boundary conditions are

$$x = 0 : w = 0, w' = 0 (\text{deflection and slope} = 0)$$

and

$$x = 1 : w'' = w''' = 0 \ (\text{zero bending moment and shear force} : M = M' = 0).$$

The well-known solution is

$$w'''' = \frac{q}{EI}, w''' = \frac{q}{EI}(x - l), w'' = \frac{M}{EI} = \frac{q}{EI}(x - l)^2, w' = \frac{q}{24EI}\left(4x^3 - 12x^2 l + 12xl^2\right)$$

$$w = \frac{q}{24EI}\left(x^2 - 4xl + 6l^2\right)$$

Thus, the maximum deflection (at $x = l$) is $w_{max} = \frac{ql^4}{8EI}$ and the maximum slope there is $w'_{max} = \frac{ql^3}{6EI}$. Therefore, the *spring constant* (= force/displacement) is evaluated as $\frac{ql}{w_{max}} = \frac{8EI}{10l^3}$ and the mass is ρ_{Al}.

Therefore, the *fundamental frequency of bending* is

$$f_0 = \frac{1}{2\pi}\sqrt{\frac{\text{spring\_constant}}{\text{mass}}} = \frac{1}{2\pi}\sqrt{\frac{8EI}{l^3 \rho Al}} = \sqrt{\frac{8}{2\pi}}\frac{i}{l^2}\sqrt{\frac{E}{\rho}} = 0.450\frac{i}{l^2}\sqrt{\frac{E}{\rho}}\left[\frac{1}{s}\right] \quad (11.8)$$

$$f_0 = \frac{1}{2\pi}\sqrt{\frac{\text{spring\_constant}}{\text{mass}}} = \frac{1}{2\pi}\sqrt{\frac{8EI}{l^3 \rho Al}} = \sqrt{\frac{8}{2\pi}}\frac{i}{l^2}\sqrt{\frac{E}{\rho}}, \left[\frac{1}{s}\right]$$

The *self-oscillation frequency of the blade* (Kruschik) is given as

$$f_{so} = \frac{17.5 k_f i}{b^2}\sqrt{\frac{E}{\rho}} \quad (11.9)$$

where k_f = a coefficient to account for deviation from a simple cantilever beam case, i = smallest radius for the moment of radius of the root profile (in cm), b = blade height (in cm), E = elasticity module (in kgf/cm^2), and ρ = density of blade material (in kg/m^3). The value of k_f depends on the tapering of the blade, the way the blade is fixed, and whether there is a damping wire, and it can be determined only experimentally. For rotors, the blade oscillation increases

somewhat due to the centrifugal force, and the corrected blade frequency is
determined from the equation

$$f_s = \sqrt{f_{SO}^2 + k_b \left(\frac{n}{60}\right)^2} \tag{11.10}$$

where k_b depends on the blade taper ratio and D_m/b and generally k_b are greater
than 1.

11.5.2   Torsion Self-Oscillation Frequency

For cylindrical steel blades, the frequency is given by the formula (Kruschik 1960)

$$f_t = 1.375 \times 10^4 m \frac{A^2}{bJ_p} \tag{11.11}$$

where

b = height (length) of blade [cm]
J_p = polar mass moment of inertia [cm^4]
A = cross section [cm^2]

$$m = \begin{cases} 4 \\ 2 \\ 3 \\ 1 \end{cases} \text{for} \begin{cases} 4/4 \\ 2/4 \\ 3/4 \\ 1/4 \end{cases} \text{wavelength.}$$

11.6   Disk Loading

The tangential force (Fig. 11.5) due to the centrifugal force of the rotating mass
dm is

$$dF_t = (r\omega^2 dm) \sin \alpha; dm = \rho h r dr d\alpha; \omega = n\pi/30$$

Hence,

$$dF_t = r \left(\frac{n\pi}{30}\right)^2 \rho h \sin \alpha dr d\alpha$$

Fig. 11.5 Toward disk
loading

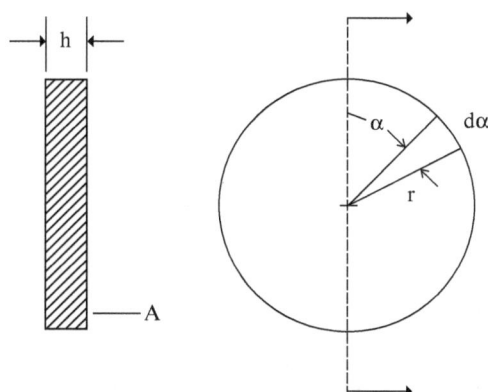

Therefore,

$$F_t = \left(\frac{n\pi}{30}\right)^2 \rho h \int_0^r \int_0^\pi r^2 \sin\alpha \, dr \, d\alpha = \left(\frac{n\pi}{30}\right)^2 \rho h \frac{2r^3}{3} \qquad (11.12)$$

Since the *area moment of inertia* is $J = hr^3/3$, $A = 2rh$ the *tangential stress* is

$$\sigma_t = \frac{F_t}{A} = 2\left(\frac{n\pi}{30}\right)^2 \rho \left(\frac{J}{A}\right) \qquad (11.13)$$

11.7   Shaft Loading

For big aircraft engines, *hollow shafts* are used in order to (1) reduce shaft weight, and (2) pass cooling air from the compressor through the shaft to cool the turbine blade disk. The usual formulas to compute the bending moment on the shaft can be computed by assessing the weight of various rotating components in a shaft. The *bending stress* is then the ratio of the bending moment and the *bending moment of resistance*; the latter, for a *hollow shaft* (Fig. 11.6), can be computed as

$$W_b = \frac{\pi}{32} D_o^3 \left[1 - \left(\frac{D_i}{D_o}\right)^4 \right] \qquad (11.14)$$

The bending stress is $\sigma_b = M_b/W_b$, in which the bending moment has to be computed by distributing weights in an elastic shaft.

For steel, the maximum tolerable bending stress is $\sigma_{b,max} = 300\text{--}600$ kgf/cm^2 $= 3 \times 10^7$ to 6×10^7 N/m^2.

Fig. 11.6 A hollow shaft

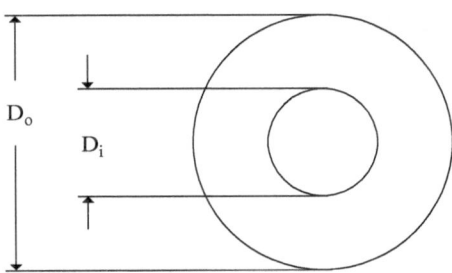

In a similar fashion, the torque moment is computed from the shaft power and radian speed of the shaft, and then it is divided by the *torsional resistance of inertia*, which, for a hollow shaft, is computed from

$$W_t = \frac{\pi}{16} D_o^3 \left[1 - \left(\frac{D_i}{D_o} \right)^4 \right]$$ (11.15a)

where the *torsional stress* is equal to the *torsional moment* [N/m] divided by the torsional resistance [m^3]:

$$\tau_t = \frac{M_t[\text{Nm}]}{W_t[\text{m}^3]}, \left[\frac{\text{N}}{\text{m}^2} \right]$$ (11.15b)

and the *torsional moment* [N/m] is given by P/ω.

Since the shaft power is given by

$$P = M_t \omega = M_t \left(\frac{n\pi}{30} \right) = W_t \tau_t \left(\frac{n\pi}{30} \right) = \tau_t \frac{\pi}{16} D_o^3 \left[1 - \left(\frac{D_i}{D_o} \right)^4 \right] \left(\frac{n\pi}{30} \right)$$ (11.16a)

by using appropriate dimensions, we can write

$$P_{\text{kW}} \times 10^3 = \tau_{t,\text{N/m}^2} \frac{\pi}{16} D_{o,\text{cm}}^3 \times 10^{-6} \left(\frac{n\pi}{30} \right) \left[1 - \left(\frac{D_i}{D_o} \right)^4 \right]$$ (11.16b)

For the *maximum allowable torsion stress* for steel shaft as $\tau_t = 1.20 \times 10^7$ N/m^2, and the shaft power given in kW, the outer shaft diameter in centimeters is given by the relation

$$D_{o,\text{cm}} = 16.943 \times \sqrt[3]{\frac{P_{\text{kW}}}{n \left[1 - \left(\frac{D_i}{D_o} \right)^4 \right]}}$$ (11.17)

Should the bending and the torsion be acting simultaneously, it is usual to compute a reference moment:

$$M_{ref} = \sqrt{M_b^2 + \frac{1}{4}M_t^2} \qquad (11.18)$$

and further calculations are done as for bending.

11.8   Critical rpm

For a rotating shaft, the critical rpm can be computed from a formula given by Kruschik (1960) as

$$n_{crit} = 300\sqrt{\frac{1}{\sum f_{cm}}}, \; [\text{min}^{-1}] \qquad (11.19)$$

where

$$\sum f = f_1 + f_2 + f_3 + \cdots,$$

in which f_1, f_2, etc. are individual bending displacements (in cm) determined for each weight.

11.9   Heat Transfer Problems

Among the various types of cooling of turbine blades (Fig. 11.7), we consider the two
While the upper two are examples of the internal cooling of turbine blades with the help of cooling air drawn from one of the compression stages through the hollow shaft, the lower two pictures are examples of cooling the combustion chamber and turbine blade surface with the help of cooler air sent through slits to form a comparatively cold surface.

Film cooling effectiveness:
This is defined as the ratio of temperatures

$$\eta_f = \frac{T_f - T_1}{T_2 - T_1}\begin{cases} T_w = T_2 : \eta_f = 1 \\ T_w = T_1 : \eta_f = 0 \end{cases} \qquad (11.20)$$

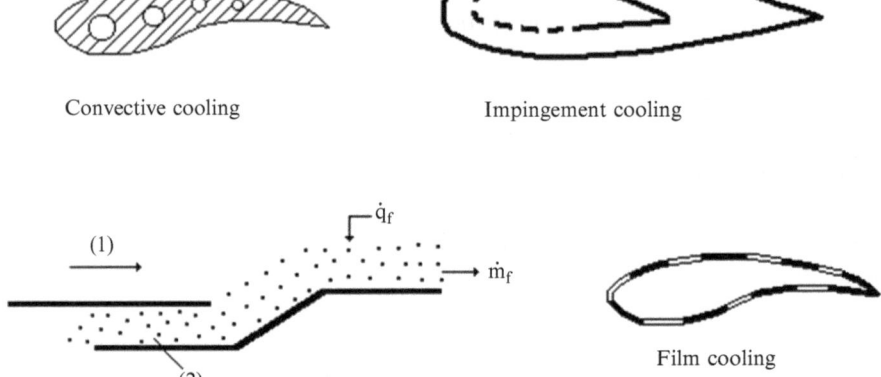

$$\xrightarrow{}\quad\begin{array}{c}\textit{convective\_cooling}\\[2pt]\textit{transpiration\_cooling}\end{array}$$

Combustion chamber cooling ➔ *Film cooling*:

Convective cooling   Impingement cooling

Film cooling

Combustion chamber cooling ⟶ Film cooling

Fig. 11.7 Various cooling arrangements in a gas turbine

where $T_w = T_f$ (wall temperature = film temperature with uniform temperature distribution across the film) is defined for a thermally insulated adiabatic wall, T_1 is the hot external gas temperature, and T_2 is the initial temperature. If the film is assumed to be of uniform temperature across the film, then heat flowing from the cold gas in a two-dimensional case is

$$\dot{q}_f = \alpha(T_1 - T_f) = \dot{m}_f c_f \frac{dT_f}{dx}, \quad \left[\frac{W}{m^2}\right]$$

where the subscript "f" stands for "film" with uniform properties across the film, c_f for *the specific heat of the film*, and α is the *heat transfer coefficient* [W/(m²K)]. Further, $\dot{m}_f = \rho_f u_f d_f$ [kg/ms] is the mass flow rate per unit length in the lateral direction, ρ_f [kg/m³] is the density of the film fluid, and d_f = film thickness [m]; initially, $d_f = d_{fo}$. First, it is assumed that the film thickness remains constant.

Now, the *Nusselt number relation for heat transfer* is defined as

$$Nu_x = \frac{\alpha(x - x_0)}{k_f}$$

where k_f is the *heat conductivity coefficient* of the external gas at the edge of the film [W/mK]

For a turbulent boundary layer, we write

$$\mathrm{Nu_x} = 0.0296\ \mathrm{Re_x^{0.8} Pr^{1/3}}$$

where

$$\mathrm{Re_x} = \frac{\rho_1 u_1 (x - x_0)}{\mu_1},\quad \mathrm{Pr} = \frac{\mu_1 c_{p1}}{k_1}$$

are the *Reynolds number* and *Prandtl number*, respectively (all dimensionless).

In addition, we write for the energy equation the length scale $\hat{x} = x - x_0$, and from the energy balance in the film, we write

$$\dot{q} = \dot{m}_f c_f \frac{dT_f}{dx} = \rho_f u_f d_f \frac{dT_f}{dx} = \alpha(T_1 - T_f) = \frac{k_1 \mathrm{Nu_x}}{\hat{x}}(T_1 - T_f)$$

$$= 0.0296 \frac{k_1}{\hat{x}} \left(\frac{\rho_1 u_1 \hat{x}}{\mu_1}\right)^{0.8} \mathrm{pr}_1^{1/3}(T_1 - T_f) = 0.0296 k_1 \left(\frac{\rho_1 u_1}{\mu_1}\right)^{0.8} \hat{x}^{-0.2}(T_1 - T_f)$$

$$(11.21a)$$

from which we can write for constant mass flow rate in the film the equation

$$\frac{dT_f}{dx} = -0.0296 \frac{k_1}{\dot{m}_f c_f}\left(\frac{\rho_1 u_1}{\mu_1}\right)^{0.8} \hat{x}^{-0.2}\ \mathrm{Pr}_1^{1/3}(T_1 - T_f)$$

$$= 0.0296 \frac{1}{\mathrm{Re_x^{0.2}\ Pr_1^{2/3}}} \frac{c_{p1}}{c_f} \frac{\rho_1 u_1}{\dot{m}_f}(T_1 - T_f)$$

By setting the film mass flow rate $\dot{m}_f = \rho_f u_f d_f$, we get the relation (since $0.0296/0.8 = 0.037$)

$$\frac{T_f - T_1}{T_2 - T_1} = \exp^{-0.037\frac{k_1}{\dot{m}_{f0}c_f}} \left(\frac{\rho_1 u_1 \hat{x}}{\mu_1}\right)^{0.8} \mathrm{Pr}^{1/3-0.037\frac{c_p}{c_f}} \frac{\rho_1 u_1}{\rho_2 \mu_2} \frac{\hat{x}}{d_2} \frac{1}{\mathrm{Pr}^{2/3}} \mathrm{Re_x^{0.2}}$$

which is the expression for the film-cooling effectiveness η_f.

It has been found, however, that the cooling effectiveness calculated from the above expression is too good to agree with the experiment, although the exponential behavior is maintained. Therefore, the mass transfer relations also need to be taken into account.

The *mass transfer Nusselt relation* is

$$\mathrm{Nu}_x' = \frac{\beta(x - x_0)}{D_f}$$

where D_f is the diffusion coefficient at the same place as for the heat transfer.

Now, for a turbulent boundary layer, we write

$$\mathrm{Nu}'_x = 0.0296\mathrm{Re}_x^{0.8}\mathrm{Sc}^{1/3}$$

where

$$\mathrm{Sc} = \frac{\mu_1}{\rho_1 D_1}$$

in which Sc is the *Schmidt number* (dimensionless).

The *mass transfer equation* is now

$$\frac{\mathrm{d}\dot{m}_f}{\mathrm{d}x} = -0.0296\frac{D_1}{\dot{m}_f}\left(\frac{\rho_1 u_1}{\mu_1}\right)^{0.8}\hat{x}^{-0.2}\mathrm{Sc}_1^{1/3}(\rho_1 - \rho_f)$$

$$= -0.0296\left(\frac{u_1}{\rho_1 \mu_1 \hat{x}}\right)^{0.2} \cdot \frac{u_1(\rho_1 - \rho_f)}{\mathrm{Sc}_1^{2/3}}$$

which can be rewritten as

$$\frac{\mathrm{d}d_f}{\mathrm{d}x} = -0.0296\frac{D_1}{(\rho_f u_f)^2 d_f}\left(\frac{\rho_1 u_1}{\mu_1}\right)^{0.8}\hat{x}^{-0.2}\mathrm{Sc}_1^{1/3}(\rho_1 - \rho_f)$$

$$= -0.0296\left(\frac{u_1}{\rho_1 \mu_1 \hat{x}}\right)^{0.2} \cdot \frac{u_1(\rho_1 - \rho_f)}{\mathrm{Sc}_1^{2/3}} \qquad (11.21b)$$

Now (11.21a) and (11.21b) have to be solved simultaneously, and one gets excellent results.

In addition to tangential slits in the combustion chamber, there are other slots to facilitate radial injection.

As mentioned earlier, this author, along with his one doctoral student (Sarkar and Bose), studied flow and heat transfer in a 2D turbine blade cascade, which have been included in the "Bibliography" of this book. The convective heat transfer distribution was shown to be strongly dependent on the relative positions of the injectant hole and orientation apart from blowing and orientation apart from blowing and temperature ratio. Details of these calculations and the results can be seen further in the references of our work. Needless to say, the formation of the coolant layer is critical on the pressure surface, and the regions between the two holes if the blades are not well protected.

11.10   Noise Pollution and Other Pollution Problems

A famous nineteenth-century book on sound, *Theory of Sound* (1895), by John Stuart, the third Baron of Rayleigh, discussed the world of church bells, tuning fork, whistles, etc. Since then, as Lighthill (1952a, b, 1962) pointed out, the development of the science of acoustics had been rather slow in comparison to the other branches of physics and remained almost static since the 1890s, when Lord Rayleigh wrote his famous book just described. However, since then, Rayleigh's world of tuning forks, violin, whispering galleries, organ pipes, church bells, singing flames and bird calls have been stormed, even though by the niceties of acoustic insulation and high-fidelity reproduction, from the air by cacophonous sequence of whines and roars and bangs, by aircrafts flying over land and sea between continents.

While the supersonic bangs occur generally in the external shocks of the supersonic flying aircrafts, which can fly only overseas and restricted uninhabited land areas, and are out of the scope of this book, the roars of the jet and the whistling sound of the compressor of the aircrafts coming to airport tarmacs are very dominating sounds.

It is enough to point out here that a human body with a good hearing capability can hear waves with *sound frequencies* from a few cycles per second (Hz) to about 10,000 Hz, with the maximum sensitivity at 1,000 Hz. In addition, a person with good hearing can barely hear a sound wave with an amplitude of 0.0002 dyn/cm^2 = 2×10^{-5} N/m^2. With air density $\rho_o = 1.16$ kg/m^3 and sonic speed = $c_o = 330$ m/s, the equivalent amplitude of fluctuation of the density is 1.83654×10^{-10} kg/m^3 and the amplitude for the fluctuation of velocity is 5.22466×10^{-8} m/s.

In principle, the propagating sound waves are caused, as shown by Lighthill (1952a, 1952b, and 1963), by an analogy of fluctuating sources.

While the sound waves from a source to a receiver (human ear) proceed in the form of pressure waves, which have wave amplitude and frequency (single frequency or broadband) noise, they are measured in a logarithmic scale compared to a standard sound level (subscript "s") in decibels (one tenth of a bel, named after the inventor of telephone, Graham Bell) by the formula

$$\text{Sound pressure level [dB]} = 20\log_{10}\sqrt{\frac{\overline{(p - p_o)^2}}{\overline{(p_s - p_o)^2}}} = 20\log_{10}\sqrt{\frac{\overline{(p')^2}}{\overline{(p'_s)^2}}}$$

where p_o is the ambient pressure. The amplitude of the standard pressure is 0.0002 dyn/cm^2.

The intensity of sound in decibels is

$$I[\text{dB}] = 10\log_{10}\left(\frac{\overline{p'^2}}{\overline{p_s'^2}}\right) = 10\log_{10}\frac{c_o^3\left(\overline{p'^2}\right)}{\rho_o \times 10^{-12}}$$

and a corresponding formula in decibels for the acoustic power is

$$P[\text{dB}] = 10\log_{10}\left(\frac{P[\text{W}]}{10^{-11}}\right)$$

Since the evaluation of the intensity of sound in dB requires a comparison of fluctuations at any frequency with standard fluctuations at a standard frequency, there are difficulties in evaluation since the human ear is not equally sensitive at all frequencies (or wavelengths). Thus, a new unit, *perceived noise level in decibels* (PNdB), has been developed, which is defined such that the PNdB rating of the complex sound should approximate the dB rating of a 1,000 Hz octave band by listening to the two sounds through two earphones. Conversion charts have been prepared in various countries, which are slightly different from each other.

For noise generated due to fluctuation of mass, force, and turbulence, Lighthill (1952a, b) showed that the acoustic equation can be derived directly from the standard flow equations of conservation of mass and momentum equations and was due to fluctuating monopole (combustion noise), dipole (oscillating blades), or quadrupole (turbulence).

Lighthill's expression for stationary fluctuating quadrupole sources (initially without convection) is obtained by dimensional analysis (and including an experimentally determined proportionality constant) for a subsonic jet and is given as

$$p[\mathrm{W}] = 10^{-4} \rho_o U_{\mathrm{jet}}^8 D_{\mathrm{jet}}^2 / c_o^5 \qquad (11.22)$$

The above expression gives hints about the ways to reduce the jet noise. The first is by reduction of the jet speed by increasing the diameter, as is done in fanjet engines. Thus, fanjet engines have not only a better propulsive efficiency (and a better overall efficiency) than the straight-jet engines, but the former are also better from an acoustic point of view. The second is by reducing the characteristic jet dimension itself. This is done by dividing the actual jet dimension into smaller multiple nozzles and multiple lobes, which has the secondary effect of reducing the turbulence in the space between the multiple structures, and thus further reducing the value of the proportionality constant in the above equation.

Besides the above jet noise, there are very disturbing shrill sounds at discrete frequencies coming from the compressor blades (due to fluctuating forces on the downstream blades coming into the wake of the blades in the previous blade row). The frequency of such a sound is obtained from the product of the number of blades and the blade rotations per second. Thus, if there is problem in a blade row, it can be found by analyzing the sound spectrum of the blades.

Reducing this type of sound load has been attempted by putting sound-absorbing materials in the engine inlet, and also by trying to change the frequency of the audible sound by blade design (for example, by having long curved blades instead of straight radial blades).

Besides the noise pollution, there are very serious problems due to chemical problems (*chemical pollution*). If there is not enough reaction time available in the combustion chamber, then there is the problem of smoke, carbon monoxide, and unburned hydrocarbons, which can cause higher specific fuel combustion rates. In addition, high temperatures can cause the formation of toxic oxides of nitrogen. There are contradictory requirements to deal with these. If there are excess air and a higher temperature, then there will be less carbon monoxide but more oxides of nitrogen. Also, for large combustion chambers, there would be fewer hydrocarbons, less carbon monoxide, and more nitrogen oxides. In order to take care of these various requirements, some of the modern types of combustion chambers are Verbix combustion (for P&W JT8D and JT9D engines) and the double annular combustion (for CFM-56 engines).

The *Verbix combustor* is an advanced low-emission combustor incorporating a two-stage fuel system. For lower-power operations, the fuel is introduced only into the pilot stage. At higher-power conditions, additional fuel is injected through four nozzles downstream of the pilot region, where the combustion air is introduced into the secondary combustion zone in the form of swirling large-velocity jets.

The *double annular combustor* (DAC) development was initialized in 1989 in response to growing airline concerns over allowable emissions standards in various countries. It reduces NO_x and CO emission levels by as much as 50% of the emission values of pre-1989 period.

NO_x is formed by the reaction of oxygen and nitrogen at very high temperatures when there is an optimum mixture of fuel and air. The amount of NO_x is determined by the residence time the burning fuel–air mixture stays in the high-temperature region. DAC reduces the flame temperature and residence time by increasing the airflow velocity in the combustion zone and physically shortening the length of the combustor. Further, it incorporates a second inner ring of the fuel nozzle.

11.10.1   Starting and Running Gas Turbines

There are various ways of staring a gas-turbine engine. For smaller gas-turbine engines, the simplest is an electric starter, which can start the engine and bring it to a minimum speed before fuel is injected and the engine is started, with one or more ignition devices like a *spark-plug,* bringing the engine to a self-sustainable speed. While the aircraft is flying through clouds or for some other reason, the combustion flame may extinguish and the engine runs *free-wheeling*, when the spark-plug operation may be started again to restart the engine.

On the ground, the first turning of the rotor shaft can also be done by some other means, such as with the help of an external electric motor, and then the external help is withdrawn after the engine is started.

11.10.2   Climate Control for Passenger Aircraft

For climate control inside a passenger cabin for the comfort of the passengers, the climate has to be maintained at a proper temperature and humidity. Air for this purpose is taken from one of the compressor stages and split into two streams. In one of the streams, the hot gas is cooled again by expansion, and then both the streams are mixed to keep the proper temperature (and humidity, if necessary) inside the passenger cabin. In many large aircrafts, there is an *auxiliary power unit* (APU), which is a small gas-turbine engine located near the tail of the aircraft separately.

11.10.3   Blade-Tip Clearance Control

Clearance control of blade tip in the compressor and turbine sections of gas turbine engines can improve efficiency, minimize leakage flow, and shorten engine development time. Tip clearance varies throughout different operating conditions, for example, startup, idle, full power, and shutdown, because of different radial forces on the blades and different thermal expansion coefficients and heat transfer. A real-time clearance control system can lead to turbine designs that eliminate the rubbing of housing and minimize leakage flow for maximum engine efficiency.

Bose and Murthy (1994) studied the change in the compressor blade clearance of a generic engine when the engine inlet was exposed to a water inlet while passing through clouds. One active clearance system for the compressor and turbine blade clearance is to utilize through-the-case eddy-current sensors for primary measurement of the tip clearance of individual blades. Earlier a high-resolution, nonintrusive stress measurement system used optical probes, but a major requirement of predrilled holes in engine casing is a big restriction for the use of such a method, and their performance is severely compromised by contaminants such as water and oil.

11.10.4   Geared Turbofans

The *geared turbofan* is a type of turbofan similar to a turbojet. It consists of a geared ducted fan with a smaller-diameter turbojet engine mounted behind it that powers the fan, so that the fan can rotate at its optimum speed. Typical of such a fan are the Honeywell TFE731 and the Pratt & Whitney PW1000G; the latter was selected for the Bombardier C Series (a family of narrow-body, twin-engine, medium-range jetliners being developed by Bombardier Aerospace of Canada), Mitsubishi Regional Jet (from Japan, for a jet aircraft for 70–80 passengers, scheduled to have its maiden flight in 2012), and Irkut MS-21 (being developed by a Russian company for twin-engine and short- and midrange Russian jet airliners with a capacity of 150–212 passengers; flight certification is planned in 2016 to replace the Tupolev Tu-154 airliners currently in service. This new aircraft is to include about one third composite materials, increasing to 40–45% provided a composite wing is added by 2015) and is an option for the Airbus A320neo. The Airbus A320neo would be the *new engine option* between GE's CFM International Leap-X and Pratt & Whitney's PW1100G. Though the new engine would burn 16% less fuel, the actual fuel gain on an A320 would be slightly less, 1–2% less typically, upon installation on an existing aircraft. Airbus is planning to schedule the planned delivery with the new engine by 2016.

Pratt & Whitney first tried to build a geared turbofan (GTF), known as PW8000, starting around 1998, upgrading essentially its earlier engines, PW6000, but with a new gearbox and a single-stage fan. In July 2008, the engine being developed along with *Motoren-Turbinen-Union* (MTU) of Germany was renamed PW1000G.

Various models of the engine will have a fan diameter of 1.4–2.1 m, a bypass ratio of 9:1 to 12:1, and a thrust per engine between 15,000–33,000 lb$_f$, equivalent to 67–150 kN.

11.11 Exercises

1. For an aircraft single-spool gas-turbine engine, for which the data can be taken from the databank given in this book, compute the power [in kW] of the compressor and turbine, and compute the hollow outside and internal shaft diameters by assuming the diameter ratio of internal to external shaft diameter equal to 0.8. Study the weight of the hollow shaft for different values of the shaft diameter ratio. Compute the *torsional moment* and *torsional stress*.
2. Compute the acoustic power in [W] decibels of a jet speed 200 m/s, density $\rho_o = 1.12$ kg/m^3, $D_{jet} = 0.8$ m/s, and $c_o = 330$ m/s, and also the power of the jet [W]. Compare the two.
3. Compute the tangential stress on a disk of 0.3-m diameter and 3-mm thickness rotating at a speed of 10,000 min^{-1}.
4. Compute the root stress of a cylindrical steel blade (density $= 8,000$ kg/m^3) of mean diameter 0.6 m and height of the blade $= 6$ cm, rotating at 10,000 min^{-1}.

Chapter 1
Introduction

T. Bose (ed.), *Airbreathing Propulsion: An Introduction*,
DOI 10.1007/978-1-4614-3532-7, pp. 1–25,
© Springer Science+Business Media LLC 2012

DOI 10.1007/978-1-4614-3532-7_12

The publisher regrets that in the print and online versions of this title, the following errors occurred. The corrected version is provided below.

Page 6:
Aircraft piston engine: weight/power 0.63-1.52 kg/kW

Page 17, after 1st line under Eq. (1.23b) add:
Maximum wing lift-drag ratio is different for different aircrafts and depends on the flight speed. The following data are given under cruise condition in the internet:

Aircraft at cruise	(L/D)-ratio
Virgin Atlantic	37
Lockheed U2	28
Boeing 747	17
Concorde at M2	7.14
Cessna 150	7.0

Page 25, Exercise 1.7:
Given that a commercial airliner with a dry mass of the aircraft 800 t and (L/D) = 21.15, aviation fuel 11200t to fly a range of 2,000 km, that is $M_{init}/M_{final} = 12000/800 = 15$ and specific impulse $I = 1560 m/s$ for Gasoline, estimate the range of the aircraft ....

The online version of the original chapter can be found at
http://dx.doi.org/10.1007/978-1-4614-3532-7_1

Appendix: Engine Data Tables

Piston

T. Bose, *Airbreathing Propulsion: An Introduction*,
Springer Aerospace Technology, DOI 10.1007/978-1-4614-3532-7,
© Springer Science+Business Media, LLC 2012

Country	Company	Model No.	No. of Cyl.	Cyl. Dia.	Stroke length	Displ. Vol.	Take-off power	R.P.M. (k-rpm)	Weight (KG)	Spec. fuel consumption (mu-g/J)	Volume compression ratio	Otto/diesel, No. of stroke	Remarks
CZ	AVIA	M337	6	105	115	5.97	170	2.6	148	72.7	6.3	04	INV INLINE
CZ	AVIA	MINR6111	6			5.97	125	2.3	127	83.1		04	
CZ	AVIA	M137	6	105	115	5.97	140	2.6	137	77.4	6.3	04	
CZ	AVIA	M332	4	105	101	3.98	100	2.4	102	77.4	6.3	04	
CZ	AVIA	M462	9	105	130	10.16	190	1.9		83.1	6.4	04	
FR	RECTIM	4AR1200	4				40	3.6	61		7	04	USES VW ENGINE
FR	RECTIM	4AR1600	4				61	3.6	64		8	04	USES VW ENGINE
DE	BMW	HORNET	9	156	162	27.5	335	1.8	350	82	5	04	
DE	JUNKER	JUMO205E	6	105	160	16.6	447	2	521	59.6	8.2	D4	DIESEL
DE	LIMBAC	SL1700E	4	88	69	1.68	61	3.2	73		8	04	HORIZ OPPOSED
DE	PIEPER	STAM1500	4			1.5	45		60			04	MODIFIED VW
DE	PULCH	003A	6	90	70	2.7	150	5	110			04	RADIAL
DE	PULCH	003B	6	94	70	2.9	180	5.5	120			04	RADIAL
PL	JANWSK	SAT500	2	70	65	0.5	25	4	27	111.5	8.5	04	HORIZ OPPOSED
PL	BORZEC	2RB	4	70	35	0.54	16	4.5	15	126.7	7.2	04	HORIZ OPPOSED
PL	WSKPZL	AI14RA	9	105	130	10.16	132	1.7	197	76.4	5.9	04	INLINE RADIAL
PL	PZLRZE	PZL-3S	7	155.5	155	20.6	550	2.1	411	105	6.4	04	RADIAL
PL	WSKPZL	ASZ62IR	9	155.5		29.87	611	2.1	579	110	6.4	04	
PL	PZLFRA	2A-120C	2	117.5	88.9	1.91	60	3.2	75		8.5	04	HORIZ OPPOSED
PL	PZLFRA	4A-235B	4	117.5	88.9	3.85	125	2.8	118		8.5	04	HORIZ OPPOSED

Country	Mfr	Model											Notes
PL	PZLFRA	6A-350C	6	117.5	88.9	5.74	220	2.8	167		10.5	O4	HORIZ OPPOSED
PL	PZLFRA	6AS-350A	6	117.5	88.9	5.74	250	2.8	189		7.4	O4	HORIZ OPPOSED
PL	PZLFRA	6A-350D	6	117.5	88.9	5.74	235	3.2	145		10.5	O4	HORIZ OPP HELE E
PL	PZLFRA	6V-350B	6	117.5	88.9	5.74	235	3.2	144		10.5	O4	VERT OPPO-HELE E
RU	ASH	ASH-62M	9	155.5		29.87	738	2	567	71.5	6.4	O4	Superch.
RU	ASH	ASH-82T	14	155.5	155	41.2	1,630	2.4	1,020		6.9	O4	
RU	ASH	ASH-82V	14	155.5	155	41.2	1,530	2.4	1,070			O4	
RU	IVCHEN	AI-14RT	9	105	130	10.2	223	2.4	217	74.5	6.2	O4	Superch.,An-14, Yak-18A
RU	IVCHEN	AI-14VF	9	105	130	10.2	208	2.4	242	78.2	6.2	O4	Heli
RU	IVCHEN	AI-26V	9	155.5	150	20.6	428	2.2	450	83.8	6.4	O4	Heli
RU	VEDEN	M14V26	9	105	130	10.2	242	2.8	245	78.2	6.2	O4	Heli;Superch.
UK	LEONID	MAJ755/1	14	122	112	18.3	593	2.9	483	95	6.8	O4	Heli
UK	LEONID	524/1	9	122	112	11.8	388	3.2	340	83.8	6.5	O4	Heli
UK	LEONID	531/8	9	122	122	12.8	71	2.6	84	87.6	7	O4	
UK	ROLASN	ARDEMMKX		83	69	1.5	53	3.6			8.5	O4	HORIZ OPPOSED
UK	ROLASN	ARDEMKXI		85.5	69	1.5	55	3.6	72		8.5	O4	HORIZ OPPOSED
UK	ROLSRO	CO0-240A	4	122.5	98.4	3.93	98	2.5	112	71	8.5	O4	HORIZ OPPOSED
UK	ROLSRO	CO10-368	4	133.4	108	6.02	145		136	74.4	8.2	O4	
UK	ROLSRO	CO-O-200	4	103	98	3.3	74	2.7	86	92.8	7	O4	
UK	ROLSRO	CO-O-240	4	112.5	98.4	3.9	97	2.8	97	84.6	8.5	O4	
UK	ROLSRO	CO-O-300	6	103	98	4.9	108	2.7	122	85.7	7	O4	
UK	WESLAK	TYP274-6	2	66	40	0.27	18	6.5	7	186		O4	SIMULT FIRING

(continued)

Country	Company	Model No.	No. of Cyl.	Cyl. Dia.	Stroke length	Displ. Vol.	Take-off power	R.P.M. (k-rpm)	Weight (KG)	Spec. fuel consumption (mu-g/J)	Volume compression ratio	Otto/diesel, No. of stroke	Remarks
UK	WESLAK	TYPE430	2	90	60	1.52	40		11	149		O2	2 STROKE
UK	WESLAK	TYPE1527	4	90	60	1.52	71	5.5	45	84.5		O4	
UK	NAPIER		12	152	187	41.1	2,668	2	1,630	59.6	27	D4	DIESEL
US	AVLYCO	0-235C	4	111	98.4	3.85	115	2.7	98		8.5	O4	
US	AVLYCO	0-235H	4	111	98.4	3.85	115	2.6	96		8.5	O4	
US	AVLYCO	O-235-L	4	111	98.4	3.85	115	2.7	98		8.5	O4	
US	AVLYCO	O290-D2C	4	124	98	4.7	104	2.8	120	78.2	7	O4	
US	AVLYCO	O-320-A	4	130	98.4	5.2	150	2.7	110		7	O4	
US	AVLYCO	O-320-D	4	130	98.4	5.2	160	2.7	114		8.5	O4	
US	AVLYCO	O-320-E	4	130	98.4	5.2	160	2.7	113		7	O4	
US	SALYCO	O-320-H	4	130	98	5.2	160	2.7	115		9	O4	
US	AVLYCO	AEO320-E	4	130	98.4	5.2	150	2.7	117		7	O4	
US	AVLYCO	O-360-A	4	130	111	5.92	180	2.7	118		8.5	O4	
US	AVLYCO	LO-360-A	4	130	111	5.92	180	2.7	120		8.5	O4	
US	AVLYCO	O-360-C	4	130	111	5.92	180	2.7	116		8.5	O4	
US	AVLYCO	O-360-E	4	130	111	5.92	180	2.7	122		9	O4	
US	AVLVCO	LO-360-E	4	130	111	5.92	180	2.7	122		9	O4	
US	AVLYCO	O-360-F	4	130	111	5.92	180	2.7	122		8.5	O4	
US	AVLYCO	IV-360-A	4	130	111	5.92	180	2.7	124		8.5	O4	
US	AVLYCO	TO-360-C	4	130	111	5.92	210	2.6	154		8.5	O4	
US	AVLYCO	IO-360-A	4	130	111	5.92	200	2.7	133		8.5	O4	
US	AVLYCO	IO-360-B	4	130	111	5.92	180		122		8.5	O4	
US	AVLYCO	IO-350-C	4	130	111	5.92	200		134		8.7	O4	
US	AVLYCO	HIO360-C	4	130	111	5.92	205		132		8.7	O4	
US	AVLYCO	HIO360-D	4	130	111	5.92	190		132		10	O4	
US	AVLYCO	HIO360-E	4	130	111,0	5.92	190		132		8	O4	

US	AVLYCO	TIO360-E	4	130	111,0	5.7	181	2.6	181	84.6	7.3	O4	
US	AVLYCO	AEIO360A	4	130	111	5.92	200		139		8.7	O4	
US	AVLYCO	AEIO360B	4	130	111	5.92	180		125		8.5	O4	
US	AVLYCO	AEIO360H	4	130	111	5.92	180		122		8.5	O4	
US	AVLYCO	O-540-B	6	130	111	8.86	235	2.6	166		7.2	O4	
US	AVLYCO	O-540-E	6	130	111	8.86	260	2.7	167		8.5	O4	
US	AVLYCO	O-540-G	6	130	111	8.86	260		174		8.5	O4	
US	AVLYCO	O-540-J	6	130	111	8.86	235	2.4	162		8.5	O4	
US	AVLYCO	VO435-23	6	124	98	7.5	194	3.4	181	93.1	7.3	O4	Mil.Heli.
US	AVLYCO	TVO-435	6	124	98	7.5	201	3.2	210	84.6	7.3	O4	Superch.
US	AVLYCO	GO-480	6	124	98	7.9	220	3.4	198	78.2	8.7	O4	Horiz.oop.
US	AVLYCO	IGSO480	6	130	98	7.9	253	3.4	218	81.2	7.9	O4	Horiz.oop.
US	AVLYCO	VO-540-B	6	130	111	8.86	305	3.2	202		7.3	O4	
US	AVLYCO	IGSO540	6	130	111	8.6	283	3.4	241	81.2	7.3	O4	Horiz.opp.
US	AVLYCO	VO-540-C	6	130	111	8.86	305	3.3	200		8.7	O4	
US	AVLYCO	IO-540-C	6	130	111	8.86	250	2.6	170		8.7	O4	
US	AVLYCO	IO-540-E	6	130	111	8.86	290	2.6	187		8.5	O4	
US	AVLYCO	IO-540-K	6	130	111	8.86	300	2.7	201		8.5	O4	
US	AVLYCO	IO-540-S	6	130	111	8.86	300	2.7	201		8.7	O4	
US	AVLYCO	IO-540-T	6	130	111	8.86	260	2.7	171		8.5	O4	
US	AVLYCO	AEIO540D	6	130	111	8.86	260	2.7	174		8.5	O4	
US	AVLYCO	TIO540-A	6	130	111	8.86	310	2.7	232		7.3	O4	
US	AVLYCO	TIO540-C	6	130	111	8.86	250	2.6	205		7.2	O4	
US	AVLYCO	TIO540-F	6	130	111	8.86	325	2.6	233		7.3	O4	
US	AVLYCO	LTIO540F	6	130	111	8.86	325	2.6	233		7.3	O4	
US	AVLYCO	TIO540-J	6	130	111	8.86	350	2.6	235		7.3	O4	
US	AVLYCO	LTIO540J	6	130	111	8.86	350	2.6	235		7.3	O4	
US	AVLYCO	TIO540-R	6	130	111	8.86	350	2.5	238		7.3	O4	
US	AVLYCO	TIO540-S	6	130	111	8.86	300	2.7	228		7.3	O4	

(continued)

Country	Company	Model No.	No. of Cyl.	Cyl. Dia.	Stroke length	Displ. Vol.	Take-off power	R.P.M. (k-rpm)	Weight (KG)	Spec. fuel consumption (mu-g/J)	Volume compression ratio	Otto/diesel, No. of stroke	Remarks
US	AVLYCO	TIO541-E	6	130	111	8.86	380	2.9	270		7.3	O4	
US	AVLYCO	TIGO541D	6	130	111	8.86	450	3.2	311		7.3	O4	
US	AVLYCO	TIGO541E	6	130	111	8.86	425	3.2	319		7.3	O4	
US	AVLYCO	IO-720-A	8	130	111	11.84	400	2.7	257		8.7	O4	
US	AVLYCO	IO-720-B	8	130	111	11.84	400	2.7	252		8.7	O4	
US	CTOTIA	460	1	88	75	0.46	20	3.5	14			O2	2 STROKE
US	FRANK	SPORT4	4	117	89	3.8	97	2.8	121	82.7	8.5	O4	Horiz.opp.
US	FRANK	4A235	4	117	89	3.8	93	2.8	117	82.7	8.5	O4	Horiz.opp.
US	FRANK	6A-335A	6	114	89	5.5	134	2.8	144	83.8	7	O4	Horiz.opp.
US	FRANK	6AS335A	6	114	89	5.5	194	3.2	157	81.2	7	O4	Horiz.opp.
US	FRANK	6A-350	6	117	89	5.6	164	2.8	145	77.5	10.5	O4	Horiz.opp.
US	FRANK	6AS350	6	117	89	5.6	186	2.8	171	84.5	7.4	O4	Horiz.opp.
US	JACOBS	R-755-A	7	133	127	12.3	300	2.2	229			O4	RADIAL
US	TELCON	O-200-A	4	103.2	98.4	3.28	100	2.8	98		7	O4	
US	TELCON	O-300-A	6	103	98	4.9	108	2.7	121	85.7	7	O4	
US	TELCON	IO-346	4	133	102	5.6	123	2.7	134	81.2	7.5	O4	Horiz.opp.
US	TELCON	IO-360-A	6	113	98	5.7	156	2.8	133	74.5	8.5	O4	Horiz.opp.
US	TELCON	IO-360-B	6	113	98	5.7	134	2.7	151		6.5	O4	Horiz.opp.
US	TELCON	IO-360-D	6	112.5	98.4	5.9	210	2.8	148		8.5	O4	Horiz.opp.
US	TELCON	TSIO360A	6	113	98	5.7	156	2.8	136	74.5	7.5	O4	Horiz.opp.
US	TELCON	TSIO360D	6	112.5	98	5.9	225	2.8	136		7.5	O4	
US	TELCON	TSIO360E	6	112.5	98.4	5.9	200	2.6	175		7.5	O4	
US	TELCON	TSIO360F	6	112.5	98.4	5.9	200	2.6	175		7.5	O4	
US	TELCON	Tia4-180	4	123	92	4.3	134	4	120	76	9	O4	Horiz.opp.
US	TELCON	Tia6-260	6	123	92	6.5	194	4	160	76	7.5	O4	Horiz.opp.
US	TELCON	TiT6-260	6	123	92	6.5	194	4	160	85.7	8	O4	Horiz.opp.

US	TELCON	TiT6-285	6	123	92	6.5	212	4	182	85.7	8	O4	Horiz.opp.
US	TELCON	TiT6-320	6	123	92	6.5	238	4	187	85.7	8	O4	Horiz.opp.
US	TELCON	TiT8-450	8	123	92	8.6	335	4.5	233	76	8	O4	Horiz.opp.
US	TELCON	IO-470-H	6	127	101.6	7.7	260	2.6	203		8.6	O4	
US	TELCON	O-470-R	6	127	101.6	7.7	230	2.6	193		7	O4	
US	TELCON	O-470-S	6	127	101	7.7	230	2.6	193		7	O4	
US	TELCON	A65	4	98	92	2.8	48	2.3	77	83.8	6.3	O4	Horiz.opp.
US	TELCON	C-90	4	103	98	3.3	71	2.6	84	87.6	7	O4	Horiz.opp.
US	TELCON	GIO470A	6	127	102	7.7	231	3.2	229	89.4	7.7	O4	Horiz.opp.
US	TELDYN	TSIO470D	6	127	101.6	7.7	260	2.6	232		7.5	O4	
US	TELDYN	IO-520-A	6	133	101.6	8.5	285	2.7	216		8.5	O4	
US	TELDYN	IO-520-B	6	133	101.6	8.5	285	2.7	207		8.5	O4	
US	TELDYN	IO-520-D	6	133	101.6	8.5	285	2.7	208		8.5	O4	
US	TELDYN	IO-520-M	6	133	101.6	8.5	285	2.7	188		8.5	O4	
US	TELDYN	TSIO520B	6	133	101.6	8.5	285	2.7	219		8.5	O4	
US	TELDYN	TSIO520C	6	133	101.6	8.5	285	2.7	209		7.5	O4	
US	TELDYN	TSIO520E	6	133	101.6	8.5	300	2.7	219		7.5	O4	
US	TELDYN	TSIO520J	6	133	101.6	8.5	310	2.7	221		7.5	O4	
US	TELDYN	TSIO50-N	6	133	101.6	8.5	310	2.7	221		7.5	O4	
US	TELDYN	TSIO520L	6	133	101.6	8.5	310	2.7	245		7.5	O4	
US	TELDYN	TSIO520M	6	133	101.6	8.5	285	2.6	198		7.5	O4	
US	TELDYN	TSIO520R	6	133	101.6	8.5	285	2.6	198		7.5	O4	
US	TELDYN	GTSI520C	6	133	101.6	8.5	340	3.2	253		7.5	O4	
US	TELDYN	GTSI520F	6	133	101.6	8.5	435	3.4	290		7.5	O4	
US	TELDYN	GTSI520K	6	133	101.6	8.5	435	3.4	290		7.5	O4	
US	TELDYN	GTSI520H	6	133	101.6	8.5	375	3.4	250		7.5	O4	
US	TELDYN	GTSI520L	6	133	101.6	8.5	375	3.4	250		7.5	O4	
US	TELDYN	GTSI520M	6	133	101.6	8.5	375	3.4	250		7.5	O4	
US	TELDYN	C90-16F	4	103.2	98.4	3.28	90	2.5	85		7	O4	

(continued)

Country	Company	Model No.	No. of Cyl.	Cyl. Dia.	Stroke length	Displ. Vol.	Take-off power	R.P.M. (k-rpm)	Weight (KG)	Spec. fuel consumption (mu-g/J)	Volume compression ratio	Otto/ diesel, No. of stroke	Remarks
US	TELDYN	O-200-B	4	103.2	98.4	3.28	100	2.8	100		7	O4	
US	TELDYN	TSIO360A	6	112.5	98.4	5.9	210	2.8	152		7.5	O4	
US	TELDYN	IO-470-D	6	127	101.6	7.7	260	2.6	193		8.6	O4	
US	TELDYN	IO-520-B	6	133	101.6	8.5	285	2.7	207		8.5	O4	
US	TELDYN	GTSI520D	6	133	101.6	8.5	375	3.4	250		7.5	O4	
US	TELDYN	GTSI520G	6	133	101.6	8.5	375	3.4	253		7.5	O4	
US	TELDYN	4-180	4	123.8	92.1	4.44	135	3.6	120		9	O4	
US	TELDYN	6-285A	6	123.8	92.1	6.65	214	3.7	161		9	O4	
US	TELDYN	T6-285	6	123.8	92.1	6.65	214	3.6	182		8	O4	
US	TELDYN	6-320	6	123.8	92.1	6.65	240	4	161		9.6	O4	
US	TELDYN	T6-320	6	123.8	92.1	6.65	240	4	187		8	O4	
US	TELDYN	T8-450	6	123.8	92.1	8.88	338	4	233		8	O4	
US	WRIGHT	R1820-76	9	156	174	30	1,077	2.7	627	82	7.2	O4	SIKORSKY
US	WRIGHT	R3350-34	18	156	160	55	2,460	2.9	1,600	68.9	6.7	O4	SUPCONSTEL G
US	WRIGHT	R3350-40	18	155	160	54.9	2,758	2.9	1,667	63.3		O4	
US	LYCOM	R1300-3	7	156	160	21.4	605	2.6	490	82	7.2	O4	
US	P&W	2800CB16	18	146	152	45.7	1,815	2.8	1,090	82		O4	

Turbojet

Country	Company	Model No.	Mass flow rate (kg/s)	Take-off thrust(kN)	Overall compression ratio	Turbine inlet temp. (Deg. C)	Diameter(m)	Weight(kg)	Specific fuel consumption (milligram/N-s)	By-pass ratio	No. of spools	No. of axial fan stages	Fan pressure	No. of compressor stages	krpm
CA	UACL	JT15D-1	34.1	9.8	10	960	0.69	230	15.29	3.3	2	1	1.5		
CA	UACL	JT15D-4	34.1	10.6	10	960			15.92	3.3	2	1	1.5		
CA	ORENDA	14	59	33.3	6		1.09	1,120	25.5		1			10	7.8
CA		ORENDA14	59	33.3	6		1.09	1,120	25.48		2			10	7.8
CA		IROQUOIS	152	98.1	8		1.19	2,270			2				
CZ	MOTORLET	M701	16.9	8.7	4.3				32.28		1			1	15.4
CZ	MOTORLET	AI25													
FR	MICROTUR	COGUA022		0.8				28	35.4		1				48.5
FR	SNECMA	ATAR101E	24.9	34.3	5.5						1			8	8.4
FR	SNECMA	ATAR 8	26	43.1	6.9						1			9	8.4
FR	SNECMA	SUPRATAR	49		6.2				27.5		1			5	
FR	SNECMA	ATAR09C3	68	42	5.7	890	0.79	1,409	28.63		1			9	8.4
FR	SNECMA	TF306	122	52	17	1,100	1.2	1,760	18.42		2	3	2	6	9.6
FR	TURBOMEC	ARBIZO3B	6	3.2	5.5			115	31.44		1			1	32
FR	MICROTUR	ECLAIR		0.8				35			1			1	47
FR	SERMEL	TRS18		1			0.32	30	41.06		1			1	
FR	TURBOMEC	ASTFAN2G		6.9				285	10.76	7					
FR	TURBOMEC	MARBORE	9.8	4.7	3.8	613		140	30.86		1			1	21.5
FR	TURBOMEC	TR281		3.6			0.41	105	27.75		1	1		1	32.3
FR	TUR-SNEC	LARZAC03	27	12.3	10		0.45	265	19.12	1.2	2	2		4	22.3
FR	TUR-SNEC	LARZAC04	27.6	13.2	10.7	1,130	0.45	290	20.1	1.1	2	2		4	22.8
FR	TURBOMEC	MARBORII	7.6	3.9	4		0.57	133	30.58		1			1	22.6
FR	TURBOMEC	PALAS	3.1	1.6	4		0.41	72	31.15		1			1	34
FR	TURBOMEC	GOURDON		6.4			0.57	104	28.28		1	1		1	
FR	TURBOMEC	GABIZO	14.8	10.8	5.1		0.67	172	29.45		1	1		1	
FR		DASSALR7	24.9	13.3	3.8		0.69	340	30.9		1			7	11.8
FR		HISPR804	26	14.7	4.8		0.69	305	30.24		1			7	12
DE	MTU	6012-C	1		3.1			46			1			4	45
DE	HEINKEL	HES053	100	63.8	7.4		1.1	1,570	26.33		1			11	6
DD		PIRNA104	50	30.9			0.98	1,000	24.07		1			12	
IN	HAL	HJE-2500	20.4	11.1	4.2		0.66	265	27.75		1			1	12.5
IT	FIAT	4002.01	6.3	3.2	4		0.57	88	34.26		1			1	25
IT	FIAT	4032	50	26.5	5.5		1.01	490	27.75		1			9	8.2
IT	FIAT	ORPH803	38.2	22.3	4.4		0.82	372	30.02		1			7	10
IT	FIAT	J79GE19	77	52.8	13.5	1,038	0.99	1,745	23.79		1			17	
IT	ALPHSROM	J85GE13A	20	18.2	7		0.53	271	35.71		1			8	
IT	PIAGGIO	VIPER500	23.9	15	5.6	892	0.49	347	28.6		1			8	13.8
JP	IHI	J3-IHI-7	25.4	13.7	4.5		0.63	430	29.74		1			8	
JP	IHI	J3-IHI-8	25.4	15.2	4.5		0.63	430	29.74		1			8	13
JP	IHI	JR100F	27.5	14	3.9	850	0.6	156	32.56		1			1	
JP	IHI	JR100H	27.5	14.9	3.9	850	0.6	156	32		1			1	
JP	IHI	JR200	37.2	17.9	4	850		127	33.13		1			5	
JP	NAL	FJR 710		49			1.52	980	9.83	6.5	2			1	
JP	NAL	JR 200	37.2	20.4	4	850		127	33.2		1			5	12.5
PL	IL	IL-SO-1		8.7	4.8			303	29.6		1			7	15.1
PL	IL	IL-SO-3		9.8	4.8			325	29.6		1			7	15.1
RU	SOLOVIEV	D-30P		68.1	18.6	1,030		1,520	17.3	1	2			4	8
RU	SOLOVIEV	D-20P	113	54	13		0.98	1,470	22.1	1	2			3	8.5
RU	SOLOVIEV	D-30K		115	20		1.56	2,150	14.1	2.3				3	
RU		VK-1		27				900							

Compression ratio	Type of compressor	No. of turbine stages	No. of compressor stages2	krpm2	Compression ratio2	Type of compressor2	No. of turbine stages2	No. of compressor stages3	krpm3	Compression ratio3	Type of compressor3	No. of turbine stages3	Combustor type	Remarks
		2	1		6.7	Centrifugal	1						Annulus	REVERSE FLOW
		2	1		6.7	Centrifugal	1						Annulus	RF CC
	Axial	2				Centrifugal								
6	Axial	2											Can	Turbojet
		2					1						Annulus	Turbofan
4.3	Centrifugal	1											Annulus	Turbojet
													Annulus	Turbojet
5.5	Axial	1											Annulus	Turbojet
6.9	Axial	2											Annulus	Turbojet
	Axial	2											Annulus	Turbojet
5.7	Axial	2											Annulus	Turbojet ABF=58.9kN,sfc=57.5.Mirage3/5.
	Axial	3	7	14.2		Axial	1						Cannular	Turbofan ABF=101kN, sfc=56.7
5.5	Axial	1											Annulus	Turbojet
	Centrifugal	1											Annulus	REV
	Centrifugal	1											Annulus	FOLD
													Annulus	ASTZ18
3.8	Axial	1											Annulus	
	Centrifugal	1											Annulus	
		1											Annulus	
		1											Annulus	
4	Centrifugal	1											Annulus	Turbojet
4	Centrifugal	1											Annulus	Turbojet
	Centrifugal	1											Annulus	Turbojet
5.1	Centrifugal	1											Annulus	Turbojet
3.8	Axial	1											Annulus	Turbojet
4.8	Axial	1											Annulus	Turbojet
3.1	Centrifugal	1											Annulus	TS
7.4	Axial	2											Annulus	Turbojet
	Axial	2											Annulus	Turbojet
4.2		1											Cannular	Turbojet
4	Centrifugal	1											Cannular	Turbojet
5.5	Axial	1											Cannular	Turbojet
4.4	Axial	1											Cannular	Turbojet
	Axial	3											Cannular	Turbojet ABF=79.7kN, sfc=55.5
7	Axial	2											Annulus	Turbojet ABF=18.1kN, sfc=62.4
5.6	Axial	1											Annulus	Turbojet
	Axial	1											Annulus	Turbojet
	Axial	1											Annulus	Turbojet
	Axial	1											Annulus	LIFTJET
	Axial	1											Annulus	LIFTJET
	Axial	1											Annulus	LIFTJET
			1				2						Annulus	Turbofan
4	Axial	1											Annulus	LIFTJET
4.8	Axial	1											Annulus	Turbojet
4.8	Axial	1											Annulus	Turbojet
	Axial	2	10	11.6		Axial	2						Cannular	Turbofan
2.6	Axial	2	8	11.7	5	Axial	1						Cannular	TurbofanAB
	Axial	4	11			Axial	2						Cannular	Turbofan,IL-62
													Cannular	Afterburner Thrust 34.5kN, Mig-17

(continued)

Country	Company	Model No.	Mass flow rate (kg/s)	Take-off thrust(kN)	Overall compression ratio	Turbine inlet temp. (Deg. C)	Diameter(m)	Weight(kg)	Specific fuel consumption (milligram/N-s)	By-pass ratio	No. of spools	No. of axial fan stages	Fan pressure	No. of compressor stages	krpm
RU		NK-8		105						1	2			3	
RU		NK-144		113	15		1.5	2,850			2			5	
RU		RD-9F	44	21.5		710	0.67	720	25.48		1			9	11.2
RU		R11-F	65.2	39		936	0.9	1,126	29		2			3	11.1
RU		R-25	67	41		1,040	0.9	1,210	29		2			3	11.1
RU		R-29B	105	80		1,130	0.99	1,772	29		2			5	
RU	IVCHENKO	AI-25	45	14.7	8	950	0.6	330	16.44	2	2	3	2.2	2	10.7
RU	KUZNETSO	NK-8-2		93.2		1,143		2,350	21.53	1	2	2	2.2	2	5.4
RU	KUZNETSO	NK-144	250	127.5	15	1,000	1.5	2,850	19.84	1	2	2		2	
RU	LOTAREV	D-36	63.7					1,080	10.62	5.3					
RU	SOLOVIEV	D-20P	113	53	13		0.98	1,468	20.4	1	2			3	8.6
RU	KUZNETSO	NK-8-4	99.1			870		2,400	22.1	1	2	2	2.2	2	5.4
RU	SOLOVIEV	D-30	125	66.7	17.4		1.05	1,550	17.56	1	2			4	7.7
RU	SOLOVIEV	D-30K	269	108	20	1,122	1.56	2,650	13.88	2.4	2			3	4.7
RU	KUZNETSO	NK-144		171.6	15		1.5	2,850		1	2			5	
SE	FLYGMOTO	RM6B	71	49	7.7		1.07	1,700			1			15	8
SE	FLYGMOTO	RM6C	79	56.4	8.4		1.07	1,770			1			16	8.1
SE	FLYGMOTO	RM8A	145	115.6	16.5		1.4	2,100		1	2	2		4	
SE	FLYGMOTO	RM8B	145	25	16.5		1.4	2,350		1	2	2		4	
SE	FLYGMOTO	RM8	146	115.8	16.5		1.34	2,100	17.57	1	2	2	2.1	4	8.6
CH	SULZER	ATAR09C	68	42	5.6	890	0.79	1,372	28.63		1			9	5.6
SE	FLYGMOTO	RM6B	71	49	7.7		1.07	1,700			1			15	8
SE	FLYGMOTO	RM6C	79	56.4	8.4		1.07	1,770			1			16	8.1
SE	FLYGMOTO	RM8A	145	115.6	16.5		1.4	2,100		1	2	2		4	
SE	FLYGMOTO	RM8B	145	25	16.5		1.4	2,350		1	2	2		4	
SE	FLYGMOTO	RM8	146	115.8	16.5		1.34	2,100	17.57	1	2	2	2.1	4	8.6
CH	SULZER	ATAR09C	68	42	5.6	890	0.79	1,372	28.63		1			9	5.6
UK	ALVIS	RO TJ125		0.5				18	37.09		1			1	
UK	BUDWORTH	PUFFIN	1.5	0.8		927	0.38	31			1			1	
UK	ARMSTROM	SAP:ASA7	70	49	8		0.95	1,375	25.06		1			13	8.6
UK	ARMSTROM	VIPEASV8	14.5	7.8	3.8		0.71	231	31.7		1			7	13.8
UK	ARMSTROM	VIPASV10	19	89	4		0.71	260	28.5		1			7	13.4
UK	ARMSTROM	VIPASV11	19	10.9	4		0.71	260	31.4		1			7	13.4
UK	ARMSTROM	PALAS600	3.3	1.7	4.1		0.43	67	33.98		1			1	
UK	BRISTOL	OLYMP104		57.9			1.02	1,600			2			8	
UK	BRISTOL	OLYMP301	131	88.3	13.1		1.13	1,946	21.26		2			6	6.8
UK	BRISTOL	OLYMP593	188	170.7	14		1.21	2,640	19.84		2			6	6.5

Compression ratio	Type of compressor	No. of turbine stages	No. of compressor stages2	krpm2	Compression ratio2	Type of compressor2	No. of turbine stages2	No. of compressor stages3	krpm3	Compression ratio3	Type of compressor3	No. of turbine stages3	Combustor type	Remarks
	Axial	2	8			Axial	1							Turbofan, Tu-154
													Annulus	Afterburner Thrust 175.kN, Tu144(SST)
7.5	Axial	2											Can	Afterburner Thrust 37.3kN, sfc=59.5.Mig-19.
			3	11.4									Cannular	Mig-21 FL/M
3.3	Axial	5	11.4	2.9		Axial								B Mig-21/Bis
			6	8.8									Annulus	Mig-27M
1.7	Axial	2	8	16.3	4.7	Axial	1						Cannular	Turbofan.Yak40
	Axial	2	6	6.8	10	Axial	1						Annulus	Turbofan REVERSER.IL62, Tu154.
	Axial	2	11			Axial	1						Annulus	Turbofan. ABF=171.7kN. Tu144(M=2.2)
2.4	Axial	2	8	11.2	5	Axial	1						Cannular	Turbofan
	Axial	2	6	7	10.8	Axial	1						Annulus	Turbofan
2.7	Axial	2	10	11.6	7.1	Axial	2						Cannular	Turbofan
	Axial	4	11	10.5			2						Cannular	Turbofan REVERSER
	Axial	2	11				1						Annulus	Turbofan
7.7	Axial	2											Cannular	Turbojet ABF=64.7kN, sfc=48.2
8.4	Axial	2											Cannular	Turbojet ABF=78.4kN, sfc=53.8
	Axial	3	7			Axial	1						Cannular	Turbofan ABsfc=70.0
	Axial	3	7			Axial	1						Cannular	Turbofan ABsfc=71.4
	Axial	3	7	11.9		Axial	1						Cannular	Turbofan ABF=115.8kN, sfc=73.7.Viggen.
8.4	Axial	2											Annulus	TurbojetABF=58.9kN, sfc=57.5
7.7	Axial	2											Cannular	Turbojet ABF=64.7kN, sfc=48.2
8.4	Axial	2											Cannular	Turbojet ABF=78.4kN, sfc=53.8
	Axial	3	7			Axial	1						Cannular	Turbofan ABsfc=70.0
	Axial	3	7			Axial	1						Cannular	Turbofan ABsfc=71.4
	Axial	3	7	11.9		Axial	1						Cannular	Turbofan ABF=115.8kN, sfc=73.7.Viggen.
8.4	Axial	2											Annulus	TurbojetABF=58.9kN, sfc=57.5
	Centrifugal	1											Annulus	Turbojet
	Centrifugal	1											Annulus	Turbojet
8	Axial	2											Annulus	Turbojet
3.8	Axial	1											Annulus	Turbojet
4	Axial	1											Annulus	Turbojet
4	Axial	1											Annulus	Turbojet
4.1	Centrifugal	1											Annulus	Turbojet
	Axial	1	7				1						Cannular	Turbofan
	Axial	1	7	8		Axial	1						Cannular	Turbojet Vulcan B2.
	Axial	1	7	8.8		Axial	1						Cannular	Turbojet ABsfc=33.59. Concorde

(continued)

Country	Company	Model No.	Mass flow rate (kg/s)	Take-off thrust(kN)	Overall compression ratio	Turbine inlet temp. (Deg. C)	Diameter(m)	Weight(kg)	Specific fuel consumption (milligram/N-s)	By-pass ratio	No. of spools	No. of axial fan stages	Fan pressure	No. of compressor stages	krpm
UK	BRISTOL	OLYMB1.7		75.5			1.06	1,630	22.65		2			7	
UK	BRISTOL	ZEPHIER	104	55.5	10		1.04	1,630	20.33		2			7	
UK	BRISTOL	ORPHBOR3	38.2	22.3	4.4		0.82	372	30.02		1			7	10
UK	DEHAVILL	GOBLIN35	28.5	15.6	3.7		1.77	726	32.28		1			1	
UK	DEHAVILL	GHOST105	40	23.5			1.34	975	33.7		1			1	
UK	DEHAVILL	GYROP442		82.3			1.26	1,900	29.4		1			7	
UK	DEHAVILL	GYDGJ10R		44.4			0.82				1			8	
UK	ROLSROYS	NENE RN6	43	24	4.5		1.25	735	30.02		1			1	12.5
UK	ROLSROYS	AVONRA21		35.6			1.07	1,116	27.04		1			12	7.9
UK	ROLSROYS	AVONRA24		50			1.05	1,305			1			15	
UK	ROLSROYS	AVONRA28	72.8	44.4	8		1.05	1,305	24.4		1			15	8
UK	ROLSROYS	AVONRA29	78.5	46.7	9.3		1.05	1,500	21.9		1			16	8
UK	ROLSROYS	AV-RB146	80	56.5	8.4		1.07	1,724	24.09		1			16	8
UK	ROLSROYS	RB108		8.9							1			5	
UK	ROLSROYS	RB145		12.3			0.51				1			7	
UK	ROLLSROY	RB162-81	38.5	26.7	4.5		0.74	188			1			6	
UK	ROLLSROY	RB162-86	38.5	23.4	4.5		0.74	236			1			6	
UK	ROLLSROY	ADOUR		19.6			0.56	600	19.27	1.1	2			2	
UK	ROLLSROY	CONWAY42	104	90.6	14.8		1.29	2,270	17	0.6	2	4		3	7.2
UK	ROLLSROY	CONWAY43	102	97	15.8		1.29	2,300	16.72	0.6	2	4		4	7.2
UK	ROLLSROY	RB21122B	626	187	25		2.17	4,171	17.7	5	3			1	
UK	ROLLSROY	RB163555	90.3	42.1	15		0.94	995	15.57	1	2			4	11.9
UK	ROLLSROY	RB163505	90.3	42.1	15		0.94	998	15.43	1	2			4	12.3
UK	ROLLSROY	RB163506	92	44.5	17.2		0.94	1,024	15.74	1	2			4	12.4
UK	ROLLSROY	RB163512	94.4	55.8	21.2		0.94	1,168	22.7	0.6	2			5	12.6
UK	ROLLSROY	RB202		57.9			1.9	392	12.76	10	2			1	
UK	ROLLSROY	SPEY201	95	55.8	20		0.96	1,633	18.14	0.7	2			5	8.6
UK	ROLLSROY	SPEY250		53.4			0.96	1,225	17.86		2			5	
UK	ROLLSROY	SPEY512	93	55.6	20.7		0.99	1,252	17.01	0.7	2			5	8.1
UK	ROLLSROY	SPEY555	92	43.8	15.4		0.94	995	21.26	1	2			5	8.6
UK	ROLLSROY	TRENT	136	44.4	16		0.98	806	20.12	3	3			1	8.7
UK	ROLLSROY	VIPER600	26.4	16.7	5.8		0.62		26.6		1			8	
UK	ROLLSROY	VIPER11	20	11.1	4.4	830	0.62	249	29.48		1			7	13.8
UK	ROLLSROY	VIPER601		16.7			0.62	358	26.6		1			8	
UK	ROLLSROY	VIPER632		17.8			0.62	358	26.6		1			8	
UK	ROLLSROY	VIPER535	23.9	15	5.6		0.62	354	28.6		1			8	13.8
UK	ROLLSROY	RB401-07	82.5	24.7				447	12.72	4.2	2	1			
UK	ROLLSROY	PEGAS104	196	95.6	14		1.22	1,429		1.4	2	3	2.3		
UK	ROLLSROY	PEGAS103	196	95.6	14		1.22	1,409		1.4	2	3	2.3		
UK	ROVER	TJ125	0.9	0.6	3.9	912	0.26	18			1			1	65

Compression ratio	Type of compressor	No. of turbine stages	No. of compressor stages2	krpm2	Compression ratio2	Type of compressor2	No. of turbine stages2	No. of compressor stages3	krpm3	Compression ratio3	Type of compressor3	No. of turbine stages3	Combustor type	Remarks
	Axial	1	5				1						Cannular	Turbofan
	Axial	1	5				1						Cannular	Turbofan
4.4	Axial	1											Cannular	
3.7	Centrifugal	1											Can	Turbojet
	Centrifugal	1											Can	Turbojet
	Axial	2											Annulus	Turbojet
	Axial	2											Annulus	Turbojet GYRON JUNIOR
4.5		1											Can	
	Axial	2											Can	Turbojet
	Axial	2											Cannular	Turbojet
8	Axial	2											Cannular	Turbojet
9.3	Axial	3											Cannular	Turbojet Caravelle/Comet4
8.4	Axial	2											Cannular	Turbojet ABF=74.5kN, sfc=56.7.Lightning/Canberra
	Axial	1											Annulus	Turbojet
	Axial	1											Annulus	Turbojet
4.5	Axial	1											Annulus	LIFTJET
4.5	Axial	1											Annulus	LIFTJET
	Axial	1	5		9.6	Axial	1						Annulus	Turbojet ABF=64.7kN, sfc=42.5.Jaguar-B
	Axial	2	9	10		Axial	1						Cannular	Turbofan Boeing-707, VC10,DC8
	Axial	2	9	10.2		Axial	1						Cannular	Turbofan
	Axial	3	7				1	6				1	Annulus	Turbofan L101
	Axial	2	12			Axial	2						Annulus	Turbofan
	Axial	2	12			Axial	2						Annulus	Turbofan
	Axial	2	12			Axial	2						Annulus	Turbofan
	Axial	2	12			Axial	2						Annulus	Turbofan
	Axial	1				Axial	1						Annulus	Turbofan, V/STOL.HS141, Do231C.
2.7	Axial	2	12	12.6	7.4	Axial	2						Cannular	Turbofan ABF=94.8kN, sfc=55.3.Phantom.
	Axial	2	12										Cannular	
2.6	Axial	2	12	12.8	8	Axial	2						Cannular	Turbofan.Trident, Gulfstream.
2.5	Axial	2	12	12	6.3	Axial	2						Cannular	
	Axial	1	4	13		Axial	1	5	15.8		A	2	Annulus	
5.8	Axial	2											Annulus	Turbojet BUISINESS JET A/C AND MILITARY TRAINERS
4.4	Axial	1											Annulus	Turbojet HJT16.
	Axial	2											Annulus	Turbojet
	Axial	2											Annulus	Turbojet
	Axial	1											Annulus	Turbojet
		2	8		11.5	Axial	1						Annulus	Turbofan CRUISE THR:5.0KN AT12.KM
		2	8		6.1	Axial	2						Annulus	Turbofan V/STOL
		2	8		6.1	Axial	2						Annulus	Turbofan V/STOL
3.9	Centrifugal	1											Annulus	Turbojet

(continued)

Country	Company	Model No.	Mass flow rate (kg/s)	Take-off thrust(kN)	Overall compression ratio	Turbine inlet temp. (Deg. C)	Diameter(m)	Weight(kg)	Specific fuel consumption (milligram/N-s)	By-pass ratio	No. of spools	No. of axial fan stages	Fan pressure	No. of compressor stages	krpm
UK	LUCAS	CT 3201	0.9	0.5		917		19	37.1		1			1	
UK	NPT	401A	2.3	1.5	4.5	955	0.34	43	28.8		1	1		1	
UK	NPT	251	2.3	1.3		977	0.34	34	33.4		1				
UK	NPT	151	1.3	0.7		897	0.28	15	34		1				
US	ALLISON	J33-A-37	40.8	20.5	4.4		1.22	793	32.27		1			1	11.8
US	ALLISON	J35-A-29	43.2	24.9	5.5		0.94	1,020	29.73		1			11	7.8
US	ALLISON	J71-A-11	72.6	45.4	8.3		0.94	1,850	22.65		1			16	6.1
US	AVCOLYCO	ALF502L		33.4	13.7		1.07	578	11.9	5	2			2	
US	AVCOLYCO	ALF502R3		29.8	13.7		1.07	565	11.9	5.9	2			2	
US	CONTINEN	J69-T-9	7.6	4.1	4		0.57	140	32		1			1	22.7
US	DREHER	TJD-76C	0.5	0.3	2.8	770	0.15	10	42.5		1			1	
US	DREHER	TJD-79A	1.1	0.5	3		0.28	16	39.64		1			1	
US	GARRET	ATF3-6	73.5	22.4	21		0.85	431	13.6	2.8	3			1	
US	GARRET	TFE731-3	53.7	16.4	14.6	1,010	0.72	329	14.33	2.8	2	1		4	
US	GARRET	TFE731		15.6	19		0.72	283	13.88		2			4	
US	GARRET	ATF3	73.5	18	23		0.81	396	12.46	3.1	3			1	
US	GEC	GE1/10		44.5		1,093	0.97	1,360	22.67	1	2			2	
US	GEC	GE4/J5P	287	229.2	12.5	1,204	1.54	5,126	29.48		1			9	
US	GEC	J79GE-17	77	52.8	13.5	987	0.99	1,745	23.81		1			17	
US	GEC	CF700-2C	39.9	18.4	8.3		0.84	330	18.4	1.6	1			9	
US	GEC	CF700-2D	39.9	18.9	8.3		0.84	330	18.4	1.6	1			9	
US	GEC	CF7002D2	39.9	20.2	8.3		0.84	334	18.4	1.6	1			9	
US	GEC	F404	63.5	71.2	25		0.88	908		0.3	2			3	
US	GEC	TF34G400	153	41.3	21	1,225	1.33	661	10.3	6.2	2			1	7.4
US	GEC	TF34G100	153	40.3	21	1,225	1.26	647	10.5	6.2	2			1	7.4
US	GEC	CF34	139	35.5	17.5			692	10.18	6.3	2			1	7.4
US	GEC	CF6-6D	591	203.5	24.3	1,330		3,679		5.7	2	1		1	
US	GEC	CF6-50A	658	258	28.4			3,956		4.3	2	1		3	
US	GEC	CF6-80B	680	240				3,769		4.3	2	1		3	
US	GEC	CF6-80C	737	241				3,951		4.7	2	1		3	
US	GEC	CF6-80C2	812	273	31.9		2.36			5.3					
US	GEC	J85-5	20	17.2	7		0.53	265			1			8	
US	GEC	J85-13	20	18.2	7		0.53	271			1			8	
US	GEC	J85-17A	20	12.6	6.5	932	0.45	181	26.93		1			8	16.5
US	GEC	J85-21	23.8	15.6	8.1	982	0.53	301	35.15		1			9	16.6
US	GEC	CJ610-1	20	12				181	28.04		1			8	
US	GEC	CJ610-4	20	12				176	28.04		1			8	

Compression ratio	Type of compressor	No. of turbine stages	No. of compressor stages2	krpm2	Compression ratio2	Type of compressor2	No. of turbine stages2	No. of compressor stages3	krpm3	Compression ratio3	Type of compressor3	No. of turbine stages3	Combustor type	Remarks
	Centrifugal	1											Annulus	Turbojet SURVELLANCE RDV
4.5	Centrifugal	1											Annulus	Turbojet
	Centrifugal	1											Annulus	Turbojet
	Centrifugal	1											Annulus	Turbojet
4.4	Centrifugal	1											Can	Turbojet
5.5	Axial	1											Can	Turbojet
8.3	Axial	3											Cannular	Turbojet
		2	7			Axial	2						Annulus	Turbofan CAN CHALLENGER
		2	7			Axial	2						Annulus	Turbofan BAE146
4	Centrifugal	1											Annulus	Turbojet
	Mixed	1											Annulus	Turbojet PRUE215A SAILPLANE
	Mixed	1											Annulus	Turbojet
		3	5			Axial	2			1	C	1	Annulus	Turbofan REVER FLOW CC
	Axial	3	1			Centrifugal	1						Annulus	Turbofan
	Axial	3	1			Centrifugal	1						Annulus	Turbofan
	Axial	3	5			Axial	2			1	C	1	Annulus	Turbofan
	Axial	2	14			Axial	1						Annulus	Turbofan ABF=75.54kN, sfc=51.0
	Axial	2											Annulus	Turbojet ABF=305.2kN, sfc=52.7. Boeing 2707 (M=2.7)
	Axial	3											Cannular	Turbojet ABF=79.7kN, sfc=55.6. F-104,F4.
	Axial												Annulus	Turbofan DER DER:J85
	Axial												Annulus	Turbofan HIGH THER:EFF
	Axial												Annulus	Turbofan NEW DESIGN TAILPIPE
	Axial	1	7			Axial	1						Annulus	Turbofan AUG
1.5	Axial	4	14	17.9	14	Axial	2						Annulus	Turbofan
1.5	Axial	4	14	17.9	14	Axial	2						Annulus	Turbofan
1.4	Axial	4	14	17.9	12.5	Axial	2						Annulus	Turbofan CAN CHALLENGER
	Axial	5	16				2						Annulus	Turbofan
	Axial	5	16				2						Annulus	Turbofan
	Axial	4	16				2						Annulus	Turbofan
	Axial	4	16				2						Annulus	Turbofan
7	Axial	2											Annulus	Turbojet ABsfc=62.28
7	Axial	2											Annulus	Turbojet ABsfc=62.87
6.5	Axial	2											Annulus	Turbojet
8.1	Axial	2											Annulus	Turbojet ABF=22.2kN, sfc=60.37. Northrop-F5.
	Axial	2											Annulus	Turbojet
	Axial	2											Annulus	Turbojet

(continued)

Country	Company	Model No.	Mass flow rate (kg/s)	Take-off thrust(kN)	Overall compression ratio	Turbine inlet temp. (Deg. C)	Diameter(m)	Weight(kg)	Specific fuel consumption (milligram/N-s)	By-pass ratio	No. of spools	No. of axial fan stages	Fan pressure	No. of compressor stages	krpm
US	GEC	CJ610-9	20	13				189	27.85		1			8	
US	GEC	CJ610-5	20	12.4				183	27.19		1			8	
US	GEC	CF700-2C	38	18.6	7		0.91	330	18.46	1.9	1	1		8	
US	GEC	CF700-2D	38	18.9	7		0.91	330	18.46	1.9	1	1		8	
US	GEC	CF7002D2	38	19.2	7		0.91		18.21	1.9	1	1		8	
US	GEC	J79-GE-2	75		12.5		0.97	1,642			1			17	
US	GEC	J79-GE3B	75		12.5			1,508			1			17	
US	GEC	J79-GE5C	75		12.5		0.97	1,671			1			17	
US	GEC	J79-GE7A	75		12.5		0.97	1,622			1			17	
US	GEC	J79-GE-8	76.5		12.5		0.97	1,666			1			17	
US	GEC	J79-GE10	75		13.5		0.99	1,749			1			17	
US	GEC	J79-GE15	76.5		12.9		0.97	1,672			1			17	
US	GEC	J79-GE17	77	52.8	13.5	987	0.99	1,740	23.81		1			17	
US	GEC	J79GE11A	75		12.5		0.97	1,615			1			17	
US	GEC	TF34	153	40		1,201	1.3			6.2	2			13	7.4
US	GEC	CF6-6D	593	174.9	28.2	1,297		3,379	10.03	5.9	2	1		1	3.5
US	GEC	CF6-50A	658	214	30.2	1,297		3,731	11.02	4.4	2	1		3	3.8
US	GEC	CF6-50B	658	218.5	30.2	1,297		3,731	11.02	4.4	2	1		3	3.8
US	GEC	CF6-50C	658	223.4	30.2	1,297		3,731	11.16	4.4	2	1		3	3.8
US	GEC	J47-G-23	45.3	25.8	5.5		1	1,200	27.78		1			12	7.2
US	GEC	J73-GE-3	70.3	40	7		0.94	1,650	25.48		1			17	8
US	GEC	J79-GE-1	72.5	49.1	12		0.88	1,450			1				
US	GEC	J85		11.1				95			1			17	
US	GEC	CJ805	72.5	46.7	12		0.84	1,220	22.7		1				
US	GEC	CFM56-C2	465.9	139	38.3		1.84			6.6					
US	GEC	CFM56-C3	474	145	38.3		1.84			6.5					
US	GEC	CFM56-C4	483.1	151	38.3		1.84			6.4					
US	GENELEC	GE404	70.6	88	26		0.7	1,048		0.3					
US	NEIGHTIN	J34WE-46	27	15.1	4.1		0.81	550	28.32		1			11	12.5
US	NEIGHTIN	J46-WE-8	35.3	20.5	6		0.82	945	27.18		1			11	12.5
US	NEIGHTIN	J54-WE-2	45.3	28.9	8		0.89	635	24.07		1			16	
US	P&W	PW2037			32					5.8	2		1.4	4	
US	P&W	PW4000		222	30.2		2.46	4,173		4.8	2				
US	P&W	F100			25					0.8			3		
US	PRAT&WHI	JT8D-1	143	62.3	15.8			1,431	16.57	1.1	2	2	1.9	6	
US	PRAT&WHI	JT8D-1A	143	62.3	15.8			1,431	16.57	1.1	2	2		6	
US	PRAT&WHI	JT8D-7	143	62.3	15.8			1,454	16.57	1.1	2	2		6	
US	PRAT&WHI	JT8D-7A	143	62.3	15.8			1,454	16.57	1.1	2	2		6	
US	PRAT&WHI	JT8D-11	146	66.7	16.2			1,537	17.56	1.1	2	2		6	
US	PRAT&WHI	JT8D-15	146	69	16.5			1,537	17.84	1	2	2		6	

Compression ratio	Type of compressor	No. of turbine stages	No. of compressor stages2	krpm2	Compression ratio2	Type of compressor2	No. of turbine stages2	No. of compressor stages3	krpm3	Compression ratio3	Type of compressor3	No. of turbine stages3	Combustor type	Remarks
	Axial	2											Annulus	Turbojet
	Axial	2											Annulus	Turbojet
	Axial	2											Annulus	Turbofan
	Axial	2											Annulus	Turbofan
	Axial	2											Annulus	Turbofan
12.5	Axial	3											Cannular	Turbojet ABF=65.9kN, sfc=57.8.F-4A PHANTOMII, NORTH AMER. A-5A
12.5	Axial	3											Cannular	Turbojet ABF=69.4kN, sfc=62.3F-104
12.5	Axial	3											Cannular	Turbojet ABF=70.3, sfc=55.8.B-58 HUSTLER
12.5	Axial	3											Cannular	FTJ ABF=75.7,sfc=54.6 F-104 C AND CANDAIR CF-104
12.9	Axial	3											Cannular	Turbojet ABF=79.7, sfc=55.6.F-413, RF-4B,PHANTOM I I,A-5A,RA-5C
13.5	Axial	3											Cannular	Turbojet ABF=75.7, sfc=55.1, F4J,RA5-C
12.9	Axial	3											Cannular	Turbojet F-4C ,RF-4C
13.5	Axial	3											Cannular	Turbojet ABF=79.7kN, sfc=55.56.F-4
12.5	Axial	3											Cannular	Turbojet ABF=75.0, sfc=55.79. F-104G
	Axial	4	1	17.9	14.1	Axial	2						Annulus	Turbofan
			1	16			2						Annulus	Turbofan
			1	16	.		2						Annulus	Turbofan
			1	16			2						Annulus	Turbofan
			1	16			2						Annulus	Turbofan
5.5	Axial	1											Can	Turbojet
7	Axial	2											Cannular	Turbojet
12	Axial	3											Cannular	Turbojet
12	Axial	3											Annulus	Turbojet
4.1	Axial	2											Annulus	Turbojet
6	Axial	2											Annulus	Turbojet
8	Axial	2											Annulus	Turbojet
	Axial	5	12			Axial	2						Annulus	Turbofan
													Annulus	Turbofan Year:1987
	Axial	3	7			Axial	1						Cannular	Turbofan
	Axial	3	7			Axial	1						Cannular	Turbofan
	Axial	3	7			Axial	1						Cannular	Turbofan
	Axial	3	7			Axial	1						Cannular	Turbofan
	Axial	3	7			Axial	1						Cannular	Turbofan
	Axial	3	7			Axial	1						Cannular	Turbofan

(continued)

Country	Company	Model No.	Mass flow rate (kg/s)	Take-off thrust(kN)	Overall compression ratio	Turbine inlet temp. (Deg. C)	Diameter(m)	Weight(kg)	Specific fuel consumption (milligram/N-s)	By-pass ratio	No. of spools	No. of axial fan stages	Fan pressure	No. of compressor stages	krpm
US	PRAT&WHI	JT8D-17R	148	72.9	17.3			1,585	18.55	1	2	2		6	
US	PRAT&WHI	JT8D-209	213	82.2	17.1			2,001		1.8	2	1		6	
US	PRAT&WHI	JT8D-217	217	89	18.6			2,025		1.7	2	1		6	
US	PRAT&WHI	JT9D-3A	684	169.9	21.5	1,243	2.42	3,905	17.84	5.2	2	1	1.6	3	3.8
US	PRAT&WHI	JT9D-7	698	206	21.5	1,243	2.43	3,902	18.01	5.2	2	1	1.6	3	3.8
US	PRAT&WHI	TF30P414	112			1,137	1.29	1,905			2	3	2.1	6	
US	PRAT&WHI	TF30P100	118			1,240	1.24	1,813			2	3	2.2	6	
US	PRAT&WHI	F100P100			25	1,399	1.18	1,371		0.7	2			3	
US	PRAT&WHI	F100P200			25	1,399	1.18	1,390		0.7	2			3	
US	PRAT&WHI	JT3C-2	82	61.2	13		0.99	1,755	26.9		2			9	
US	PRAT&WHI	JT3C-6	82	60.1	13		0.99	1,922	25.7		2			9	
US	PRAT&WHI	JT3C-8	82	61.2	13		0.99	1,959	26.9		2			9	
US	PRAT&WHI	JT3C-26	82	86.2	13		1.02	2,156	21.54		2			9	
US	PRAT&WHI	JT3D-2	196	75.7	13		1.35	1,770	14.78		2			8	
US	PRAT&WHI	JT3D-3A	196	80.2	13		1.35	1,891	15.15		2			8	
US	PRAT&WHI	JT3D-3B	196	86.2	13		1.35	1,950	15.15		2			8	
US	PRAT&WHI	JT3D-7	196	84.5	13		1.35	1,950	15.58		2			8	
US	PRAT&WHI	JT3D-8A	196	93.4	16.1		1.34	2,109	15.86		2			8	
US	PRAT&WHI	JT4A-9	118	74.8	12.5		1.09	2,290	22.94		2			9	
US	PRAT&WHI	JT4A-11	118	77.9	12.5		1.09	2,315	23.79		2			9	
US	PRAT&WHI	JT4A-28	118	109	12.5		1.09	2,665	60.88		2			9	
US	PRAT&WHI	JT4A-29	118	118	12.5		1.09	2,706	62.3		2			9	
US	PRAT&WHI	JT8B-1	118	37.8	14.5		0.77	933	22.22		2			9	
US	PRAT&WHI	JT8B-3	118	41.4	14.5		0.77	961	24.35		2			9	
US	PRAT&WHI	JT8B-5	118	49.8	14.5		0.77	1,052	25.2		2			9	
US	PRAT&WHI	JT8D-5	118	54.5	14.5		1.08	1,431	16		2			9	
US	PRAT&WHI	JT8D-9	146	64.5	16.1		1.08	1,431	16.85		2			9	
US	PRAT&WHI	JT9D-15	687	209.1	22		2.43	3,833	10.05	5.1	2	1		3	
US	PRAT&WHI	JTF10A20		82.3			1.22	1,755	70.8		2				
US	PRAT&WHI	JTF10A-8		50.5			1.07	1,232	17.56		2				
US	PRAT&WHI	JTF10A-9		54.3			1.07	1,146	17.84		2				
US	PRAT&WHI	JTF10A16		59.6			1.07	1,178	18.12		2				
US	PRAT&WHI	JTF10A21	106	55.6	17		0.96	1,843	17.84		2	3	1.8	6	9.4
US	PRAT&WHI	JF10A27A		90.1			1.29	1,827			2				
US	PRAT&WHI	JT10A27D		90.6			1.28	1,869			2				
US	PRAT&WHI	JT10A27F		89							2				
US	PRAT&WHI	JTF10A36		87.2			1.25	1,846	73.91		2				
US	PRAT&WHI	JT11D20B	133.42								1			9	
US	PRAT&WHI	JT12A-5	13.4				0.56	203	27.19		1			9	
US	PRAT&WHI	JT12A-6A	13.4				0.56	206	27.19		1			9	
US	PRAT&WHI	JT12A-8	14.7				0.56	212	28.18		1			9	
US	PRAT&WHI	J57P20	82	50.7	13		1.02	2,150	21.82		2			9	6.5
US	PRAT&WHI	J85-P-2		44.4			0.86		22.7		2			9	
US	PRAT&WHI	JT3C-7	82	53.5	13		0.99	1,580	22.23		2			9	8
US	PRAT&WHI	JT3D-1	196	71.1	13		1.35	1,830	16.99		2			8	
US	PRAT&WHI	JT4A-10	118	73.6	12.5		1.12	1,910			2			9	

Compression ratio	Type of compressor	No. of turbine stages	No. of compressor stages2	krpm2	Compression ratio2	Type of compressor2	No. of turbine stages2	No. of compressor stages3	krpm3	Compression ratio3	Type of compressor3	No. of turbine stages3	Combustor type	Remarks
	Axial	3	7			Axial	1						Cannular	Turbofan
	Axial	3	7			Axial	1						Cannular	Turbofan AX:FLOW
	Axial	3	7			Axial	1						Cannular	Turbofan AX:FLOW
		4	11	7.6			2						Annulus	Turbofan
		4	11	8			2						Annulus	Turbofan
		3	7				1						Cannular	Turbofan ABF=93.0kN, sfc=78.7
		3	7				1						Cannular	Turbofan ABF=111.7kN, sfc=69.4
		2	10	13.5	8		2						Annulus	Turbofan AB
		2	10	13.5	8		2						Annulus	Turbofan AB
	Axial	2	7			Axial	1						Cannular	Turbojet J57-P-43WB
	Axial	2	7			Axial	1						Cannular	Turbojet
	Axial	2	7			Axial	1						Cannular	Turbojet J57-P-59W
	Axial	2	7			Axial	1						Cannular	Turbojet J57-P-20,-20A
	Axial	3	7			Axial	1						Cannular	Turbofan TF33-P-3
	Axial	3	7			Axial	1						Cannular	Turbofan TF33-P-5,9
	Axial	3	7			Axial	1						Cannular	Turbofan
	Axial	3	7			Axial	1						Cannular	Turbofan
	Axial	3	7			Axial	1						Cannular	Turbofan TF33-P-7
	Axial	2	7			Axial	1						Cannular	Turbofan
	Axial	2	7			Axial	1						Cannular	Turbofan
	Axial	2	7			Axial	1						Cannular	Turbofan J75-P-17
	Axial	2	7			Axial	1						Cannular	Turbofan J75-19W
	Axial	2	7			Axial	1						Cannular	Turbofan J52-P-6A
	Axial	2	7			Axial	1						Cannular	Turbofan J52-P-8A
	Axial	2	7			Axial	1						Cannular	Turbofan J52-P-408
	Axial	2	7			Axial	1						Cannular	Turbofan
	Axial	2	7			Axial	1						Cannular	Turbofan
	Axial	4	11			Axial	2						Annulus	Turbofan
														Turbofan TF30-P-1-1A
														Turbofan TF30-P-6
														Turbofan TF30-P-8
														Turbofan TF30-P-408
	Axial	3	7	13.8		Axial	1						Cannular	Turbofan ABF=89.0kN, sfc=70.8.TF30-P-3
														Turbofan ABF=90.1kN, sfc=86.1.TF 30-P-12
														Turbofan ABF=90.6kN. sfc=85.32. TF 30-P-7
														Turbofan TF30-P-412
														Turbofan TF TF
	Axial	2											Annulus	Turbojet J60-P-3,5
	Axial	2											Annulus	Turbojet J60-P-6
	Axial	2											Annulus	Turbojet
	Axial	2											Annulus	Turbojet
	Axial	2	7	9.5	13	Axial	1						Cannular	Turbojet Afterburner F=80.0kN, sfc=79.36. Crusader.
	Axial	2	7			Axial	1						Annulus	Turbofan
	Axial	2	7			Axial	1						Annulus	Turbofan
	Axial	3	7			Axial	1						Annulus	Turbofan
	Axial	2	7			Axial	1						Annulus	Turbofan

(continued)

Country	Company	Model No.	Mass flow rate (kg/s)	Take-off thrust(kN)	Overall compression ratio	Turbine inlet temp. (Deg. C)	Diameter(m)	Weight(kg)	Specific fuel consumption (milligram/N-s)	By-pass ratio	No. of spools	No. of axial fan stages	Fan pressure	No. of compressor stages	krpm
US	PRAT&WHI	JT11		133.4			1.27	2,500			1			9	
US	PRAT&WHI	JT12		13.3			0.56	195	26.3		1				
US	TELEDYNE	J69T-41A	13	8.5	6	954	0.57	159	30.89		1	1		1	22
US	TELEDYNE	J69T-406	13.8	8.5	5.5	993	0.57	163	31.18		1	1		1	22.1
US	TELEDYNE	J100C101	4.4		5.8			159	31.16		1	1		1	
US	TELEDYNE	CAJ69T25	9	4.6	3.8	788	0.57	165	32.28		1			1	21.7
US	TELEDYNE	CAJ69T29	13	7.6	5.5	871	0.57	154	30.73		1			1	22
US	TELEDYNE	J69-T406		8.6				765	31.15		1			1	22.2
US	TELEDYNE	CAEJ100-	20.4	12	6.3			193	31.15		1	2		1	20.7
US	TELEDYNE	CA35628C	24.5	15.6	8.1			240			1	2		1	
US	TELEDYNE	CA35628D	29.4	18.7	8.1			231			1	2		1	
US	TELEDYNE	CA35628E	29.4	18.7	8.1			220			1	2		1	
US	WILLIAMS	WR2-6	1	0.6	4.1	955	0.27	14			1			1	60
US	WILLIAMS	WR24-6	1.4	0.5	5.3	955	0.27	14			1			1	60
US	WILLIAMS	WR19	2	3.2	8.1	955		64		1.1	2	2		2	60
W	ROLS-MTU	RB199	75	80	25		0.75	1,028		0.9					
W	ALL-ROLS	TF41A-1	117	64	20	1,182		1,470	18.23	0.8	2	3		2	
W	ROLS-MTU	RB-193	93	45.2	16.5			1,050	18.41	1.1	2	3		2	
W	ROLS-SNE	M45H-01	106	32.3	18			673	12.91	2.8	2	1		5	
W	ROLS-SNE	OLYMP593		169.3			1.22	2,628			2			7	
W	ROLS-SNE	OLYMP593		170.9			1.22	2,628			2			7	
W	ROLS-SNE	OLYMP593		177.7			1.22	2,628			2			7	
W	ROLS-TUR	ADOUR102		29.4	11			704	27	0.8	2	2			
W	TUR-UNIO	RB19934R	70	71	23	1,327	0.87	900		1.1	3			3	
W		ATAR09C	68	42	5.5	890	0.79	1,409	28.63		1			9	8.4

Compression ratio	Type of compressor	No. of turbine stages	No. of compressor stages2	krpm2	Compression ratio2	Type of compressor2	No. of turbine stages2	No. of compressor stages3	krpm3	Compression ratio3	Type of compressor3	No. of turbine stages3	Combustor type	Remarks
	Axial	2											Annulus	
	Centrifugal	1											Annulus	Turbojet
	Centrifugal	1											Annulus	Turbojet
	Centrifugal	1											Annulus	Turbojet
3.8	Centrifugal	1											Annulus	Turbojet Cessna T37B
5.5	Centrifugal	1											Annulus	Turbojet
	Centrifugal	1											Annulus	Turbojet
6.3	Centrifugal	2											Annulus	Turbojet CA-100
8.1	Centrifugal	2											Annulus	Turbojet
8.1	Centrifugal	2											Annulus	Turbojet
8.1	Centrifugal	2											Annulus	Turbojet
4.1	Centrifugal	1											Annulus	Turbojet
5.3	Centrifugal	1											Annulus	Turbojet
	Centrifugal	2	1			Centrifugal	1						Annulus	Turbofan
	Axial	2	11		6.2	Axial	2						Annulus	Turbofan NORH
		3	6				1						Annulus	VETF
	Axial	3	7			Axial	1						Annulus	TFVFW614
	Axial	1	7			Axial	1						Annulus	Turbojet 602
	Axial	1	7			Axial	1						Annulus	Turbojet 612
	Axial	1	7			Axial	1						Annulus	Turbojet 621
		1	5			Axial	1						Annulus	Turbofan RH JAGUAR
	Axial	2	3			Axial	1			6	A	1	Annulus	Turbofan AB
5.5	Axial	2											Annulus	Turbojet ABF=58.9kN, sfc=57.5. Mirage3D.

Turbprop

Country	Company	Model No.	Mass flow rate (kg/s)	Take-off thrust(kN)	Shaft power(kW)	Overall compression ratio	Turbine inlet temp. (Deg. C)	Diameter(m)	Weight(kg)	Specific fuel consumption (micro-g/J)	No. of spools
CA	UAC	PT6B	2.5		514	6.3		0.48	111	115.6	2
CA	UAC	ST6L	2.9		536	7.1		0.48	135	94.8	2
CA	P&W	PW100	6.7			14.8					3
CZ	MOTORLET	M601	3.1	0.4	484	6.2		0.43	142	182	2
FR	SOCEMA	TGA1		5.5	1,764	3.6					1
FR	SNECMA	TYNE20	21	5	4,213	13.6		1.1	975	74.2	2
FR	TURBOMEC	ARTOUST3	4		483	5.1		0.45	151	125	1
FR	TURBOMEC	ARTOUS3B	4.3	0.4	410	5.2			154	128.6	1
FR	TURBOMEC	TURMO3	4.8		559	5.1		0.56	240	131.8	2
FR	TURBOMEC	TURMO3C	5.9		968	5.8			226	108.1	2
FR	TURBOMEC	TURMO16			1,490						
FR	TURBOMEC	ASTAZON			239			0.46	111	122.7	1
FR	TURBOMEC	ASTAZ3N	2.5	0.3	447	6		0.46	147	107.4	1
FR	TURBOMEC	ASTAZ14A	3.4	0.6	447	7.6		0.46	160	89.5	1
FR	TURBOMEC	ASTAZ16	3.4	0.4	765	8.1		0.55	206	83.9	1
FR	TURBOMEC	BASTAN			485			0.55	180	124.6	1
FR	TURBOMEC	BASTAN6	4.5	0.8	745	5.7		0.55	230	105.5	1
FR	TURBOMEC	BASTAN7	5.6	0.9	790	6.8		0.55	295	102.1	1
FR	TURBOMEC	TURMAS14	3.3	0.4	670	8		0.46	160	89.5	2
DE	KHD	T112	0.9		106	5			36	156.9	1
DE	KHD	T53	5.8	0.5	1,043	7.4		0.58	249	111.1	2
DE	MTU	6022	1.9		279	6.4			90	125.3	1
JP	IH	CT58-110	5.6	0.6	931	8.3		0.4	143		2
JP	IH	T58-8B	14		931	1.8					
JP	IH	T64-10	11	0.9	2,064	12.6					2
JP	KAWASAKI	KT5311A	5	0.6	819	6.1		0.58	225	116.3	2
JP	MITSUBI	CT63	1.4	0.1	236	6.2	996		64	117.8	2
RU		MO22	30	5.9	4,191		757	1.05	1,400	92.5	1
RU		NK12	62	11.8	8,823	13	877	1.15	2,300	98.2	1
UK	ARMSTRNG	P181	5.6	0.7	596			0.69	250	114.8	2
UK	ARMSTRNG	P182	5.6	0.9	820			0.69	272	111.4	2
UK	ARMSTRNG	MAMBA5	8.2	1.3	1,103	5.4		0.84	370	120.8	1
UK	ARMSTRNG	MAMBA8	9.6		1,454						1
UK	ARMSTRNG	DMAMBA3	16.4	3.6	2,044	5.4		1.34	1,098	128	1
UK	ARMSTRNG	DMAMBA8	19.1	3.2	2,684			1.47	1,110	114.8	1
UK	ARMSTRNG	PYTHON		5.2	3,020	5					1
UK	AUTODIES	GT15	0.1		11	3	680	0.23	10		1
UK	BLACKBRN	ART600	3.2		354	4.1		0.48	126	169.2	1
UK	BLACKBRN	TURM600	3.2		354	4.1		0.48	127	177.9	2
UK	BRISTOL	PROT755	20	5.4	2,720	7.2		1.04	1,300	102	2
UK	BRISTOL	PROT765	20.1	5.6	2,952	7.2		1.02	1,315	102	2
UK	BRISTOL	PROT720	20.1	5.6	3,077	7.2		1.02	1,315	99.3	2
UK	BRISTOL	ORION2	37.1	8.6	3,280	10	762	1.06	1,430	109.5	2
UK	NAPIER	ELAND1	14		2,004	7		0.91	715	105.7	1
UK	NAPIER	ELAND6	14		2,353	7		0.91	735	102	1
UK	NAPIER	ELAND4	14		2,809	7		0.91	818	94.8	1

Type of compressor	No. of compressor stages	kilo-rpm	No. of turbine stages	Type of compressor2	No. of compressor stages2	kilo-rpm2	No. of turbine stages2	Combustor type	Remarks
Axial	3	37.5	1	Centrifugal	1	33	1	Annulus	Helicopter
Axial	3	37.5	1	Centrifugal	1	33	1	Annulus	APU
Centrifugal	2		1	Axial	1	20	1	Annulus	
Axial	2	37.8	1	Centrifugal	1	31	1	Annulus	L410
Axial	15	6.5	1					Can	
Axial	6		3	Axial	9	15.2	1	Cannular	
Mixed	2	34.5	2					Annulus	
Mixed	2	33.5	3					Annulus	Helicopter
Axial	1	34.5	1	Centrifugal	1		1	Annulus	
Mixed	2	33.5	2			33.5	1	Annulus	HELICOPTER
Mixed	2							Annulus	
Mixed	2	43.5	3					Annulus	Helicopter
Mixed	3	89.5	3					Annulus	Helicopter
Mixed	3	43	3					Annulus	
Mixed	2	33						Annulus	
Mixed	2	33.5	3					Annulus	
Mixed	3	32	3					Annulus	
Mixed	3	43	2			29	2	Annulus	
Mixed	2	64	2					Annulus	APU
Axial	5	25.4	2	Centrifugal	1	20.1	2	Annulus	Mil.Heli.
Centrifugal	2	42	3					Cannular	MBB Heli.
Axial	10	26.3	2			19.5	1	Annulus	Helicopter
		19.5	1						
Axial	14	16.9	2			15.6	2	Annulus	
Axial	5	25.2	1	Centrifugal	1	21.2	1	Annulus	Helicopter
Axial	6	51.6	2	Centrifugal	1	35	1	Annulus	Helicopter
Axial	14	7.6	3						
Axial	14	8.2	5					Cannular	Tu-114D,An-22
Axial	2	14	1	Centrifugal	1	20	2	Annulus	
Axial	2		1	Centrifugal	1	20	2	Annulus	
Axial	10	15	3					Annulus	
Axial	11	15	3					Annulus	
Axial	10	15	3					Annulus	
Axial	11	15	3					Annulus	
Centrifugal	14	8	1					Can	
Centrifugal	1	85	1					Can	APU
Centrifugal	1		2					Annulus	
			1	Centrifugal	1		1		
Axial	12	11.6	2	Centrifugal	1		2	Can	
Axial	12		2	Centrifugal	1		2	Can	
Axial	12		2	Centrifugal	1		2	Can	
Axial	7	10	3	Axial	5		1	Cannular	
Axial	10	12.5	3					Can	
Axial	10	12.5	3					Can	
Axial	10	12.5	3					Can	

(continued)

Country	Company	Model No.	Mass flow rate (kg/s)	Take-off thrust(kN)	Shaft power(kW)	Overall compression ratio	Turbine inlet temp. (Deg. C)	Diameter(m)	Weight(kg)	Specific fuel consumption (micro-g/J)	No. of spools
UK	NAPIER	ELAND5	14		2,809	7		0.91	818	94.8	1
UK	NAPIER	GAZ2	7.2		1,228	6.4		0.85	376	115.6	2
UK	NAPIER	GAZ3	7.2		1,338	6.4		0.85	392	113.7	2
UK	NAPIER	GAZ4	7.2		1,485	6.4		0.85	408	108.8	2
UK	ROLLS	DART510	9.1	1.6	1,193	5.5	860	0.96	526	117.8	1
UK	ROLLS	DART526	10	2.2	1,423	5.7	890	0.96	565	114	1
UK	ROLLS	DART529	10.6	2.2	1,512	5.6		0.96	555	96.9	1
UK	ROLLS	DART542	12	3.2	2,049	6.3		0.96	625	92.8	1
UK	ROLLS	DART545		2.2	2,206			0.96	570	108.8	1
UK	ROLLS	TYNE1	20.8	5.7	3,354	13		1.03	940	85.3	2
UK	ROLLS	TYNE3		5	3,951			1.03		81.9	2
UK	ROLLS	TYNE12	21	4.7	3,796	13.5		1.1	987	66	2
UK	ROLLS	RS360-07	3.3		670	12.1	968	0.56	136	83.9	2
UK	ROLLS	GN-H1200	5.6	0.7	1,006	8.1	898		142	104.8	2
UK	ROLLS	GN-H1400	6	0.9	1,117	8.4	725		151	104.8	2
UK	ROLLS	GN-H1800	7		1,304	9.9			212		2
UK	ROLLS	NIM103	5.1		529	5.9		0.58	304	121.5	2
UK	ROVER	1S/60	0.6		51	3	860		63	224.8	1
UK	ROVER	2S/150A	0.9		108	3.9	927		73	154.3	2
US	ALLISON	501D13	17.7		2,581	9.2	971	0.68	793	92.5	1
US	ALLISON	250B2			184			0.49	48	120.1	2
US	ALLISON	250B17	1.6		311	7.2			86	103.3	2
US	ALLISON	250C20	1.6		298	7.2			70	108.1	2
US	ALLISON	T56A15	15	3.4	3,658	9.3	1,080	1.12	828	88	1
US	ALLISON	T56A18	14.5	3.6	3,967	9.7	1,132		705	88	1
US	BOEING	502	1.6	0.2	147	3				192	1
US	CONTIN	T51	2		209	3.7		0.53	120	171.1	1
US	GARRETT	TPE331	3.5	0.7	627	10.4	1,004	0.53	161	92.8	1
US	GARRETT	TSE36	1.4		179	4.3	912	0.71	81	141.6	1
US	GARRETT	TSE231	1.9		354	8.6	1,038	0.39	77	101.3	2
US	GARRETT	TSE331	3.5		597	10.3	1,004	0.55	161	101.3	1
US	GARRETT	GTP30	0.9		75	3	898	0.41	39		1
US	GARRETT	GTCP36	0.9		145	3	843	0.58	66		1
US	GARRETT	GTP85	2.6		149	3.7	898	0.65	104		1
US	GARRETT	GTCP85	2.5		261	3.4	871	0.66	129		1
US	GARRETT	GTCP95	0.8		276	3.6	871	0.71	135		1
US	GARRETT	GTCP105	5.2		142	5.2	843	0.66	176		1
US	GARRETT	GTCP165	2.7		95	3.7	871	0.61	103		1
US	GARRETT	GTP331	2.7		474	9	960	0.57	145		1
US	GARRETT	GTCP660	11		597	4.3	913	1.03	252		1
US	GARRETT	TSCP700					974	0.95	247		2
US	GE	T58GE2	5.6	0.3	765	8.3	880	0.4	147	113.3	2
US	GE	T58GE10	6.2	0.6	1,044	8.4		0.53	159	104.7	2
US	GE	T58GE16	6.2		1,395	8.4	1,073	0.61	200	94.6	2
US	GE	T64GE14	11	0.9	2,066	12.5	971	0.74	512	84.6	2

Type of compressor	No. of compressor stages	kilo-rpm	No. of turbine stages	Type of compressor2	No. of compressor stages2	kilo-rpm2	No. of turbine stages2	Combustor type	Remarks
Axial	10	12.5	3					Can	
	11	20.4	1	Axial	11		2	Cannular	
	11	20.4	1	Axial	11		2	Cannular	
	11	20.4	1	Axial	11		2	Cannular	
Centrifugal	2	14.5	2					Can	
Centrifugal	2	15	3					Can	
Centrifugal	2	15	3					Can	
Centrifugal	2	15	3					Can	
Centrifugal	2	15	3					Can	
Axial	6	15.2	3	Axial	9		1	Cannular	
Axial	6		3	Axial	9		1	Cannular	
Axial	6	15.2	3	Axial	9	15.2	1	Cannular	
Axial	4	40	1	Centrifugal	1	40	1	Annulus	Helicopter
Axial	10	26.3	2			26.7	1	Annulus	Helicopter
Axial	10	26.3	2			26.8	1	Annulus	Helicopter
Axial	11	26.3	2			26.3	2	Annulus	Helicopter
Axial	2	35	2	Centrifugal	1	34.2	1	Annulus	Helicopter
Centrifugal	1	46	1					Annulus	APU
Centrifugal	1	64.5	1			64.5	1	Annulus	APU
Axial	14	13.8	4					Cannular	
Axial	7		2	Centrifugal	1		1	Can	
Axial	6	52	2	Centrifugal	1	33.3	2	Annulus	
Axial	6	52	2	Centrifugal	1	33.3	2	Annulus	Helicopter
Axial	14	13.8	4					Cannular	
Axial	14	13.8	4					Annulus	
Centrifugal	1	38	1					Can	
Centrifugal	1	34	1					Annulus	
Centrifugal	2	41.7	1					Annulus	
Centrifugal	1	58	1					Annulus	Heli
Centrifugal	2	44.8	1			51	1	Annulus	Heli
Centrifugal	2	41.7	3						Heli
Centrifugal	1	59.2	1						APU
Centrifugal	1	58	1						APU
Centrifugal	2	42.2	1						APU U
Centrifugal	2	40.8	1						APU
Centrifugal	1	42	1						APU
Centrifugal	2	35.1	2						APU
Centrifugal	1	38	1					Annulus	APU
Centrifugal	2	41.7	3						APU
Axial	4	20	2					Annulus	APU
Axial	3		2	Centrifugal	1	35.3	1	Annulus	APU
Axial		19.5	1	Axial	10		2	Annulus	
		19.5	1	Axial	10	27.3	2	Annulus	Mil.Heli
			1	Axial	10	26.8	2	Annulus	Mil.Heli
Axial		15.6	2	Axial	14	17.8	2	Annulus	

(continued)

Country	Company	Model No.	Mass flow rate (kg/s)	Take-off thrust(kN)	Shaft power(kW)	Overall compression ratio	Turbine inlet temp. (Deg. C)	Diameter(m)	Weight(kg)	Specific fuel consumption (micro-g/J)	No. of spools
US	GE	T64G716	12		2,536	13	1,093	0.6	317	81.2	2
US	GE	GE12			1,119		1,093	0.4	122	149	2
US	GE	T700	4.6			15					1
US	LYCOMING	T53L3	4.9	0.5	713	5.7		0.59	225	112.2	2
US	LYCOMING	T53L11	5	0.5	820	6.1		0.58	224	116.2	2
US	LYCOMING	T53L13	5.8	0.5	1,044	7.4		0.58	249	111	2
US	LYCOMING	T55L11	11.3	1.1	2,794	8		0.61	304	94.4	2
US	LYCOMING	T55L13	5.8	0.5	1,043	7.4		0.58	249	111	2
US	LYCOMING	T53L701	5.8	0.6	1,043	7.4		0.58	312	99.9	2
US	LYCOMING	T5319	5.4	0.7	1,341	8		0.58	256	99.9	2
US	LYCOMING	T5321A	5.4	0.7	1,368	8		0.58	306	99.9	2
US	LYCOMING	LTC4B12	12	1.5	3,432	8.5		0.61	308	86.6	2
US	LYCOMING	LTC4V1	11.8	1.1	3,730			0.56	258	69.3	3
US	P&W	T34	30.5	5.5	4,117	6.7		0.86	1,200	112.5	1
US	P&W	JFTD12A	23		3,633	6.7		0.51	444	116.6	2
US	P&W	PT6A-65	4.5	0.8	900	9.2	1,127	0.22			1
US	SO-TITAN	T62T12	1		78	3.5	788	0.32	33		1
US	SO-TITAN	T62T25	0.6		60	3.5	788	0.32	32		1
US	SO-TITAN	T62T39	0.9		30	2.9	788	0.47	41		1
US	TELEDYNE	TS120G6	0.8		134	5.6	649		97		1
US	TELEDYNE	T65T1	1.5		249	6	538		59	124.9	
US	TELEDYNE	T67T1	3.2		1,266	7.8			250	98	
US	WILLIAMS	WR9-7C	1		23	4	690		43	457	1

Type of compressor	No. of compressor stages	kilo-rpm	No. of turbine stages	Type of compressor2	No. of compressor stages2	kilo-rpm2	No. of turbine stages2	Combustor type	Remarks
Axial		13.6	2	Axial	14	18.2	2	Annulus	Heli
Centrifugal	1		2	Axial	5		2	Annulus	Heli
Mixed	4		2	Axial				Annulus	Heli
Axial	5	21.5	1	Centrifugal	1		1	Annulus	
Axial	5	21.2	1	Centrifugal	1		1	Annulus	Heli
Axial	5	21.2	1	Centrifugal	1		1	Annulus	Heli
Axial	7	19	1	Centrifugal	1	16	2	Annulus	Heli
Axial	5	25.4	1	Centrifugal	1		1	Annulus	Heli
Axial	5	24.4	1	Centrifugal	1		1	Annulus	Heli
Axial	5	26.4	1	Centrifugal	1		1	Annulus	Heli
Axial	5	26.4	1	Centrifugal	1	21.3	1	Annulus	Heli
Axial	7	16	2	Centrifugal	1	19.8	2	Annulus	Heli,V/STOL
Axial	12		3	Axial		17	2	Annulus	Reverse Flow.Heli
Axial	13	11	3					Cannular	
			2	Axial	9	16.7	2	Cannular	Sikorsky
Mixed	4	17	3	Axial				Annulus	
Centrifugal	1	56.7	1					Annulus	APU
Centrifugal	1	56.7	1					Annulus	APU
Centrifugal	1	56.7	1					Annulus	APU
Mixed	2	67	1					Cannular	APU
Centrifugal	1	59.6	2					Annulus	APU

List of Symbols

A	Flow area
A_e	Exhaust area
A_w	Wall surface area
A^*	Flow area corresponding to sonic speed
a	Sonic speed
b	Bypass ratio
b	Blade width/height
$\mathbf{b}$	Acceleration vector
C_F	Force coefficient
C_M	Torque coefficient
C_P	Power coefficient
$c, \mathbf{c}$	Absolute fluid velocity
c_f	Friction coefficient
c_L	Lift coefficient
c_p	Specific heat at constant pressure
c_u	Azimuthal component of $\mathbf{c}$
c_v	Specific heat at constant volume
D	Diameter
$\mathbf{D}$	Drag force
d	Stroke length
d_f	Film thickness
F	Thrust, force
f	Frequency of oscillation
f	Displacement
f_c	Heating factor
f_m	Maximum camber
H	Work done
H^*	Isentropic work done
h, h^o	Enthalpy (static, stagnation)
h_w	Enthalpy of gas at wall
I	Specific impulse

T. Bose, *Airbreathing Propulsion: An Introduction*,
Springer Aerospace Technology, DOI 10.1007/978-1-4614-3532-7,
© Springer Science+Business Media, LLC 2012

I	Mass moment of inertia
I	Intensity of sound
J	Advance ratio
J	Area moment of inertia
K_n	Dimensionless rpm
k	Blockage factor
$\mathbf{L}$	Lift force
L, l	Lengths
l	Chord length
l	Connecting rod length
M	Mach number
M	Bending/torque moment
$\dot{m}$	Fluid mass flow rate
$\dot{m}_\mathbf{a}$	Mass flow rate of air
$\dot{m}_c$	Mass flow rate of cold stream
$\dot{m}_f$	Mass flow rate of fuel
$\dot{m}_H$	Mass flow rate of hot stream
$\dot{m}_L$	Mass flow rate through labyrinth
N	Number
N_B	Number of blades
n	rpm
P	Power
P_D	Dissipative power
P_E	Increase in kinetic energy
p_E	Power developed from energy
P_F	Developed power
p, p°	Pressure (static, stagnation)
p_m	Mean indicated pressure
Q	Heat added per unit time
q	Heat per unit mass
q_a	Heat added
q_r	Heat rejected
q_w	Wall heat flux
$\dot{q}$	Heat added per unit mass and time
R	Radius of curvature
R	Gas constant
$\mathbf{R}$	Resultant force
r	Radius
$\hat{r}$	Reaction or degree of reaction
s	A length
s	Entropy
s	Labyrinth gap
SFC	Specific fuel consumption
SFC*	Nondimensional SFC

T, T°	Temperature (static, stagnation)
t	Time
t	Pitch
u	Azimuthal velocity
u	Flow velocity
u_c	Cold stream flow velocity
u_e	Exhaust velocity
u_H	Hot stream flow velocity
u_∞	Approaching flow velocity
V	Volume
V_c	Clearance volume
V_d	Displacement volume
V_t	Total volume
W	Weight of the engine
W_b	Moment of resistance against bending
W_t	Moment of resistance against torsion
$w, \mathbf{w}$	Relative flow velocity
w	Deflection
w	Work per unit mass
w_f	Friction work
w_t	Technical work
w_u	Azimuthal component of $\mathbf{w}$
α	Heat transfer coefficient
α, β	Blade angles
β	Mass transfer coefficient
χ	Width-to-diameter ratio
χ	Complex velocity potential
χ_M	Correction factor due to Mach number
χ_R	Correction factor due to Reynolds number
χ_δ	Correction factor due to blade tail thickness
γ	Specific heat ratio
Δ	Difference, gap
ΔH_p	Heat of reaction
Δh	Difference in enthalpy
δ	Boundary-layer thickness
δ	Pressure ratio
ε	Compression ratio
φ	Hot jet to approaching flow speed ratio
φ	Discharge coefficient
φ	Angle
φ	Mass flow coefficient
η	Efficiency
η_{ad}	Adiabatic efficiency
η_{comb}	Combustor efficiency

η^k	Kinetic energy efficiency
η_p	Propulsive efficiency
η_{pol}	Polytropic efficiency
η_{rot}	Rotor efficiency
η_{stat}	Stator efficiency
η_{th}	Thermodynamic efficiency
π_c	Compressor pressure ratio (based on p)
π_c^o	Compressor pressure ratio (based on p^o)
π_τ	Turbine pressure ratio (based on p)
π_t^o	Turbine pressure ratio (based on p^o)
μ	Dynamic viscosity of fluid
μ	Slip factor or power-lowering factor
v	Blade root-to-tip diameter ratio
σ	Solidity
σ	Stress
σ_b	Bending stress
σ_c	Stress due to centrifugal force
σ_t	Torsional stress
Θ	Temperature ratio
θ	Crank angle
θ	Included angle
ρ	Mass density
τ_c	Temperature ratio for compressor
τ_t	Temperature ratio for turbine
τ_w	Wall shear stress
ψ	Work coefficient
Ω	Cross-sectional area
Ω	Combustion volume ratio
ω	Azimuthal speed
ζ	Cold jet-to-hot jet speed ratio
ζ	Profile loss coefficient

Bibliography

T. Arts, Aerothermal performance of a two-dimensional highly loaded transonic turbine nozzle guide vane: a test case for inviscid and viscous flow computations. ASME J. Turbomachine. **114**, 147–154 (1992)

S. Bock, W. Horn, G. Wilfert, J. Sieber, Active case technology with NEWAC research program for cleaner and more efficient aeroengines, in *European Workshop on New Engines Concept (NEWAC)*, Munich, 30 June to 1 July 2010

T.K. Bose, *Numerical Fluid Dynamics* (NAROSA, India, 1997)

T.K. Bose, S.N.B. Murthy, Blade clearance estimation in a generic compressor with air-water mixture operation, in *30th AIAA/ASME/SAE/ASEE Joint Propulsion Conference*, Indianapolis, AIAA-94-2693, June 1994

B. Eckert, *Axial- und Radial Kompressoren* (Springer, Berlin, 1953)

P. Hill, C. Peterson, *Mechanics and Thermodynamics of Propulsion*, 2nd edn. (Addison-Wesley, Reading, 1992)

J.L. Kerrebrock, *Aircraft Engines and Gas Turbines*, 2nd edn. (Cambridge University Press, New York, 1992)

S.G. Koff, J.P. Nikkanen, R.S. Mazzowy, Rotor case treatment. U.S. Patent 5,282,718, Aug 1994

J. Kruschik, *Gasturbine* (Springer, Vienna, 1960)

M.J. Lighthill, On sound generated aerodynamically. I. General theory. Proc. Roy. Soc. Lond. **211a**, 1 (1952a)

M.J. Lighthill, On sound generated aerodynamically. II. Turbulence as a source of sound. Proc. Roy. Soc. Lond. **222A**, 1 (1952b)

M.J. Lighthill, Jet noise. AIAA J. **1**, 1507 (1963)

E.M. Murman, J.D. Cole, Calculation of plane steady transonic flow. AIAA J. **9**, 114–121 (1971)

J.H. Nicholson, A.E. Forest, M.L.G. Oldfield, D.L. Schultzm, Heat transfer optimized turbine rotor blades – an experimental study using transient techniques. ASME J. Eng. Gas Turbine. Power **106**, 173–181 (1984)

G.C. Oates, *Aerodynamics of Aircraft Engine Components* (Pandora, 1985)

G.C. Oates, *Aircraft Propulsion System Technology* (AIAA, 1989)

C. Pfleiderer, *Stroemungsmaschinen* (Springer, Berlin, 1964)

S. Sarkar, T.K. Bose, Comparison of different turbulence models for prediction of slot film cooling: Flow and temperature field. J. Numer. Heat Transf. **28**(Part B), 217–238 (1995a)

S. Sarkar, T.K. Bose, Numerical simulation of a 2D jet-cross flow interaction related to film cooling applications: effect of blowing rates, injection angle and free-stream turbulence. Sadhana (J. Indian Acad. Sci.) **20**(Part 6), 915–935 (1995b)

S. Sarkar, T.K. Bose, Numerical analysis of slot-film cooling: a parametric study. J. Aeron. Soc. India **48**, 80–90 (1996)

T. Bose, *Airbreathing Propulsion: An Introduction*,
Springer Aerospace Technology, DOI 10.1007/978-1-4614-3532-7,
© Springer Science+Business Media, LLC 2012

S. Sarkar, T.K. Bose, Numerical study of film cooling: a three-dimensional calculation. J. Energy
    Heat Mass Transf. **19**, 199–206 (1997)
S. Sarkar, Numerical study of film cooling in a turbine blade cascade. PhD thesis, IITM (Aero),
    1997
J. Stuart (Lord Rayleigh), *Theory of Sound*, 2 vols. (Paperback edition, Dover, 1895)
J.F. Thompson, N.P. Weatherhill (eds.), *Handbook of Grid Generation* (CRC Press, Boca Raton,
    1999)
J.F. Thompson, Z.U.A. Warsi, C.W. Mastin, *Numerical Grid Generation*. (North Holland,
    New York, 1985)
W. Traupel, *Thermische Turbomaschinen*, vol. 1 (Springer, Berlin, 1966)

Index

A

Absolute velocity, 39, 112, 115, 171, 204
Adiabatic efficiency, 100–101, 105, 107,
  110, 121, 190
Adiabatic flame temperature, 12, 25
Advance ratio, 233, 239, 241, 244
Afterburner, 19, 54–80, 84, 94, 98, 277,
  279, 287
Alternative fuels, 16
Angle of attack, 168, 218, 233, 241
Annular combustor, 24, 95
APU. *See* Auxiliary power unit (APU)
Area moment of inertia, 251, 252, 255
Auxiliary power unit (APU), 232, 263, 291,
  293, 295
Average friction coefficient, 110, 137, 139
Average volume flow rate, 185, 187, 190
Axial inflow factor, 237
Axial momentum theory, 233–240,
  244, 249
Azimuthal Mach number, 123, 124,
  132, 161
Azimuthal speeds, 91, 114, 132, 139, 153, 166,
  180–181, 188, 189, 194, 240
Azimuthal velocity, 112, 121, 127, 163, 238

B

Barometric pressure formula, 25
Bending moment, 249, 250, 252, 253, 255
Bending moment of resistance, 255
Bending stress, 248–251, 255
Bernoulli's equation, 235
Blade angle, 112, 113, 121, 126, 128, 129, 144,
  154, 163, 166, 173, 189, 203, 240, 241,
  243–244
Blade arc length, 137

Blade damping wires, 138, 142
Bladed diffuser, 204, 215
Blade distance, 137, 169
Blade element theory, 168, 233, 240–244
Blockage factor, 150–151
Bore, 27
Buckingham pi theorem, 160
Bypass ratio, 3, 8, 9, 17, 19, 21–23, 25, 64, 66,
  68, 74, 80, 84, 86, 265

C

Can combustors, 18, 19, 95
Cannular combustor, 21, 95
Carnot cycle, 1, 44, 46, 50, 57
Carnot cycle efficiency, 1, 44, 46, 50, 57
Case treatment, 230–232
Centre of gravity, 249
Centrifugal acceleration, 113
Centrifugal force, 114, 246, 251–254
Change in total specific enthalpy, 88, 91,
  133, 153
Characteristic azimuthal speed, 153
Chemical pollution, 262
Choked flow, 148
Circular arc blade, 179–180, 203, 211
Circulation of the flow, 135
Clearance control, 264
Clearance volume, 27, 28
Coefficient of losses, 134, 137–140, 144, 145,
  198, 220
Collision loss, 219, 220
Combustion chamber temperature ratio, 18, 51,
  64, 66, 74, 85
Combustion volume ratio, 33, 37
Complex potential, 171, 172
Composite wing, 264

T. Bose, *Airbreathing Propulsion: An Introduction*,
Springer Aerospace Technology, DOI 10.1007/978-1-4614-3532-7,
© Springer Science+Business Media, LLC 2012